RSMeans
Residential
Repair &
Remodeling Costs

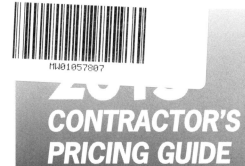

CONTRACTOR'S PRICING GUIDE

Senior Editor
Bob Mewis, CCC

Contributing Editors
Christopher Babbitt
Ted Baker
Adrian C. Charest, PE
Gary W. Christensen
David G. Drain, PE
Cheryl Elsmore
Robert J. Kuchta
Robert C. McNichols
Melville J. Mossman, PE
Jeannene D. Murphy
Marilyn Phelan, AIA
Stephen C. Plotner
Eugene R. Spencer
Phillip R. Waier, PE

Cover Design
Wendy McBay
Paul Robertson
Wayne Engebretson

President & CEO
Iain Melville

Chief Operating Officer
Richard Remington

Chief Customer Officer
Lisa Fiondella

Vice President, Marketing & Business Development
Steven Ritchie

Vice President, Finance & Operations
Renée Pinczes

Engineering Director
Bob Mewis, CCC

Director, Product Management
Jennifer Johnson

Product Manager
Andrea Sillah

Product Marketing Manager
Ryan Cummings

Senior Director, Project & Process Management
Rodney L. Sharples

Production Manager
Michael Kokernak

Editing Supervisor
Jill Goodman

Production Supervisor
Paula Reale-Camelio

Production
Jonathan Forgit
Mary Lou Geary
Debbie Panarelli
Caroline Sullivan

Technical Support
Christopher Anderson
Gary L. Hoitt
Kathryn S. Rodriguez

This book is printed using soy-based printing ink.
This book is recyclable.

PRINTED WITH SOY INK

RSMeans®

Contractor's Pricing Guide
Residential Repair &
Remodeling Costs 2013

Quick costs to help you estimate

Easy-to-use format . . . organized the way you build!

Hundreds of work items with material, labor, and equipment

Built for the small residential contractor/insurance repair specialist

$39.95 per copy (in United States)
Price subject to change without prior notice.

RS**Means**

Copyright 2012
A division of Reed Construction Data
Construction Publishers & Consultants
700 Longwater Drive
Norwell, MA 02061
www.rsmeans.com

Printed in the United States of America

ISSN 1542-9156

ISBN 978-1-936335-75-6

This book is printed on recycled paper, using soy-based printing ink. This book is recyclable.

Foreword

Our Mission

Since 1942, RSMeans has been actively engaged in construction cost publishing and consulting throughout North America.

Today, more than 70 years after RSMeans began, our primary objective remains the same: to provide you, the construction and facilities professional, with the most current and comprehensive construction cost data possible.

Whether you are a contractor, owner, architect, engineer, facilities manager, or anyone else who needs a reliable construction cost estimate, you'll find this publication to be a highly useful and necessary tool.

With the constant flow of new construction methods and materials today, it's difficult to find the time to look at and evaluate all the different construction cost possibilities. In addition, because labor and material costs keep changing, last year's cost information is not a reliable basis for today's estimate or budget.

That's why so many construction professionals turn to RSMeans. We keep track of the costs for you, along with a wide range of other key information, from city cost indexes . . . to productivity rates . . . to crew composition . . . to contractors' overhead and profit rates.

RSMeans performs these functions by collecting data from all facets of the industry and organizing it in a format that is instantly accessible to you. From the preliminary budget to the detailed unit price estimate, you'll find the data in this book useful for all phases of construction cost determination.

The Staff, the Organization, and Our Services

When you purchase one of RSMeans' publications, you are, in effect, hiring the services of a full-time staff of construction and engineering professionals.

Our thoroughly experienced and highly qualified staff works daily at collecting, analyzing, and disseminating comprehensive cost information for your needs. These staff members have years of practical construction experience and engineering training prior to joining the firm. As a result, you can count on them not only for accurate cost figures, but also for additional background reference information that will help you create a realistic estimate.

The RSMeans organization is always prepared to help you solve construction problems through its variety of data solutions, including online, CD, and print book formats, as well as cost estimating expertise available via our Marketing Intelligence, training, and seminars.

Besides a full array of construction cost estimating books, RSMeans also publishes a number of other reference works for the construction industry. Subjects include construction estimating and project and business management, special topics such as green building and job order contracting, and a library of facility management references.

In addition, you can access all of our construction cost data electronically in convenient CD format or on the Web. Visit **RSMeans Online.com** for more information on our 24/7 online cost data.

What's more, you can increase your knowledge and improve your construction estimating and management performance with an RSMeans construction seminar or in-house training program. These two-day seminar programs offer unparalleled opportunities for everyone in your organization to become updated on a wide variety of construction-related issues.

RSMeans is also a worldwide provider of construction cost management and analysis services for commercial and government owners.

In short, RSMeans can provide you with the tools and expertise for constructing accurate and dependable construction estimates and budgets in a variety of ways.

Robert Snow Means Established a Tradition of Quality That Continues Today

Robert Snow Means spent years building RSMeans, making certain he always delivered a quality product.

Today, at RSMeans, we do more than talk about the quality of our data and the usefulness of our books. We stand behind all of our data, from historical cost indexes to construction materials and techniques to current costs.

If you have any questions about our products or services, please call us toll-free at 1-800-334-3509. Our customer service representatives will be happy to assist you. You can also visit our website at **www.rsmeans.com**.

Table of Contents

How the Book Is Built: An Overview

A Powerful Construction Tool

You have in your hands one of the most powerful construction tools available today. A successful project is built on the foundation of an accurate and dependable estimate. This book will enable you to construct just such an estimate.

For the casual user the book is designed to be:

- quickly and easily understood so you can get right to your estimate.
- filled with valuable information so you can understand the necessary factors that go into the cost estimate.

For the regular user, the book is designed to be:

- a handy desk reference that can be quickly referred to for key costs.
- a comprehensive, fully reliable source of current construction costs, so you'll be prepared to estimate any project.
- a source book for project cost, product selections, and alternate materials and methods.

To meet all of these requirements we have organized the book into the following clearly defined sections.

Unit Price Section

The cost data has been divided into 23 sections representing the order of construction. Within each section is a listing of components and variations to those components applicable to that section. Costs are shown for the various actions that must be done to that component.

Reference Section

This section includes information on Location Factors and a listing of Abbreviations.

Location Factors: You can adjust total project costs to over 900 locations throughout the U.S. and Canada by using the data in this section.

Abbreviations: A listing of the abbreviations used throughout this book, along with the terms they represent, is included.

Index

A comprehensive listing of all terms and subjects in this book will help you quickly find what you need when you are not sure where it falls in the order of construction.

The Scope of This Book

This book is designed to be as comprehensive and as easy to use as possible. To that end we have made certain assumptions and limited its scope in two key ways:

1. We have established material prices based on a national average.
2. We have computed labor costs based on a seven major-region average of open shop wage rates.

Project Size

This book is intended for use by those involved primarily in residential repair and remodeling construction costing between $10,000–$100,000.

With reasonable exercise of judgment, the figures can be used for any building work. For other types of projects, such as new home construction or commercial buildings, consult the appropriate RSMeans publication for more information.

How to Use the Book: The Details

What's Behind the Numbers? The Development of Cost Data

The staff at RSMeans continually monitors developments in the construction industry in order to ensure reliable, thorough, and up-to-date cost information.

While *overall* construction costs may vary relative to general economic conditions, price fluctuations within the industry are dependent upon many factors. Individual price variations may, in fact, be opposite to overall economic trends. Therefore, costs are constantly tracked and complete updates are published yearly. Also, new items are frequently added in response to changes in materials and methods.

Costs–$ (U.S.)

All costs represent U.S. national averages and are given in U.S. dollars. The RSMeans Location Factors can be used to adjust costs to a particular location. The Location Factors for Canada can be used to adjust U.S. national averages to local costs in Canadian dollars. No exchange rate conversion is necessary.

Material Costs

The RSMeans staff contacts manufacturers, dealers, distributors, and contractors all across the U.S. and Canada to determine national average material costs. If you have access to current material costs for your specific location, you may wish to make adjustments to reflect differences from the national average.

Included within material costs are fasteners for a normal installation. RSMeans engineers use manufacturers' recommendations, written specifications, and/or standard construction practice for size and spacing of fasteners. Adjustments to material costs may be required for your specific application or location. Material costs do not include sales tax.

Labor Costs

Labor costs are based on the average of residential wages from across the U.S. for the current year. Rates, along with overhead and profit markups, are listed on the inside back cover of this book.

- If wage rates in your area vary from those used in this book, or if rate increases are expected within a given year, labor costs should be adjusted accordingly.

Labor costs reflect productivity based on actual working conditions. These figures include time spent during a normal workday on tasks other than actual installation, such as material receiving and handling, mobilization at site, site movement, breaks, and cleanup.

Productivity data is developed over an extended period so as not to be influenced by abnormal variations and reflects a typical average.

Equipment Costs

Equipment costs include not only rental costs, but also operating costs for equipment under normal use. Equipment and rental rates are obtained from industry sources throughout North America—contractors, suppliers, dealers, manufacturers, and distributors.

Factors Affecting Costs

Costs can vary depending upon a number of variables. Here's how we have handled the main factors affecting costs.

Quality—The prices for materials and the workmanship upon which productivity is based represent sound construction work. They are also in line with U.S. government specifications.

Overtime—We have made no allowance for overtime. If you anticipate premium time or work beyond normal working hours, be sure to make an appropriate adjustment to your labor costs.

Productivity—The labor costs for each line item are based on working an eight-hour day in daylight hours in moderate temperatures. For work that extends beyond normal work hours or is performed under adverse conditions, productivity may decrease.

Size of Project—The size, scope of work, and type of construction project will have a significant impact on cost. Economies of scale can reduce costs for large projects. Unit costs can often run higher for small projects. Costs in this book are intended for the size and type of project as previously described in "How the Book Is Built: An Overview." Costs for projects of a significantly different size or type should be adjusted accordingly.

Location—Material prices in this book are for metropolitan areas. However, in dense urban areas, traffic and site storage limitations may increase costs. Beyond a 20-mile radius of large cities, extra trucking or transportation charges may also increase the material costs slightly. On the other hand, lower wage rates may be in effect. Be sure to consider both of these factors when preparing an estimate, particularly if the job site is located in a central city or remote rural location.

In addition, highly specialized subcontract items may require travel and per diem expenses for mechanics.

Other Factors—

- season of year
- contractor management
- weather conditions
- building code requirements
- availability of:
    - adequate energy
    - skilled labor
    - building materials
- owner's special requirements/ restrictions
- safety requirements
- environmental considerations

Overhead & Profit—The extreme right-hand column of each chart gives the "Totals." These figures contain the installing contractor's O&P. Therefore, it is necessary for a general contractor to add a percentage of all subcontracted items. For a detailed breakdown of O&P see the inside back cover of this book.

Unpredictable Factors—General business conditions influence "in-place" costs of all items. Substitute materials and construction methods may have to be employed. These may affect the installed cost and/or life cycle costs. Such factors may be difficult to evaluate and cannot necessarily be predicted on the basis of the job's location in a particular section of the country. Thus, where these factors apply, you may find significant but unavoidable cost variations for which you will have to apply a measure of judgment to your estimate.

Final Checklist

Estimating can be a straightforward process provided you remember the basics. Here's a checklist of some of the steps you should remember to complete before finalizing your estimate.

Did you remember to . . .

- factor in the Location Factor for your locale?
- take into consideration which items have been marked up and by how much?
- mark up the entire estimate sufficiently for your purposes?
- include all components of your project in the final estimate?
- double check your figures to be sure of your accuracy?
- call RSMeans if you have any questions about your estimate or the data you've found in our publications?

Remember, RSMeans stands behind its publications. If you have any questions about your estimate . . . about the costs you've used from our books . . . or even about the technical aspects of the job that may affect your estimate, feel free to call the RSMeans editors at 1-800-334-3509.

Unit Price Section

Table of Contents

Table of Contents (cont.)

How to Use the Unit Price Pages

The prices in this publication include all overhead and profit mark-ups for the installing contractor. If a general contractor is involved, an additional 10% should be added to the cost shown in the "Total" column.

The costs for material, labor and any equipment are shown here. Material costs include delivery and a 10% mark-up. No sales taxes are included. Labor costs include all mark-ups as shown on the inside back cover of this book. Equipment costs also include a 10% mark-up.

The cost in the total column should be multiplied by the quantity of each item to determine the total cost for that item. The sum of all item's total costs becomes the project's total cost.

The book is arranged into 23 categories that follow the order of construction for residential projects. The category is shown as white letters on a black background at the top of each page.

A brief specification of the cost item is shown here.

Graphics representing some of the cost items are found at the beginning of each category.

The cost data is arranged in a tree structure format. Major cost sections are left justified and are shown as bold headings within a grey background. Cost items and sub-cost items are found under the bold headings. The complete description should read as "bold heading, cost item, sub-cost item."
For example, the identified item's description is "Beam/Girder, Solid Wood Beam, 2″ x 8″."

Each action performed to the cost item is conveniently listed to the right of the description. See the explanation below for the scope of work included as part of the action.

The unit of measure upon which all the costs are based is shown here. An abbreviations list is included within the reference section found at the back of this book.

Rough Frame / Structure

Rough Framing

Beam / Girder	Unit	Material	Labor	Equip.	Total	Specification
Solid Wood Beam						
2″ x 8″						
Demolish	L.F.		.83		.83	Cost includes material and labor to
Install	L.F.	.87	1.30		2.17	install 2″ x 8″ beam.
Demolish and Install	L.F.	.87	2.13		3	
Clean	L.F.	.11	.28		.39	
Paint	L.F.	.14	1.95		2.09	
Minimum Charge	Job		211		211	
2″ x 10″						
Demolish	L.F.		1.03		1.03	Cost includes material and labor to
Install	L.F.	1.19	1.41		2.60	install 2″ x 10″ beam.
Demolish and Install	L.F.	1.19	2.44		3.63	
Clean	L.F.	.13	.33		.46	
Paint	L.F.	.19	1.20		1.39	
Minimum Charge	Job		211		211	
2″ x 12″						
Demolish	L.F.		1.25		1.25	Cost includes material and labor to
Install	L.F.	1.45	1.54		2.99	install 2″ x 12″ beam.
Demolish and Install	L.F.	1.45	2.79		4.24	
Clean	L.F.	.14	.39		.53	
Paint	L.F.	.22	2.89		3.11	
Minimum Charge	Job		211		211	
4″ x 8″						
Demolish	L.F.		2.57		2.57	Cost includes material and labor to
Install	L.F.	3.74	1.91	.71	6.36	install 4″ x 8″ beam.
Demolish and Install	L.F.	3.74	4.48	.71	8.93	
Clean	L.F.	.13	.33		.46	
Paint	L.F.	.19	1.20		1.39	
Minimum Charge	Job		211		211	
4″ x 10″						
Demolish	L.F.		3.22		3.22	Cost includes material and labor to
Install	L.F.	4.68	2.02	.75	7.45	install 4″ x 10″ beam.
Demolish and Install	L.F.	4.68	5.24	.75	10.67	
Clean	L.F.	.14	.39		.53	
Paint	L.F.	.21	1.40		1.61	
Minimum Charge	Job		211		211	
4″ x 12″						
Demolish	L.F.		3.86		3.86	Cost includes material and labor to
Install	L.F.	5.60	2.13	.79	8.52	install 4″ x 12″ beam.
Demolish and Install	L.F.	5.60	5.99	.79	12.38	
Clean	L.F.	.18	.44		.62	
Paint	L.F.	.22	2.89		3.11	
Minimum Charge	Job		211		211	

Action	Scope
Demolish	Includes removal of the item and hauling the debris to a truck, dumpster, or storage area.
Install	Includes the installation of the item with normal placement and spacing of fasteners.
Demolish and Install	Includes the demolish action and the install action.
Reinstall	Includes the reinstallation of a removed item.
Clean	Includes cleaning of the item.
Paint	Includes up to two coats of finish on the item (unless otherwise indicated).
Minimum Charge	Includes the minimum labor or the minimum labor and equipment cost. Use this charge for small quantities of material.

Job Costs

Scaffolding	Temporary Electrical Service	Air Compressor

Construction Fees		Unit	Material	Labor	Equip.	Total	Specification
Asbestos Test	Minimum Charge	Job		247		247	Minimum charge for testing.
Architectural	Minimum Charge	Day	1975			1975	Minimum labor charge for architect.
Engineering	Minimum Charge	Day	1850			1850	Minimum labor charge for field engineer.

Construction Permits		Unit	Material	Labor	Equip.	Total	Specification
Job (less than 5K)	Minimum Charge	Job	65			65	Permit fees, job less than 5K.
Job (5K to 10K)	Minimum Charge	Job	109			109	Permit fees, job more than 5K, but less than 10K.
Job (10K to 25K)	Minimum Charge	Job	241			241	Permit fees, job more than 10K, but less than 25K.
Job (25K to 50K)	Minimum Charge	Job	460			460	Permit fees, job more than 25K, but less than 50K.
Job (50K to 100K)	Minimum Charge	Job	845			845	Permit fees, job more than 50K, but less than 100K.
Job (100K to 250K)	Minimum Charge	Job	1750			1750	Permit fees, job more than 100K, but less than 250K.
Job (250K to 500K)	Minimum Charge	Job	3300			3300	Permit fees, job more than 250K, but less than 500K.
Job (500K to 1M)	Minimum Charge	Job	5925			5925	Permit fees, job more than 500K, but less than 1 million.

Job Costs

Trade Labor		Unit	Material	Labor	Equip.	Total	Specification
Carpenter							
Daily (8 hours)							
	Install	Day		425		425	Daily labor rate for one carpenter.
	Minimum Charge	Job		211		211	
Weekly (40 hours)							
	Install	Week		2125		2125	Weekly labor rate for one carpenter.
	Minimum Charge	Job		211		211	
Drywaller							
Daily (8 hours)							
	Install	Day		425		425	Daily labor rate for one carpenter.
	Minimum Charge	Job		211		211	
Weekly (40 hours)							
	Install	Week		2125		2125	Weekly labor rate for one carpenter.
	Minimum Charge	Job		211		211	
Roofer							
Daily (8 hours)							
	Install	Day		385		385	Daily labor rate for one skilled roofer.
	Minimum Charge	Job		192		192	
Weekly (40 hours)							
	Install	Week		1925		1925	Weekly labor rate for one skilled
	Minimum Charge	Job		192		192	roofer.
Painter							
Daily (8 hours)							
	Install	Day		360		360	Daily labor rate for one painter.
	Minimum Charge	Job		180		180	
Weekly (40 hours)							
	Install	Week		1800		1800	Weekly labor rate for one painter.
	Minimum Charge	Job		180		180	
Mason							
Daily (8 hours)							
	Install	Day		415		415	Daily labor rate for one skilled mason.
	Minimum Charge	Job		207		207	
Weekly (40 hours)							
	Install	Week		2075		2075	Weekly labor rate for one skilled
	Minimum Charge	Job		207		207	mason.
Electrician							
Daily (8 hours)							
	Install	Day		460		460	Daily labor rate for an electrician.
	Minimum Charge	Job		230		230	
Weekly (40 hours)							
	Install	Week		2300		2300	Weekly labor rate for an electrician.
	Minimum Charge	Job		230		230	
Plumber							
Daily (8 hours)							
	Install	Day		485		485	Daily labor rate for a plumber.
	Minimum Charge	Job		243		243	
Weekly (40 hours)							
	Install	Week		2425		2425	Weekly labor rate for a plumber.
	Minimum Charge	Job		243		243	
Common Laborer							
Daily (8 hours)							
	Install	Day		310		310	Daily labor rate for a common
	Minimum Charge	Job		155		155	laborer.

Job Costs

Trade Labor		Unit	Material	Labor	Equip.	Total	Specification
Weekly (40 hours)		Week					
	Install	Week		1550		1550	Weekly labor rate for a common
	Minimum Charge	Job		155		155	laborer.

Temporary Utilities		Unit	Material	Labor	Equip.	Total	Specification
Power Pole							
Job Site Electricity							
	Install	Month	55			55	Includes average cost of electricity per
	Minimum Charge	Job		230		230	month.
Water Service							
Job Site Water							
	Install	Month	69.50			69.50	Includes average costs for water for
	Minimum Charge	Month	69.50			69.50	one month.
Telephone							
	Install	Month	89			89	Includes average cost for local and
	Minimum Charge	Month	89			89	long distance telephone service for
							one month.
Chemical Toilet							
	Install	Month			290	290	Monthly rental of portable toilet.
	Minimum Charge	Day			14.50	14.50	
Office Trailer							
Place and Remove							
	Install	Job	375			375	Includes cost to a contractor to deliver,
	Minimum Charge	Job		174		174	place and level an office trailer and
							remove at end of job.
Storage Container							
	Install	Job	209			209	Includes cost to a contractor to deliver
	Minimum Charge	Job		174		174	and place a storage container and
							remove at end of job.
Fencing							
	Install	L.F.	4.72			4.72	Includes material costs for temporary
	Minimum Charge	L.F.	275			275	fencing.
Railing							
	Install	L.F.	16.60	11.90	.90	29.40	Includes material and labor to install
	Minimum Charge	L.F.	275			275	primed steel pipe.

Equipment Rental		Unit	Material	Labor	Equip.	Total	Specification
Flatbed Truck							
Daily Rental							
	Install	Day			292	292	Daily rental of flatbed truck.
	Minimum Charge	Job			292	292	
Weekly Rental							
	Install	Week			1450	1450	Weekly rental of flatbed truck.
	Minimum Charge	Job			292	292	
Monthly Rental							
	Install	Month			5850	5850	Monthly rental of flatbed truck.
	Minimum Charge	Job			292	292	

Job Costs

Equipment Rental		Unit	Material	Labor	Equip.	Total	Specification
Dump Truck							
Daily Rental							
	Install	Day			460	460	Daily rental of dump truck.
	Minimum Charge	Day			460	460	
Weekly Rental							
	Install	Week			2300	2300	Weekly rental of dump truck.
	Minimum Charge	Day			460	460	
Monthly Rental							
	Install	Month			9175	9175	Monthly rental of dump truck.
	Minimum Charge	Day			460	460	
Forklift							
Daily Rental							
	Install	Day			360	360	Daily rental of forklift.
	Minimum Charge	Day			360	360	
Weekly Rental							
	Install	Week			1800	1800	Weekly rental of forklift.
	Minimum Charge	Day			360	360	
Monthly Rental							
	Install	Month			7150	7150	Monthly rental of forklift.
	Minimum Charge	Day			360	360	
Bobcat							
Daily Rental							
	Install	Day			189	189	Daily rental of wheeled, skid steer loader.
	Minimum Charge	Day			189	189	
Weekly Rental							
	Install	Week			945	945	Weekly rental of wheeled, skid steer loader.
	Minimum Charge	Day			189	189	
Monthly Rental							
	Install	Month			3775	3775	Monthly rental of wheeled, skid steer loader.
	Minimum Charge	Day			189	189	
Air Compressor							
Daily Rental							
	Install	Day			151	151	Daily rental of air compressor.
	Minimum Charge	Day			151	151	
Weekly Rental							
	Install	Week			755	755	Weekly rental of air compressor.
	Minimum Charge	Day			151	151	
Monthly Rental							
	Install	Month			3025	3025	Monthly rental of air compressor.
	Minimum Charge	Day			151	151	
Jackhammer							
Daily Rental							
	Install	Day			10.80	10.80	Daily rental of air tool, jackhammer.
	Minimum Charge	Day			10.80	10.80	
Weekly Rental							
	Install	Week			54	54	Weekly rental of air tool, jackhammer.
	Minimum Charge	Day			10.80	10.80	
Monthly Rental							
	Install	Month			216	216	Monthly rental of air tool, jackhammer.
	Minimum Charge	Day			10.80	10.80	
Concrete Mixer							
Daily Rental							
	Install	Day			157	157	Daily rental of concrete mixer.
	Minimum Charge	Day			157	157	

Job Costs

Equipment Rental		Unit	Material	Labor	Equip.	Total	Specification
Weekly Rental							
	Install	Week			785	785	Weekly rental of concrete mixer.
	Minimum Charge	Day			157	157	
Monthly Rental							
	Install	Month			3125	3125	Monthly rental of concrete mixer.
	Minimum Charge	Day			157	157	
Concrete Bucket							
Daily Rental							
	Install	Day			22.50	22.50	Daily rental of concrete bucket.
	Minimum Charge	Day			22.50	22.50	
Weekly Rental							
	Install	Week			112	112	Weekly rental of concrete bucket.
	Minimum Charge	Day			22.50	22.50	
Monthly Rental							
	Install	Month			450	450	Monthly rental of concrete bucket.
	Minimum Charge	Day			22.50	22.50	
Concrete Pump							
Daily Rental							
	Install	Day			785	785	Daily rental of concrete pump.
	Minimum Charge	Day			785	785	
Weekly Rental							
	Install	Week			3925	3925	Weekly rental of concrete pump.
	Minimum Charge	Day			785	785	
Monthly Rental							
	Install	Month			15700	15700	Monthly rental of concrete pump.
	Minimum Charge	Day			785	785	
Generator							
Daily Rental							
	Install	Day			39	39	Daily rental of generator.
	Minimum Charge	Day			39	39	
Weekly Rental							
	Install	Week			196	196	Weekly rental of generator.
	Minimum Charge	Day			39	39	
Monthly Rental							
	Install	Month			785	785	Monthly rental of generator.
	Minimum Charge	Day			39	39	
Welder							
Daily Rental							
	Install	Day			41.50	41.50	Daily rental of welder.
	Minimum Charge	Day			41.50	41.50	
Weekly Rental							
	Install	Week			207	207	Weekly rental of welder.
	Minimum Charge	Day			41.50	41.50	
Monthly Rental							
	Install	Month			830	830	Monthly rental of welder.
	Minimum Charge	Day			41.50	41.50	
Sandblaster							
Daily Rental							
	Install	Day			22	22	Daily rental of sandblaster.
	Minimum Charge	Day			22	22	
Weekly Rental							
	Install	Week			110	110	Weekly rental of sandblaster.
	Minimum Charge	Day			22	22	

Job Costs

Equipment Rental

		Unit	Material	Labor	Equip.	Total	Specification
Monthly Rental							
	Install	Month			440	440	Monthly rental of sandblaster.
	Minimum Charge	Day			22	22	
Space Heater							
Daily Rental							
	Install	Day			22.50	22.50	Daily rental of space heater.
	Minimum Charge	Day			11.35	11.35	
Weekly Rental							
	Install	Day			112	112	Weekly rental of space heater.
	Minimum Charge	Day			11.35	11.35	
Monthly Rental							
	Install	Day			450	450	Monthly rental of space heater.
	Minimum Charge	Day			11.35	11.35	

Move / Reset Contents

		Unit	Material	Labor	Equip.	Total	Specification
Small Room							
	Install	Room		37.50		37.50	Includes cost to remove and reset the
	Minimum Charge	Job		155		155	contents of a small room.
Average Room							
	Install	Room		45.50		45.50	Includes cost to remove and reset the
	Minimum Charge	Job		155		155	contents of an average room.
Large Room							
	Install	Room		62.50		62.50	Includes cost to remove and reset the
	Minimum Charge	Job		155		155	contents of a large room.
Extra Large Room							
	Install	Room		99		99	Includes cost to remove and reset the
	Minimum Charge	Job		155		155	contents of a extra large room.
Store Contents							
Weekly							
	Install	Week	275			275	Storage fee, per week.
Monthly							
	Install	Month	825			825	Storage fee, per month.

Cover and Protect

		Unit	Material	Labor	Equip.	Total	Specification
Cover / Protect Walls							
	Install	SF Wall		.12		.12	Includes labor costs to cover and
	Minimum Charge	Job		155		155	protect walls.
Cover / Protect Floors							
	Install	SF Flr.		.12		.12	Includes labor costs to cover and
	Minimum Charge	Job		155		155	protect floors.

Temporary Bracing

		Unit	Material	Labor	Equip.	Total	Specification
Shim Support Piers							
	Install	Ea.	4.35	122		126.35	Includes labor and material to install
	Minimum Charge	Job		365		365	shims support for piers.

Job Costs

Jack and Relevel		Unit	Material	Labor	Equip.	Total	Specification
Single Story							
	Install	S.F.	1.82	1.47		3.29	Includes labor and equipment to jack
	Minimum Charge	Job		365		365	and relevel single-story building.
Two-story							
	Install	S.F.	1.82	2.45		4.27	Includes labor and equipment to jack
	Minimum Charge	Job		365		365	and relevel two-story building.

Dumpster Rental		Unit	Material	Labor	Equip.	Total	Specification
Small Load							
	Install	Week			685	685	Dumpster, weekly rental, 2 dumps, 5
	Minimum Charge	Week			685	685	yard container.
Medium Load							
	Install	Week			795	795	Dumpster, weekly rental, 2 dumps, 10
	Minimum Charge	Week			795	795	yard container.
Large Load							
	Install	Week			1225	1225	Dumpster, weekly rental, 2 dumps, 30
	Minimum Charge	Week			1225	1225	yard container.

Debris Hauling		Unit	Material	Labor	Equip.	Total	Specification
Per Ton							
	Install	Ton		96.50	56.50	153	Includes hauling of debris by trailer or
	Minimum Charge	Ton		725	425	1150	dump truck.
Per Cubic Yard							
	Install	C.Y.		32	18.80	50.80	Includes hauling of debris by trailer or
	Minimum Charge	Job		725	425	1150	dump truck.
Per Pick-up Truck Load							
	Install	Ea.		172		172	Includes hauling of debris by pick-up
	Minimum Charge	Ea.		172		172	truck.

Dump Fees		Unit	Material	Labor	Equip.	Total	Specification
Per Cubic Yard							
	Install	C.Y.	35			35	Dump charges, building materials.
	Minimum Charge	C.Y.	35			35	
Per Ton							
	Install	Ton	132			132	Dump charges, building materials.
	Minimum Charge	Ton	132			132	

Scaffolding		Unit	Material	Labor	Equip.	Total	Specification
4' to 6' High							
	Install	Month	67			67	Includes monthly rental of scaffolding,
	Minimum Charge	Month	67			67	steel tubular, 30" wide, 7' long, 5' high.
7' to 11' High							
	Install	Month	80.50			80.50	Includes monthly rental of scaffolding,
	Minimum Charge	Month	80.50			80.50	steel tubular, 30" wide, 7' long, 10' high.

Job Costs

Scaffolding		Unit	Material	Labor	Equip.	Total	Specification
12' to 16' High							
	Install	Month	93.50			93.50	Includes monthly rental of scaffolding, steel tubular, 30" wide, 7' long, 15' high.
	Minimum Charge	Month	93.50			93.50	
17' to 21' High							
	Install	Month	122			122	Includes monthly rental of scaffolding, steel tubular, 30" wide, 7' long, 20' high.
	Minimum Charge	Month	122			122	
22' to 26' High							
	Install	Month	135			135	Includes monthly rental of scaffolding, steel tubular, 30" wide, 7' long, 25' high.
	Minimum Charge	Month	135			135	
27' to 30' High							
	Install	Month	149			149	Includes monthly rental of scaffolding, steel tubular, 30" wide, 7' long, 30' high.
	Minimum Charge	Month	149			149	

Construction Clean-up		Unit	Material	Labor	Equip.	Total	Specification
Initial							
	Clean	S.F.		.39		.39	Includes labor for general clean-up.
	Minimum Charge	Job		155		155	
Progressive							
	Clean	S.F.		.19		.19	Includes labor for progressive site/job clean-up on a weekly basis.
	Minimum Charge	Job		155		155	

Building Demolition		Unit	Material	Labor	Equip.	Total	Specification
Wood-framed Building							
1 Story Demolition							
	Demolish	S.F.		1.46	1.99	3.45	Includes demolition of 1 floor only
	Minimum Charge	Job		2325	3150	5475	light wood-framed building, including hauling of debris. Dump fees not included.
2nd Story Demolition							
	Demolish	S.F.		1.49	2.02	3.51	Includes demolition of 2nd floor only
	Minimum Charge	Job		2325	3150	5475	light wood-framed building, including hauling of debris. Dump fees not included.
3rd Story Demolition							
	Demolish	S.F.		1.94	2.63	4.57	Includes demolition of 3rd floor only
	Minimum Charge	Job		2325	3150	5475	light wood-framed building, including hauling of debris. Dump fees not included.
Masonry Building							
	Demolish	S.F.		2.08	2.83	4.91	Includes demolition of one story
	Minimum Charge	Job		2325	3150	5475	masonry building, including hauling of debris. Dump fees not included.
Concrete Building							
Unreinforced							
	Demolish	S.F.		4.28	1.29	5.57	Includes demolition of a one story
	Minimum Charge	Job		2325	3150	5475	non-reinforced concrete building, including hauling of debris. Dump fees not included.
Reinforced							
	Demolish	S.F.		4.91	1.48	6.39	Includes demolition of a one story
	Minimum Charge	Job		2325	3150	5475	reinforced concrete building, including hauling of debris. Dump fees not included.
Steel Building							
	Demolish	S.F.		2.78	3.77	6.55	Includes demolition of a metal
	Minimum Charge	Job		2325	3150	5475	building, including hauling of debris. Dump fees not included. No salvage value assumed for scrap metal.

Job Preparation

General Clean-up		Unit	Material	Labor	Equip.	Total	Specification
Remove Debris							
	Clean	S.F.		.96	.29	1.25	Includes removal of debris caused by
	Minimum Charge	Job		480	146	626	the loss including contents items and
							loose debris.
Muck-out							
	Clean	S.F.		.60		.60	Includes labor to remove 2" of muck
	Minimum Charge	Job		155		155	and mud after a flood loss.
Deodorize / Disinfect							
	Clean	SF Flr.		.21		.21	Includes deodorizing building with
	Minimum Charge	Job		155		155	application of mildicide and
							disinfectant to floors and walls - will
							vary depending on the extent of
							deodorization required.
Clean Walls							
Heavy							
	Clean	S.F.		.14		.14	Includes labor and material for heavy
	Minimum Charge	Job		155		155	cleaning (multiple applications) with
							detergent and solvent.
Clean Ceiling							
Light							
	Clean	S.F.		.24		.24	Includes labor and material to clean
	Minimum Charge	Job		155		155	light to moderate smoke and soot.
Heavy							
	Clean	S.F.		.36		.36	Includes labor and material to clean
	Minimum Charge	Job		155		155	heavy smoke and soot.

Flood Clean-up		Unit	Material	Labor	Equip.	Total	Specification
Emergency Service Call							
After Hours / Weekend							
	Minimum Charge	Job		206		206	Includes minimum charges for
							deflooding clean-up after hours or on
							weekends.
Grey Water / Sewage							
	Minimum Charge	Job		155		155	Includes minimum charges for
							deflooding clean-up.
Water Extraction							
Clean (non-grey)							
	Clean	S.F.	.06	.62		.68	Includes labor to remove clean
	Minimum Charge	Job		155		155	uncontaminated water from the loss.
Grey (non-solids)							
	Clean	S.F.	.06	1.24		1.30	Includes labor to remove non-solid
	Minimum Charge	Job		155		155	grey water from the loss.
Sewage / Solids							
	Clean	SF Flr.	.10	1.55		1.65	Includes labor to remove water from
	Minimum Charge	Job		155		155	the loss that may contain sewage or
							solids.

Job Preparation

Flood Clean-up		Unit	Material	Labor	Equip.	Total	Specification
Mildicide Walls							
Topographical							
	Clean	SF Wall		.28		.28	Clean-up of deflooding including
	Minimum Charge	Job		155		155	topographical mildicide application.
Injection Treatments							
	Clean	SF Flr.	32	4.76		36.76	Clean-up of deflooding including
	Minimum Charge	Job		155		155	mildicide injection.
Carpet Cleaning							
Uncontaminated							
	Clean	SF Flr.		.19		.19	Includes labor to clean carpet.
	Minimum Charge	Job		155		155	
Contaminated							
	Clean	SF Flr.		.31		.31	Includes labor to clean contaminated
	Minimum Charge	Job		155		155	carpet.
Stairway							
	Clean	Ea.	.11	3.64		3.75	Includes labor to clean contaminated
	Minimum Charge	Job		155		155	carpet on a stairway.
Carpet Treatment							
Disinfect / Deodorize							
	Clean	SF Flr.		.19		.19	Includes disinfecting and deodorizing
	Minimum Charge	Job		155		155	of carpet.
Disinfect For Sewage							
	Clean	SF Flr.		.50		.50	Includes labor to disinfect and
	Minimum Charge	Job		155		155	deodorize carpet exposed to sewage.
Mildicide							
	Clean	SF Flr.	.01	.19		.20	Includes labor and materials to apply
	Minimum Charge	Job		155		155	mildicide to carpet.
Thermal Fog Area							
	Clean	SF Flr.	.03	.48		.51	Includes labor and materials to
	Minimum Charge	Job		155		155	perform thermal fogging during
							deflooding clean-up.
Wet Fog / ULV Area							
	Clean	SF Flr.	.02	.27		.29	Includes labor and materials to
	Minimum Charge	Job		155		155	perform wet fogging / ULV during
							deflooding clean-up.
Steam Clean Fixtures							
	Clean	Ea.	.01	31		31.01	Includes labor and materials to steam
	Minimum Charge	Job		155		155	clean fixtures during deflooding
							clean-up.
Pressure Wash Sewage							
	Clean	SF Flr.		2.81		2.81	Includes labor to pressure wash
	Minimum Charge	Job		155		155	sewage contamination.
Air Mover							
	Minimum Charge	Day			22.50	22.50	Daily rental of air mover.

Job Preparation

Flood Clean-up

Flood Clean-up		Unit	Material	Labor	Equip.	Total	Specification
Dehumidifier							
Medium Size							
	Minimum Charge	Day			48	48	Daily rental of dehumidifier, medium.
Large Size							
	Minimum Charge	Day			97.50	97.50	Daily rental of dehumidifier, large.
Air Purifier							
HEPA Filter-Bacteria							
	Minimum Charge	Ea.	345			345	HEPA filter for air purifier.

Asbestos Removal

Asbestos Removal		Unit	Material	Labor	Equip.	Total	Specification
Asbestos Analysis							
	Minimum Charge	Job		247		247	Minimum charge for testing.
Asbestos Pipe Insulation							
1/2" to 3/4" Diameter							
	Demolish	L.F.	.45	6.40		6.85	Minimum charge to remove asbestos.
	Minimum Charge	Job		3950		3950	Includes full Tyvek suits for workers changed four times per eight-hour shift with respirators, air monitoring and supervision by qualified professionals.
1" to 3" Diameter							
	Demolish	L.F.	1.41	19.80		21.21	Minimum charge to remove asbestos.
	Minimum Charge	Job		3950		3950	Includes full Tyvek suits for workers changed four times per eight-hour shift with respirators, air monitoring and supervision by qualified professionals.
Asbestos-based Siding							
	Demolish	S.F.	3.12	44		47.12	Minimum charge to remove asbestos.
	Minimum Charge	Job		3950		3950	Includes full Tyvek suits for workers changed four times per eight-hour shift with respirators, air monitoring and supervision by qualified professionals.
Asbestos-based Plaster							
	Demolish	S.F.	2.40	34		36.40	Minimum charge to remove asbestos.
	Minimum Charge	Job		3950		3950	Includes full Tyvek suits for workers changed four times per eight-hour shift with respirators, air monitoring and supervision by qualified professionals.
Encapsulate Asbestos Ceiling							
	Install	S.F.	.31	.19		.50	Includes encapsulation with penetrating sealant sprayed onto ceiling and walls with airless sprayer.
	Minimum Charge	Job		3950		3950	
Scrape Acoustical Ceiling							
	Demolish	S.F.	.09	1.11		1.20	Minimum charge to remove asbestos.
	Minimum Charge	Job		3950		3950	Includes full Tyvek suits for workers changed four times per eight-hour shift with respirators, air monitoring and supervision by qualified professionals.

Job Preparation

Asbestos Removal

		Unit	Material	Labor	Equip.	Total	Specification
Seal Ceiling Asbestos							
	Install	S.F.	2.30	3.72		6.02	Includes attaching plastic cover to ceiling area and sealing where asbestos exists, taping all seams, caulk or foam insulation in cracks and negative air system.
	Minimum Charge	Job		3950		3950	
Seal Wall Asbestos							
	Install	S.F.	2.30	3.31		5.61	Includes attaching plastic cover to wall area and sealing where asbestos exists, taping all seams, caulk or foam insulation in cracks and negative air system.
	Minimum Charge	Job		3950		3950	
Seal Floor Asbestos							
	Install	S.F.	3.67	6.60		10.27	Includes attaching plastic cover to floor area and sealing where asbestos exists, taping all seams, caulk or foam insulation in cracks and negative air system.
	Minimum Charge	Job		3950		3950	
Negative Air Vent System							
	Install	Day			46	46	Daily rental of negative air vent system.
	Minimum Charge	Week			229	229	
HEPA Vacuum Cleaner							
	Install	Day			19.10	19.10	Daily rental of HEPA vacuum, 16 gallon.
	Minimum Charge	Week			95.50	95.50	
Airless Sprayer							
	Install	Day			64.50	64.50	Daily rental of airless sprayer.
	Minimum Charge	Week			325	325	
Decontamination Unit							
	Install	Day			143	143	Daily rental of decontamination unit.
	Minimum Charge	Week			715	715	
Light Stand							
	Install	Day			15.85	15.85	Daily rental of floodlight w/tripod.
	Minimum Charge	Week			79	79	

Room Demolition

		Unit	Material	Labor	Equip.	Total	Specification
Strip Typical Room							
	Demolish	S.F.		3.10		3.10	Includes 4 labor hours for general demolition.
	Minimum Charge	Job		172		172	
Strip Typical Bathroom							
	Demolish	S.F.		4.13		4.13	Includes 4 labor hours for general demolition.
	Minimum Charge	Job		172		172	
Strip Typical Kitchen							
	Demolish	S.F.		3.97		3.97	Includes 4 labor hours for general demolition.
	Minimum Charge	Job		172		172	
Strip Typical Laundry							
	Demolish	S.F.		3.54		3.54	Includes 4 labor hours for general demolition.
	Minimum Charge	Job		172		172	

Job Preparation

Selective Demolition		Unit	Material	Labor	Equip.	Total	Specification
Concrete Slab							
4″ Unreinforced							
	Demolish	S.F.		2.06	.33	2.39	Includes labor and equipment to
	Minimum Charge	Job		875	139	1014	remove unreinforced concrete slab and haul debris to truck or dumpster.
6″ Unreinforced							
	Demolish	S.F.		2.92	.46	3.38	Includes labor and equipment to
	Minimum Charge	Job		875	139	1014	remove unreinforced concrete slab and haul debris to truck or dumpster.
8″ Unreinforced							
	Demolish	S.F.		4.80	.76	5.56	Includes labor and equipment to
	Minimum Charge	Job		875	139	1014	remove unreinforced concrete slab and haul debris to truck or dumpster.
4″ Reinforced							
	Demolish	S.F.		2.59	.41	3	Includes labor and equipment to
	Minimum Charge	Job		875	139	1014	remove reinforced concrete slab and haul debris to truck or dumpster.
6″ Reinforced							
	Demolish	S.F.		3.89	.62	4.51	Includes labor and equipment to
	Minimum Charge	Job		875	139	1014	remove reinforced concrete slab and haul debris to truck or dumpster.
8″ Reinforced							
	Demolish	S.F.		6.35	1.01	7.36	Includes labor and equipment to
	Minimum Charge	Job		875	139	1014	remove reinforced concrete slab and haul debris to truck or dumpster.
Brick Wall							
4″ Thick							
	Demolish	S.F.		2.86	.45	3.31	Includes labor and equipment to
	Minimum Charge	Job		875	139	1014	remove brick wall and haul debris to a central location for disposal.
8″ Thick							
	Demolish	S.F.		3.94	.62	4.56	Includes labor and equipment to
	Minimum Charge	Job		875	139	1014	remove brick wall and haul debris to a central location for disposal.
12″ Thick							
	Demolish	S.F.		5.25	.83	6.08	Includes labor and equipment to
	Minimum Charge	Job		875	139	1014	remove brick wall and haul debris to a central location for disposal.
Concrete Masonry Wall							
4″ Thick							
	Demolish	S.F.		3.18	.50	3.68	Includes minimum labor and
	Minimum Charge	Job		875	139	1014	equipment to remove 4″ block wall and haul debris to a central location for disposal.
6″ Thick							
	Demolish	S.F.		3.94	.62	4.56	Includes minimum labor and
	Minimum Charge	Job		875	139	1014	equipment to remove 6″ block wall and haul debris to a central location for disposal.

Job Preparation

Selective Demolition		Unit	Material	Labor	Equip.	Total	Specification
8″ Thick							
	Demolish	S.F.		4.85	.77	5.62	Includes minimum labor and equipment to remove 8″ block wall and haul debris to a central location for disposal.
	Minimum Charge	Job		875	139	1014	
12″ Thick							
	Demolish	S.F.		7	1.11	8.11	Includes minimum labor and equipment to remove 12″ block wall and haul debris to a central location for disposal.
	Minimum Charge	Job		875	139	1014	
Tree Removal							
Small, 4″ - 8″							
	Demolish	Ea.		112	102	214	Minimum charge, tree removal.
	Minimum Charge	Job		155		155	
Medium, 9″ - 12″							
	Demolish	Ea.		167	153	320	Minimum charge, tree removal.
	Minimum Charge	Job		155		155	
Large, 13″ - 18″							
	Demolish	Ea.		201	184	385	Minimum charge, tree removal.
	Minimum Charge	Job		520	225	745	
Very Large, 19″ - 24″							
	Demolish	Ea.		251	230	481	Minimum charge, tree removal.
	Minimum Charge	Job		520	225	745	
Stump Removal							
	Demolish	Ea.		70.50	166	236.50	Includes minimum charges for a site demolition crew and backhoe / loader.
	Minimum Charge	Job		288	112	400	

Backhoe/Loader - Wheel Type   Tractor Loader - Wheel Type   Backhoe - Crawler Type

Excavation		Unit	Material	Labor	Equip.	Total	Specification
Footings							
Hand Trenching	Install	L.F.		1.94		1.94	Includes hand excavation for footing or trench up to 10" deep and 12" wide.
	Minimum Charge	Job		77.50		77.50	
Machine Trenching	Install	L.F.		.74	.40	1.14	Includes machine excavation for footing or trench up to 10" deep and 12" wide.
	Minimum Charge	Job		370	202	572	
Trenching w / Backfill							
Hand	Install	L.F.		3.10		3.10	Includes hand excavation, backfill and compaction.
	Minimum Charge	Job		77.50		77.50	
Machine	Install	L.F.		3.71	18.45	22.16	Includes machine excavation, backfill and compaction.
	Minimum Charge	Job		155	38.50	193.50	

Backfill		Unit	Material	Labor	Equip.	Total	Specification
Hand (no compaction)	Install	L.C.Y.		26		26	Includes backfill by hand in average soil without compaction.
	Minimum Charge	Job		155		155	
Hand (w / compaction)	Install	E.C.Y.		36.50		36.50	Includes backfill by hand in average soil with compaction.
	Minimum Charge	Job		155		155	
Machine (no compaction)	Install	L.C.Y.		3.71	2.02	5.73	Includes backfill, by 55 HP wheel loader, of trenches from loose material piled adjacent without compaction.
	Minimum Charge	Job		370	202	572	
Machine (w / compaction)	Install	L.C.Y.		9.90	5.40	15.30	Includes backfill, by 55 HP wheel loader, of trenches from loose material piled adjacent with compaction by vibrating tampers.
	Minimum Charge	Job		370	202	572	

Foundation

Concrete Footing and Formwork

Wood Pile

Piles	Unit	Material	Labor	Equip.	Total	Specification
Treated Wood, 12″ Butt						
To 30′						
Demolish	L.F.		5	1.95	6.95	Includes material, labor and
Install	V.L.F.	21	4.60	3.33	28.93	equipment to install up to 30′ long
Demolish and Install	V.L.F.	21	9.60	5.28	35.88	treated wood piling.
Paint	L.F.	.42	2.06		2.48	
Minimum Charge	Job		2875	2075	4950	
30′ to 40′						
Demolish	L.F.		5	1.95	6.95	Includes material, labor and
Install	V.L.F.	22.50	4.10	2.98	29.58	equipment to install 30′ to 40′ long
Demolish and Install	V.L.F.	22.50	9.10	4.93	36.53	treated wood piling.
Paint	L.F.	.42	2.06		2.48	
Minimum Charge	Job		2875	2075	4950	
40′ to 50′						
Demolish	L.F.		5	1.95	6.95	Includes material, labor and
Install	V.L.F.	22.50	3.99	2.90	29.39	equipment to install 40′ to 50′ long
Demolish and Install	V.L.F.	22.50	8.99	4.85	36.34	treated wood piling.
Paint	L.F.	.42	2.06		2.48	
Minimum Charge	Job		2875	2075	4950	
50′ to 60′						
Demolish	L.F.		5	1.95	6.95	Includes material, labor and
Install	V.L.F.	23.50	3.59	2.61	29.70	equipment to install 50′ to 60′ long
Demolish and Install	V.L.F.	23.50	8.59	4.56	36.65	treated wood piling.
Paint	L.F.	.42	2.06		2.48	
Minimum Charge	Job		2875	2075	4950	
60′ to 80′						
Demolish	L.F.		5	1.95	6.95	Includes material, labor and
Install	V.L.F.	36	3.42	2.99	42.41	equipment to install 60′ to 80′ long
Demolish and Install	V.L.F.	36	8.42	4.94	49.36	treated wood piling.
Paint	L.F.	.42	2.06		2.48	
Minimum Charge	Job		2875	2075	4950	
Untreated Wood, 12″ Butt						
To 30′						
Demolish	L.F.		5	1.95	6.95	Includes material, labor and
Install	V.L.F.	22.50	4.60	3.33	30.43	equipment to install up to 30′ long
Demolish and Install	V.L.F.	22.50	9.60	5.28	37.38	untreated wood piling.
Paint	L.F.	.42	2.06		2.48	
Minimum Charge	Job		2875	2075	4950	

Foundation

Piles

	Unit	Material	Labor	Equip.	Total	Specification
30' to 40'						
Demolish	L.F.		5	1.95	6.95	Includes material, labor and
Install	V.L.F.	22.50	4.10	2.98	29.58	equipment to install 30' to 40' long
Demolish and Install	V.L.F.	22.50	9.10	4.93	36.53	untreated wood piling.
Paint	L.F.	.42	2.06		2.48	
Minimum Charge	Job		2875	2075	4950	
40' to 50'						
Demolish	L.F.		5	1.95	6.95	Includes material, labor and
Install	V.L.F.	22.50	3.99	2.90	29.39	equipment to install 40' to 50' long
Demolish and Install	V.L.F.	22.50	8.99	4.85	36.34	untreated wood piling.
Paint	L.F.	.42	2.06		2.48	
Minimum Charge	Job		2875	2075	4950	
50' to 60'						
Demolish	L.F.		5	1.95	6.95	Includes material, labor and
Install	V.L.F.	22.50	3.59	2.61	28.70	equipment to install 50' to 60' long
Demolish and Install	V.L.F.	22.50	8.59	4.56	35.65	untreated wood piling.
Paint	L.F.	.42	2.06		2.48	
Minimum Charge	Job		2875	2075	4950	
60' to 80'						
Demolish	L.F.		5	1.95	6.95	Includes material, labor and
Install	V.L.F.	28	3.42	2.48	33.90	equipment to install 60' to 80' long
Demolish and Install	V.L.F.	28	8.42	4.43	40.85	untreated wood piling.
Paint	L.F.	.42	2.06		2.48	
Minimum Charge	Job		2875	2075	4950	
Precast Concrete						
10" Square						
Demolish	L.F.		5.10	2	7.10	Includes material, labor and
Install	V.L.F.	19.20	4.10	2.98	26.28	equipment to install a precast,
Demolish and Install	V.L.F.	19.20	9.20	4.98	33.38	prestressed, 40' long, 10" square
Paint	L.F.	.42	2.06		2.48	concrete pile.
Minimum Charge	Job		2875	2075	4950	
12" Square						
Demolish	L.F.		5.50	2.14	7.64	Includes material, labor and
Install	V.L.F.	24	4.23	3.07	31.30	equipment to install a precast,
Demolish and Install	V.L.F.	24	9.73	5.21	38.94	prestressed, 40' long, 12" square
Paint	L.F.	.42	2.06		2.48	concrete pile.
Minimum Charge	Job		2875	2075	4950	
14" Square						
Demolish	L.F.		5.60	2.19	7.79	Includes material, labor and
Install	V.L.F.	28	4.79	3.47	36.26	equipment to install a precast,
Demolish and Install	V.L.F.	28	10.39	5.66	44.05	prestressed, 40' long, 14" square
Paint	L.F.	.42	2.06		2.48	concrete pile.
Minimum Charge	Job		2875	2075	4950	
16" Square						
Demolish	L.F.		5.90	2.30	8.20	Includes material, labor and
Install	V.L.F.	44.50	5.15	3.72	53.37	equipment to install a precast,
Demolish and Install	V.L.F.	44.50	11.05	6.02	61.57	prestressed, 40' long, 16" square
Paint	L.F.	.42	2.06		2.48	concrete pile.
Minimum Charge	Job		2875	2075	4950	
18" Square						
Demolish	L.F.		6.40	2.49	8.89	Includes material, labor and
Install	V.L.F.	54	5.55	4.84	64.39	equipment to install a precast,
Demolish and Install	V.L.F.	54	11.95	7.33	73.28	prestressed, 40' long, 18" square
Paint	L.F.	.42	2.06		2.48	concrete pile.
Minimum Charge	Job		2875	2075	4950	

Foundation

Piles

		Unit	Material	Labor	Equip.	Total	Specification
Cross Bracing							
Treated 2″ x 6″							
	Demolish	L.F.		1.01		1.01	Cost includes material and labor to
	Install	L.F.	.70	5.95		6.65	install 2″ x 6″ lumber for cross-bracing
	Demolish and Install	L.F.	.70	6.96		7.66	of foundation piling or posts.
	Reinstall	L.F.		4.74		4.74	
	Clean	S.F.		.39		.39	
	Paint	L.F.	.37	1.39		1.76	
	Minimum Charge	Job		735		735	
Treated 2″ x 8″							
	Demolish	L.F.		1.01		1.01	Cost includes material and labor to
	Install	L.F.	.95	6.10		7.05	install 2″ x 8″ lumber for cross-bracing
	Demolish and Install	L.F.	.95	7.11		8.06	of foundation piling or posts.
	Reinstall	L.F.		4.90		4.90	
	Clean	S.F.		.39		.39	
	Paint	L.F.	.50	1.44		1.94	
	Minimum Charge	Job		735		735	
Treated 2″ x 10″							
	Demolish	L.F.		1.04		1.04	Cost includes material and labor to
	Install	L.F.	1.24	6.35		7.59	install 2″ x 10″ lumber for
	Demolish and Install	L.F.	1.24	7.39		8.63	cross-bracing of foundation piling or
	Reinstall	L.F.		5.07		5.07	posts.
	Clean	S.F.		.39		.39	
	Paint	L.F.	.61	1.47		2.08	
	Minimum Charge	Job		735		735	
Treated 2″ x 12″							
	Demolish	L.F.		1.25		1.25	Cost includes material and labor to
	Install	L.F.	1.72	6.55		8.27	install 2″ x 12″ lumber for
	Demolish and Install	L.F.	1.72	7.80		9.52	cross-bracing of foundation piling or
	Reinstall	L.F.		5.25		5.25	posts.
	Clean	S.F.		.39		.39	
	Paint	L.F.	.72	1.50		2.22	
	Minimum Charge	Job		735		735	
Metal Pipe							
6″ Concrete Filled							
	Demolish	L.F.		7.10	2.77	9.87	Includes material, labor and
	Install	V.L.F.	27	6	4.34	37.34	equipment to install concrete filled
	Demolish and Install	V.L.F.	27	13.10	7.11	47.21	steel pipe pile.
	Paint	L.F.	.32	2.06		2.38	
	Minimum Charge	Job		2875	2075	4950	
6″ Unfilled							
	Demolish	L.F.		7.10	2.77	9.87	Includes material, labor and
	Install	V.L.F.	23	5.45	3.97	32.42	equipment to install unfilled steel pipe
	Demolish and Install	V.L.F.	23	12.55	6.74	42.29	pile.
	Paint	L.F.	.32	2.06		2.38	
	Minimum Charge	Job		2875	2075	4950	
8″ Concrete Filled							
	Demolish	L.F.		7.10	2.77	9.87	Includes material, labor and
	Install	V.L.F.	25	6.25	4.53	35.78	equipment to install concrete filled
	Demolish and Install	V.L.F.	25	13.35	7.30	45.65	steel pipe pile.
	Paint	L.F.	.32	2.06		2.38	
	Minimum Charge	Job		2875	2075	4950	
8″ Unfilled							
	Demolish	L.F.		7.10	2.77	9.87	Includes material, labor and
	Install	V.L.F.	23.50	5.75	4.17	33.42	equipment to install unfilled steel pipe
	Demolish and Install	V.L.F.	23.50	12.85	6.94	43.29	pile.
	Paint	L.F.	.32	2.06		2.38	
	Minimum Charge	Job		2875	2075	4950	

Foundation

Piles	Unit	Material	Labor	Equip.	Total	Specification
10" Concrete Filled						
Demolish	L.F.		7.35	2.86	10.21	Includes material, labor and
Install	V.L.F.	31.50	6.40	4.63	42.53	equipment to install concrete filled
Demolish and Install	V.L.F.	31.50	13.75	7.49	52.74	steel pipe pile.
Paint	L.F.	.32	2.06		2.38	
Minimum Charge	Job		2875	2075	4950	
10" Unfilled						
Demolish	L.F.		7.35	2.86	10.21	Includes material, labor and
Install	V.L.F.	29	5.75	4.17	38.92	equipment to install unfilled steel pipe
Demolish and Install	V.L.F.	29	13.10	7.03	49.13	pile.
Paint	L.F.	.32	2.06		2.38	
Minimum Charge	Job		2875	2075	4950	
12" Concrete Filled						
Demolish	L.F.		7.35	2.86	10.21	Includes material, labor and
Install	V.L.F.	37	6.90	5	48.90	equipment to install concrete filled
Demolish and Install	V.L.F.	37	14.25	7.86	59.11	steel pipe pile.
Paint	L.F.	.32	2.06		2.38	
Minimum Charge	Job		2875	2075	4950	
12" Unfilled						
Demolish	L.F.		7.35	2.86	10.21	Includes material, labor and
Install	V.L.F.	36	6.05	4.39	46.44	equipment to install unfilled steel pipe
Demolish and Install	V.L.F.	36	13.40	7.25	56.65	pile.
Paint	L.F.	.32	2.06		2.38	
Minimum Charge	Job		2875	2075	4950	
Steel H Section						
H Section 8 x 8 x 36#						
Demolish	L.F.		4.96	1.94	6.90	Includes material, labor and
Install	V.L.F.	20.50	4.49	3.26	28.25	equipment to install steel H section
Demolish and Install	V.L.F.	20.50	9.45	5.20	35.15	pile.
Paint	L.F.	.32	2.06		2.38	
Minimum Charge	Job		2875	2075	4950	
H Section 10 x 10 x 57#						
Demolish	L.F.		5.45	2.12	7.57	Includes material, labor and
Install	V.L.F.	32.50	4.71	3.42	40.63	equipment to install steel H section
Demolish and Install	V.L.F.	32.50	10.16	5.54	48.20	pile.
Paint	L.F.	.32	2.06		2.38	
Minimum Charge	Job		2875	2075	4950	
H Section 12 x 12 x 74#						
Demolish	L.F.		5.60	2.19	7.79	Includes material, labor and
Install	V.L.F.	43	4.87	4.26	52.13	equipment to install steel H section
Demolish and Install	V.L.F.	43	10.47	6.45	59.92	pile.
Paint	L.F.	.32	2.06		2.38	
Minimum Charge	Job		2875	2075	4950	
H Section 14 x 14 x 89#						
Demolish	L.F.		5.90	2.30	8.20	Includes material, labor and
Install	V.L.F.	51.50	5.30	4.66	61.46	equipment to install steel H section
Demolish and Install	V.L.F.	51.50	11.20	6.96	69.66	pile.
Paint	L.F.	.32	2.06		2.38	
Minimum Charge	Job		2875	2075	4950	

Foundation

Foundation Post

Foundation Post	Unit	Material	Labor	Equip.	Total	Specification
Wood Foundation Post						
4″ x 4″						
Demolish	L.F.		.81	.31	1.12	Cost includes material and labor to
Install	L.F.	1.27	5.15		6.42	install 4″ x 4″ treated wood post by
Demolish and Install	L.F.	1.27	5.96	.31	7.54	hand in normal soil conditions.
Reinstall	L.F.		4.13		4.13	
Minimum Charge	Job		735		735	
4″ x 6″						
Demolish	L.F.		.81	.31	1.12	Cost includes material and labor to
Install	L.F.	1.88	5.15		7.03	install 4″ x 6″ treated wood post by
Demolish and Install	L.F.	1.88	5.96	.31	8.15	hand in normal soil conditions.
Reinstall	L.F.		4.13		4.13	
Minimum Charge	Job		735		735	
6″ x 6″						
Demolish	L.F.		.81	.31	1.12	Cost includes material and labor to
Install	L.F.	2.81	5.15		7.96	install 6″ x 6″ treated wood post by
Demolish and Install	L.F.	2.81	5.96	.31	9.08	hand in normal soil conditions.
Reinstall	L.F.		4.13		4.13	
Minimum Charge	Job		735		735	
Water Jet						
Demolish	Ea.		91.50		91.50	Includes material, labor and
Install	Ea.	73.50	710		783.50	equipment to water jet a 6″ x 6″ wood
Demolish and Install	Ea.	73.50	801.50		875	post under an existing building. Cost
Reinstall	Ea.		707.66		707.66	may vary depending upon location
Minimum Charge	Job		620		620	and access.
Shim						
Install	Ea.	4.11	73.50		77.61	Includes labor and material to trim or
Minimum Charge	Job		735		735	readjust existing foundation posts
						under an existing building. Access of
						2 - 4 foot assumed.
Cross-bracing						
Treated 2″ x 6″						
Demolish	L.F.		1.01		1.01	Cost includes material and labor to
Install	L.F.	.70	5.95		6.65	install 2″ x 6″ lumber for cross-bracing
Demolish and Install	L.F.	.70	6.96		7.66	of foundation piling or posts.
Reinstall	L.F.		4.74		4.74	
Clean	S.F.		.39		.39	
Paint	L.F.	.37	1.39		1.76	
Minimum Charge	Job		735		735	
Treated 2″ x 8″						
Demolish	L.F.		1.01		1.01	Cost includes material and labor to
Install	L.F.	.95	6.10		7.05	install 2″ x 8″ lumber for cross-bracing
Demolish and Install	L.F.	.95	7.11		8.06	of foundation piling or posts.
Reinstall	L.F.		4.90		4.90	
Clean	S.F.		.39		.39	
Paint	L.F.	.50	1.44		1.94	
Minimum Charge	Job		735		735	
Treated 2″ x 10″						
Demolish	L.F.		1.04		1.04	Cost includes material and labor to
Install	L.F.	1.24	6.35		7.59	install 2″ x 10″ lumber for
Demolish and Install	L.F.	1.24	7.39		8.63	cross-bracing of foundation piling or
Reinstall	L.F.		5.07		5.07	posts.
Clean	S.F.		.39		.39	
Paint	L.F.	.61	1.47		2.08	
Minimum Charge	Job		735		735	

Foundation

Foundation Post

	Unit	Material	Labor	Equip.	Total	Specification
Treated 2″ x 12″						Cost includes material and labor to
Demolish	L.F.		1.25		1.25	install 2″ x 12″ lumber for
Install	L.F.	1.72	6.55		8.27	cross-bracing of foundation piling or
Demolish and Install	L.F.	1.72	7.80		9.52	posts.
Reinstall	L.F.		5.25		5.25	
Clean	S.F.		.39		.39	
Paint	L.F.	.72	1.50		2.22	
Minimum Charge	Job		735		735	

Concrete Footing

	Unit	Material	Labor	Equip.	Total	Specification
Continuous						
12″ w x 6″ d						
Demolish	L.F.		1.51	1.38	2.89	Includes material and labor to install
Install	L.F.	4.20	3.46	.02	7.68	continuous reinforced concrete footing
Demolish and Install	L.F.	4.20	4.97	1.40	10.57	poured by chute cast against earth
Minimum Charge	Job		845		845	including reinforcing, forming and
						finishing.
12″ w x 12″ d						
Demolish	L.F.		2.90	2.66	5.56	Includes material and labor to install
Install	L.F.	8.65	6.90	.04	15.59	continuous reinforced concrete footing
Demolish and Install	L.F.	8.65	9.80	2.70	21.15	poured by chute cast against earth
Minimum Charge	Job		845		845	including reinforcing, forming and
						finishing.
18″ w x 10″ d						
Demolish	L.F.		3.78	3.46	7.24	Includes material and labor to install
Install	L.F.	8.15	8.50	.06	16.71	continuous reinforced concrete footing
Demolish and Install	L.F.	8.15	12.28	3.52	23.95	poured by chute cast against earth
Minimum Charge	Job		845		845	including reinforcing, forming and
						finishing.
24″ w x 24″ d						
Demolish	L.F.		6.30	5.75	12.05	Includes material and labor to install
Install	L.F.	24	27.50	.18	51.68	continuous reinforced concrete footing
Demolish and Install	L.F.	24	33.80	5.93	63.73	poured by chute cast against earth
Minimum Charge	Job		845		845	including reinforcing, forming and
						finishing.
Stem Wall						
Single Story						
Demolish	L.F.		14	12.80	26.80	Includes material and labor to install
Install	L.F.	22	74	67.50	163.50	stem wall 24″ above and 18″ below
Demolish and Install	L.F.	22	88	80.30	190.30	grade for a single story structure.
Minimum Charge	Job		845		845	Includes 12″ wide and 8″ deep
						footing.
Two Story						
Demolish	L.F.		14	12.80	26.80	Includes material and labor to install
Install	L.F.	26	106	97	229	stem wall 24″ above and 18″ below
Demolish and Install	L.F.	26	120	109.80	255.80	grade typical for a two story structure,
Minimum Charge	Job		845		845	includes 18″ wide and 10″ deep
						footing.

Foundation

Concrete Slab		Unit	Material	Labor	Equip.	Total	Specification
Reinforced							
4″							
	Demolish	S.F.		2.59	.41	3	Cost includes material and labor to
	Install	S.F.	2.13	.40		2.53	install 4″ reinforced concrete
	Demolish and Install	S.F.	2.13	2.99	.41	5.53	slab-on-grade poured by chute
	Clean	S.F.	.11	.34		.45	including forms, vapor barrier, wire
	Paint	S.F.	.19	.78		.97	mesh, 3000 PSI concrete, float finish,
	Minimum Charge	Job		195		195	and curing.
6″							
	Demolish	S.F.		3.89	.62	4.51	Cost includes material and labor to
	Install	S.F.	2.90	.60	.01	3.51	install 6″ reinforced concrete
	Demolish and Install	S.F.	2.90	4.49	.63	8.02	slab-on-grade poured by chute
	Clean	S.F.	.11	.34		.45	including forms, vapor barrier, wire
	Paint	S.F.	.19	.78		.97	mesh, 3000 PSI concrete, float finish,
	Minimum Charge	Job		195		195	and curing.
8″							
	Demolish	S.F.		6.35	1.01	7.36	Cost includes material and labor to
	Install	S.F.	3.71	.78	.01	4.50	install 8″ reinforced concrete
	Demolish and Install	S.F.	3.71	7.13	1.02	11.86	slab-on-grade poured by chute
	Clean	S.F.	.11	.34		.45	including forms, vapor barrier, wire
	Paint	S.F.	.19	.78		.97	mesh, 3000 PSI concrete, float finish,
	Minimum Charge	Job		195		195	and curing.
Unreinforced							
4″							
	Demolish	S.F.		2.06	.33	2.39	Cost includes material and labor to
	Install	S.F.	1.42	.96	.01	2.39	install 4″ concrete slab-on-grade
	Demolish and Install	S.F.	1.42	3.02	.34	4.78	poured by chute, including forms,
	Clean	S.F.	.11	.34		.45	vapor barrier, 3000 PSI concrete, float
	Paint	S.F.	.19	.78		.97	finish, and curing.
	Minimum Charge	Job		195		195	
6″							
	Demolish	S.F.		2.92	.46	3.38	Cost includes material and labor to
	Install	S.F.	2.07	.98	.01	3.06	install 6″ concrete slab-on-grade
	Demolish and Install	S.F.	2.07	3.90	.47	6.44	poured by chute, including forms,
	Clean	S.F.	.11	.34		.45	vapor barrier, 3000 PSI concrete, float
	Paint	S.F.	.19	.78		.97	finish, and curing.
	Minimum Charge	Job		195		195	
8″							
	Demolish	S.F.		4.80	.76	5.56	Cost includes material and labor to
	Install	S.F.	2.84	1.03	.01	3.88	install 8″ concrete slab-on-grade
	Demolish and Install	S.F.	2.84	5.83	.77	9.44	poured by chute, including forms,
	Clean	S.F.	.11	.34		.45	vapor barrier, 3000 PSI concrete, float
	Paint	S.F.	.19	.78		.97	finish, and curing.
	Minimum Charge	Job		195		195	
Lightweight							
	Demolish	S.F.		.92	.27	1.19	Includes material and labor to install
	Install	S.F.	1.80	2.90		4.70	lightweight concrete, 3″ - 4″ thick.
	Demolish and Install	S.F.	1.80	3.82	.27	5.89	
	Clean	S.F.	.11	.34		.45	
	Paint	S.F.	.19	.78		.97	
	Minimum Charge	Job		195		195	

Foundation

Concrete Slab

	Unit	Material	Labor	Equip.	Total	Specification
Gunite						
Demolish	S.F.		1.15	.45	1.60	Includes material and labor to install
Install	S.F.	.36	1.24	.22	1.82	gunite on flat plane per inch of
Demolish and Install	S.F.	.36	2.39	.67	3.42	thickness using 1″ as base price. No
Clean	S.F.	.11	.34		.45	forms or reinforcing are included.
Paint	S.F.	.19	.78		.97	
Minimum Charge	Job		1100		1100	
Epoxy Inject						
Install	S.F.	5.15	6.50		11.65	Includes labor and material to inject
Minimum Charge	Job		195		195	epoxy into cracks.
Slabjacking						
Install	S.F.	.47	1.44	.12	2.03	Includes labor, material and
Minimum Charge	Job		195		195	equipment to install slabjacking under
						an existing building.
Scrape and Paint						
Paint	S.F.	.24	.90		1.14	Includes labor and materials to scrape
Minimum Charge	Job		180		180	and paint a concrete floor.
Vapor Barrier						
Install	S.F.	.03	.11		.14	Cost includes material and labor to
Minimum Charge	Job		211		211	install 4 mil polyethylene vapor barrier
						10′ wide sheets with 6″ overlaps.

Rough Framing

Beam / Girder		Unit	Material	Labor	Equip.	Total	Specification
Solid Wood Beam							
2″ x 8″							
	Demolish	L.F.		.83		.83	Cost includes material and labor to
	Install	L.F.	.87	1.30		2.17	install 2″ x 8″ beam.
	Demolish and Install	L.F.	.87	2.13		3	
	Clean	L.F.	.11	.28		.39	
	Paint	L.F.	.14	1.95		2.09	
	Minimum Charge	Job		211		211	
2″ x 10″							
	Demolish	L.F.		1.03		1.03	Cost includes material and labor to
	Install	L.F.	1.19	1.41		2.60	install 2″ x 10″ beam.
	Demolish and Install	L.F.	1.19	2.44		3.63	
	Clean	L.F.	.13	.33		.46	
	Paint	L.F.	.19	1.20		1.39	
	Minimum Charge	Job		211		211	
2″ x 12″							
	Demolish	L.F.		1.25		1.25	Cost includes material and labor to
	Install	L.F.	1.45	1.54		2.99	install 2″ x 12″ beam.
	Demolish and Install	L.F.	1.45	2.79		4.24	
	Clean	L.F.	.14	.39		.53	
	Paint	L.F.	.22	2.89		3.11	
	Minimum Charge	Job		211		211	
4″ x 8″							
	Demolish	L.F.		2.57		2.57	Cost includes material and labor to
	Install	L.F.	3.74	1.91	.71	6.36	install 4″ x 8″ beam.
	Demolish and Install	L.F.	3.74	4.48	.71	8.93	
	Clean	L.F.	.13	.33		.46	
	Paint	L.F.	.19	1.20		1.39	
	Minimum Charge	Job		211		211	
4″ x 10″							
	Demolish	L.F.		3.22		3.22	Cost includes material and labor to
	Install	L.F.	4.68	2.02	.75	7.45	install 4″ x 10″ beam.
	Demolish and Install	L.F.	4.68	5.24	.75	10.67	
	Clean	L.F.	.14	.39		.53	
	Paint	L.F.	.21	1.40		1.61	
	Minimum Charge	Job		211		211	
4″ x 12″							
	Demolish	L.F.		3.86		3.86	Cost includes material and labor to
	Install	L.F.	5.60	2.13	.79	8.52	install 4″ x 12″ beam.
	Demolish and Install	L.F.	5.60	5.99	.79	12.38	
	Clean	L.F.	.18	.44		.62	
	Paint	L.F.	.22	2.89		3.11	
	Minimum Charge	Job		211		211	

Rough Frame / Structure

Beam / Girder		Unit	Material	Labor	Equip.	Total	Specification
6″ x 8″							
	Demolish	L.F.		3.86		3.86	Cost includes material and labor to
	Install	L.F.	5.40	3.65	1.35	10.40	install 6″ x 8″ beam.
	Demolish and Install	L.F.	5.40	7.51	1.35	14.26	
	Clean	L.F.	.14	.39		.53	
	Paint	L.F.	.21	1.40		1.61	
	Minimum Charge	Job		211		211	
6″ x 10″							
	Demolish	L.F.		4.83		4.83	Cost includes material and labor to
	Install	L.F.	3.58	1.69		5.27	install 6″ x 10″ triple beam.
	Demolish and Install	L.F.	3.58	6.52		10.10	
	Clean	L.F.	.18	.44		.62	
	Paint	L.F.	.24	1.60		1.84	
	Minimum Charge	Job		211		211	
6″ x 12″							
	Demolish	L.F.		5.75		5.75	Cost includes material and labor to
	Install	L.F.	4.37	1.78		6.15	install 6″ x 12″ triple beam.
	Demolish and Install	L.F.	4.37	7.53		11.90	
	Clean	L.F.	.19	.50		.69	
	Paint	L.F.	.26	1.80		2.06	
	Minimum Charge	Job		211		211	
Steel I-type							
W8 x 31							
	Demolish	L.F.		9	3.56	12.56	Includes material and labor to install
	Install	L.F.	49	2.83	1.55	53.38	W shaped steel beam / girder.
	Demolish and Install	L.F.	49	11.83	5.11	65.94	
	Minimum Charge	Job		515		515	
W8 x 48							
	Demolish	L.F.		9	3.56	12.56	Includes material and labor to install
	Install	L.F.	75.50	2.96	1.63	80.09	W shaped steel beam / girder.
	Demolish and Install	L.F.	75.50	11.96	5.19	92.65	
	Minimum Charge	Job		515		515	
W8 x 67							
	Demolish	L.F.		9	3.56	12.56	Includes material and labor to install
	Install	L.F.	106	3.11	1.70	110.81	W shaped steel beam / girder.
	Demolish and Install	L.F.	106	12.11	5.26	123.37	
	Minimum Charge	Job		515		515	
W10 x 45							
	Demolish	L.F.		9	3.56	12.56	Includes material and labor to install
	Install	L.F.	71	2.96	1.63	75.59	W shaped steel beam / girder.
	Demolish and Install	L.F.	71	11.96	5.19	88.15	
	Minimum Charge	Job		515		515	
W10 x 68							
	Demolish	L.F.		9	3.56	12.56	Includes material and labor to install
	Install	L.F.	107	3.11	1.70	111.81	W shaped steel beam / girder.
	Demolish and Install	L.F.	107	12.11	5.26	124.37	
	Minimum Charge	Job		515		515	
W12 x 50							
	Demolish	L.F.		9	3.56	12.56	Includes material and labor to install
	Install	L.F.	78.50	2.96	1.63	83.09	W shaped steel beam / girder.
	Demolish and Install	L.F.	78.50	11.96	5.19	95.65	
	Minimum Charge	Job		515		515	
W12 x 87							
	Demolish	L.F.		13.20	5.25	18.45	Includes material and labor to install
	Install	L.F.	136	3.11	1.70	140.81	W shaped steel beam / girder.
	Demolish and Install	L.F.	136	16.31	6.95	159.26	
	Minimum Charge	Job		515		515	

Rough Frame / Structure

Beam / Girder

Beam / Girder		Unit	Material	Labor	Equip.	Total	Specification
W12 x 120							
	Demolish	L.F.		13.20	5.25	18.45	Includes material and labor to install
	Install	L.F.	189	3.18	1.75	193.93	W shaped steel beam / girder.
	Demolish and Install	L.F.	189	16.38	7	212.38	
	Minimum Charge	Job		515		515	
W14 x 74							
	Demolish	L.F.		13.20	5.25	18.45	Includes material and labor to install
	Install	L.F.	117	3.11	1.70	121.81	W shaped steel beam / girder.
	Demolish and Install	L.F.	117	16.31	6.95	140.26	
	Minimum Charge	Job		515		515	
W14 x 120							
	Demolish	L.F.		13.20	5.25	18.45	Includes material and labor to install
	Install	L.F.	189	3.18	1.75	193.93	W shaped steel beam / girder.
	Demolish and Install	L.F.	189	16.38	7	212.38	
	Minimum Charge	Job		515		515	

Column

Column		Unit	Material	Labor	Equip.	Total	Specification
Concrete							
Small Diameter							
	Demolish	L.F.		14.60	2.31	16.91	Includes material and labor to install
	Install	L.F.	154	13.45	2.67	170.12	small diameter concrete column.
	Demolish and Install	L.F.	154	28.05	4.98	187.03	
	Minimum Charge	Job		675	134	809	
Large Diameter							
	Demolish	L.F.		44	6.95	50.95	Includes material and labor to install
	Install	L.F.	275	15.85	3.14	293.99	large diameter concrete column.
	Demolish and Install	L.F.	275	59.85	10.09	344.94	
	Minimum Charge	Job		675	134	809	

Floor Framing System

Floor Framing System		Unit	Material	Labor	Equip.	Total	Specification
12″ O.C.							
2″ x 6″ Joists							
	Demolish	S.F.		.71		.71	Cost includes material and labor to
	Install	S.F.	.73	.68		1.41	install 2″ x 6″ joists, 12″ O.C.
	Demolish and Install	S.F.	.73	1.39		2.12	including box or band joist. Blocking
	Reinstall	S.F.		.54		.54	or bridging not included.
	Minimum Charge	Job		211		211	
2″ x 8″ Joists							
	Demolish	S.F.		.73		.73	Cost includes material and labor to
	Install	S.F.	1.02	.77		1.79	install 2″ x 8″ joists, 12″ O.C.
	Demolish and Install	S.F.	1.02	1.50		2.52	including box or band joist. Blocking
	Reinstall	S.F.		.62		.62	or bridging not included.
	Minimum Charge	Job		211		211	
2″ x 10″ Joists							
	Demolish	S.F.		.76		.76	Cost includes material and labor to
	Install	S.F.	1.40	.94		2.34	install 2″ x 10″ joists, 12″ O.C.
	Demolish and Install	S.F.	1.40	1.70		3.10	including box or band joist. Blocking
	Reinstall	S.F.		.75		.75	or bridging not included.
	Minimum Charge	Job		211		211	

Rough Frame / Structure

Floor Framing System		Unit	Material	Labor	Equip.	Total	Specification
2" x 12" Joists							
	Demolish	S.F.		.78		.78	Cost includes material and labor to
	Install	S.F.	1.71	.97		2.68	install 2" x 12" joists, 12" O.C.
	Demolish and Install	S.F.	1.71	1.75		3.46	including box or band joist. Blocking
	Reinstall	S.F.		.77		.77	or bridging not included.
	Minimum Charge	Job		211		211	
Block / Bridge							
	Demolish	Ea.		1.08		1.08	Includes material and labor to install
	Install	Ea.	1.60	2.38		3.98	set of cross bridging or per block of
	Demolish and Install	Ea.	1.60	3.46		5.06	solid bridging for 2" x 10" joists cut to
	Minimum Charge	Job		211		211	size on site.

16" O.C.
2" x 6" Joists

		Unit	Material	Labor	Equip.	Total	Specification
	Demolish	S.F.		.53		.53	Cost includes material and labor to
	Install	S.F.	.55	.51		1.06	install 2" x 6" joists including box or
	Demolish and Install	S.F.	.55	1.04		1.59	band joist installed 16" O.C. Does not
	Reinstall	S.F.		.41		.41	include beams, blocking or bridging.
	Minimum Charge	Job		211		211	

2" x 8" Joists

		Unit	Material	Labor	Equip.	Total	Specification
	Demolish	S.F.		.55		.55	Cost includes material and labor to
	Install	S.F.	.78	.58		1.36	install 2" x 8" joists including box or
	Demolish and Install	S.F.	.78	1.13		1.91	band joist installed 16" O.C. Does not
	Reinstall	S.F.		.46		.46	include beams, blocking or bridging.
	Minimum Charge	Job		211		211	

2" x 10" Joists

		Unit	Material	Labor	Equip.	Total	Specification
	Demolish	S.F.		.57		.57	Cost includes material and labor to
	Install	S.F.	1.07	.70		1.77	install 2" x 10" joists including box or
	Demolish and Install	S.F.	1.07	1.27		2.34	band joist 16" O.C. Does not include
	Reinstall	S.F.		.56		.56	beams, blocking or bridging.
	Minimum Charge	Job		211		211	

2" x 12" Joists

		Unit	Material	Labor	Equip.	Total	Specification
	Demolish	S.F.		.59		.59	Cost includes material and labor to
	Install	S.F.	1.31	.73		2.04	install 2" x 12" joists including box or
	Demolish and Install	S.F.	1.31	1.32		2.63	band joist installed 16" O.C. Does not
	Reinstall	S.F.		.58		.58	include beams, blocking or bridging.
	Minimum Charge	Job		211		211	

Block / Bridge

		Unit	Material	Labor	Equip.	Total	Specification
	Demolish	Ea.		1.08		1.08	Includes material and labor to install
	Install	Ea.	1.60	2.38		3.98	set of cross bridging or per block of
	Demolish and Install	Ea.	1.60	3.46		5.06	solid bridging for 2" x 10" joists cut to
	Minimum Charge	Job		211		211	size on site.

24" O.C.
2" x 6" Joists

		Unit	Material	Labor	Equip.	Total	Specification
	Demolish	S.F.		.35		.35	Cost includes material and labor to
	Install	S.F.	.40	.34		.74	install 2" x 6" joists including box or
	Demolish and Install	S.F.	.40	.69		1.09	band joist installed 24" O.C. Does not
	Reinstall	S.F.		.27		.27	include beams, blocking or bridging.
	Minimum Charge	Job		211		211	

2" x 8" Joists

		Unit	Material	Labor	Equip.	Total	Specification
	Demolish	S.F.		.37		.37	Cost includes material and labor to
	Install	S.F.	.56	.38		.94	install 2" x 8" joists including box or
	Demolish and Install	S.F.	.56	.75		1.31	band joist installed 24" O.C. Does not
	Reinstall	S.F.		.31		.31	include beams, blocking or bridging.
	Minimum Charge	Job		211		211	

Rough Frame / Structure

Floor Framing System		Unit	Material	Labor	Equip.	Total	Specification
2" x 10" Joists							
	Demolish	S.F.		.38		.38	Cost includes material and labor to
	Install	S.F.	.76	.47		1.23	install 2" x 10" joists including box or
	Demolish and Install	S.F.	.76	.85		1.61	band joist installed 24" O.C. Does not
	Reinstall	S.F.		.38		.38	include beams, blocking or bridging.
	Minimum Charge	Job		211		211	
2" x 12" Joists							
	Demolish	S.F.		.39		.39	Cost includes material and labor to
	Install	S.F.	.94	.48		1.42	install 2" x 12" joists including box or
	Demolish and Install	S.F.	.94	.87		1.81	band joist installed 24" O.C. Does not
	Reinstall	S.F.		.39		.39	include beams, blocking or bridging.
	Minimum Charge	Job		211		211	
Block / Bridge							
	Demolish	Ea.		1.08		1.08	Includes material and labor to install
	Install	Ea.	1.60	2.38		3.98	set of cross bridging or per block of
	Demolish and Install	Ea.	1.60	3.46		5.06	solid bridging for 2" x 10" joists cut to
	Minimum Charge	Job		211		211	size on site.
Engineered Lumber, Joist							
9-1/2"							
	Demolish	S.F.		.57		.57	Includes material and labor to install
	Install	S.F.	1.65	.42		2.07	engineered lumber truss / joists per
	Demolish and Install	S.F.	1.65	.99		2.64	S.F. of floor area based on joists 16"
	Reinstall	S.F.		.34		.34	O.C. Beams, supports and bridging
	Minimum Charge	Job		211		211	are not included.
11-7/8"							
	Demolish	S.F.		.59		.59	Includes material and labor to install
	Install	S.F.	1.86	.43		2.29	engineered lumber truss / joists per
	Demolish and Install	S.F.	1.86	1.02		2.88	S.F. of floor area based on joists 16"
	Reinstall	S.F.		.35		.35	O.C. Beams, supports and bridging
	Minimum Charge	Job		211		211	are not included.
14"							
	Demolish	S.F.		.61		.61	Includes material and labor to install
	Install	S.F.	2.49	.46		2.95	engineered lumber truss / joists per
	Demolish and Install	S.F.	2.49	1.07		3.56	S.F. of floor area based on joists 16"
	Reinstall	S.F.		.37		.37	O.C. Beams, supports and bridging
	Minimum Charge	Job		211		211	are not included.
16"							
	Demolish	S.F.		.63		.63	Includes material and labor to install
	Install	S.F.	3.09	.49		3.58	engineered lumber truss / joists per
	Demolish and Install	S.F.	3.09	1.12		4.21	S.F. of floor area based on joists 16"
	Reinstall	S.F.		.39		.39	O.C. Beams, supports and bridging
	Minimum Charge	Job		211		211	are not included.
Block / Bridge							
	Demolish	Pr.		1.08		1.08	Includes material and labor to install
	Install	Pr.	2.04	3.13		5.17	set of steel, one-nail type cross
	Demolish and Install	Pr.	2.04	4.21		6.25	bridging for trusses placed 16" O.C.
	Minimum Charge	Job		211		211	
Block / Bridge							
2" x 6"							
	Demolish	Ea.		1.08		1.08	Includes material and labor to install
	Install	Ea.	.83	1.90		2.73	set of cross bridging or per block of
	Demolish and Install	Ea.	.83	2.98		3.81	solid bridging for 2" x 6" joists cut to
	Minimum Charge	Job		211		211	size on site.

Floor Framing System		Unit	Material	Labor	Equip.	Total	Specification
2″ x 8″							
	Demolish	Ea.		1.08		1.08	Includes material and labor to install
	Install	Ea.	1.17	2.11		3.28	set of cross bridging or per block of
	Demolish and Install	Ea.	1.17	3.19		4.36	solid bridging for 2″ x 8″ joists cut to
	Minimum Charge	Job		211		211	size on site.
2″ x 10″							
	Demolish	Ea.		1.08		1.08	Includes material and labor to install
	Install	Ea.	1.60	2.38		3.98	set of cross bridging or per block of
	Demolish and Install	Ea.	1.60	3.46		5.06	solid bridging for 2″ x 10″ joists cut to
	Minimum Charge	Job		211		211	size on site.
2″ x 12″							
	Demolish	Ea.		1.08		1.08	Includes material and labor to install
	Install	Ea.	1.95	2.80		4.75	set of cross bridging or per block of
	Demolish and Install	Ea.	1.95	3.88		5.83	solid bridging for 2″ x 12″ joists cut to
	Minimum Charge	Job		211		211	size on site.
Ledger Strips							
1″ x 2″							
	Demolish	L.F.		.29		.29	Cost includes material and labor to
	Install	L.F.	.28	1.23		1.51	install 1″ x 2″ ledger strip nailed to the
	Demolish and Install	L.F.	.28	1.52		1.80	face of studs, beams or joist.
	Minimum Charge	Job		211		211	
1″ x 3″							
	Demolish	L.F.		.29		.29	Includes material and labor to install
	Install	L.F.	.45	1.41		1.86	up to 1″ x 4″ ledger strip nailed to the
	Demolish and Install	L.F.	.45	1.70		2.15	face of studs, beams or joist.
	Minimum Charge	Job		211		211	
1″ x 4″							
	Demolish	L.F.		.29		.29	Includes material and labor to install
	Install	L.F.	.45	1.41		1.86	up to 1″ x 4″ ledger strip nailed to the
	Demolish and Install	L.F.	.45	1.70		2.15	face of studs, beams or joist.
	Minimum Charge	Job		211		211	
2″ x 2″							
	Demolish	L.F.		.31		.31	Cost includes material and labor to
	Install	L.F.	.36	1.28		1.64	install 2″ x 2″ ledger strip nailed to the
	Demolish and Install	L.F.	.36	1.59		1.95	face of studs, beams or joist.
	Minimum Charge	Job		211		211	
2″ x 4″							
	Demolish	L.F.		.34		.34	Cost includes material and labor to
	Install	L.F.	.40	1.69		2.09	install 2″ x 4″ ledger strip nailed to the
	Demolish and Install	L.F.	.40	2.03		2.43	face of studs, beams or joist.
	Reinstall	L.F.		1.35		1.35	
	Minimum Charge	Job		211		211	
Ledger Boards							
2″ x 6″							
	Demolish	L.F.		.34		.34	Cost includes material and labor to
	Install	L.F.	4.65	2.64		7.29	install 2″ x 6″ ledger board fastened to
	Demolish and Install	L.F.	4.65	2.98		7.63	a wall, joists or studs.
	Reinstall	L.F.		2.11		2.11	
	Minimum Charge	Job		211		211	
4″ x 6″							
	Demolish	L.F.		.57		.57	Cost includes material and labor to
	Install	L.F.	7.90	3.02		10.92	install 4″ x 6″ ledger board fastened to
	Demolish and Install	L.F.	7.90	3.59		11.49	a wall, joists or studs.
	Reinstall	L.F.		2.42		2.42	
	Minimum Charge	Job		211		211	

Rough Frame / Structure

Floor Framing System

Floor Framing System		Unit	Material	Labor	Equip.	Total	Specification
4″ x 8″							
	Demolish	L.F.		.76		.76	Cost includes material and labor to
	Install	L.F.	9	3.52		12.52	install 4″ x 8″ ledger board fastened to
	Demolish and Install	L.F.	9	4.28		13.28	a wall, joists or studs.
	Reinstall	L.F.		2.82		2.82	
	Minimum Charge	Job		211		211	
Sill Plate Per L.F.							
2″ x 4″							
	Demolish	L.F.		.29		.29	Cost includes material and labor to
	Install	L.F.	.46	1.54		2	install 2″ x 4″ pressure treated lumber,
	Demolish and Install	L.F.	.46	1.83		2.29	drilled and installed on foundation
	Reinstall	L.F.		1.23		1.23	bolts 48″ O.C. Bolts, nuts and washers
	Minimum Charge	Job		211		211	are not included.
2″ x 6″							
	Demolish	L.F.		.44		.44	Cost includes material and labor to
	Install	L.F.	.70	1.69		2.39	install 2″ x 6″ pressure treated lumber,
	Demolish and Install	L.F.	.70	2.13		2.83	drilled and installed on foundation
	Reinstall	L.F.		1.35		1.35	bolts 48″ O.C. Bolts, nuts and washers
	Minimum Charge	Job		211		211	not included.
2″ x 8″							
	Demolish	L.F.		.59		.59	Cost includes material and labor to
	Install	L.F.	.95	1.88		2.83	install 2″ x 8″ pressure treated lumber,
	Demolish and Install	L.F.	.95	2.47		3.42	drilled and installed on foundation
	Reinstall	L.F.		1.50		1.50	bolts 48″ O.C. Bolts, nuts and washers
	Minimum Charge	Job		211		211	not included.
Earthquake Strapping							
	Install	Ea.	2.99	2.64		5.63	Includes labor and material to replace
	Minimum Charge	Job		211		211	an earthquake strap.

Subflooring

Subflooring		Unit	Material	Labor	Equip.	Total	Specification
Plywood							
1/2″							
	Demolish	S.F.		.90		.90	Cost includes material and labor to
	Install	SF Flr.	.48	.45		.93	install 1/2″ plywood subfloor, CD
	Demolish and Install	SF Flr.	.48	1.35		1.83	standard interior grade, plugged and
	Minimum Charge	Job		211		211	touch sanded.
5/8″							
	Demolish	S.F.		.91		.91	Cost includes material and labor to
	Install	SF Flr.	.61	.51		1.12	install 5/8″ plywood subfloor, CD
	Demolish and Install	SF Flr.	.61	1.42		2.03	standard interior grade, plugged and
	Reinstall	SF Flr.		.40		.40	touch sanded.
	Minimum Charge	Job		211		211	
3/4″							
	Demolish	S.F.		.92		.92	Cost includes material and labor to
	Install	SF Flr.	.64	.55		1.19	install 3/4″ plywood subfloor, CD
	Demolish and Install	SF Flr.	.64	1.47		2.11	standard interior grade, plugged and
	Reinstall	SF Flr.		.44		.44	touch sanded.
	Minimum Charge	Job		211		211	

Rough Frame / Structure

Subflooring		Unit	Material	Labor	Equip.	Total	Specification
Particle Board							
1/2"							
	Demolish	S.F.		.45		.45	Cost includes material and labor to
	Install	SF Flr.	.45	.56		1.01	install 1/2" particle board subfloor.
	Demolish and Install	SF Flr.	.45	1.01		1.46	
	Minimum Charge	Job		211		211	
5/8"							
	Demolish	S.F.		.45		.45	Cost includes material and labor to
	Install	SF Flr.	.55	.62		1.17	install 5/8" particle board subfloor.
	Demolish and Install	SF Flr.	.55	1.07		1.62	
	Minimum Charge	Job		211		211	
3/4"							
	Demolish	S.F.		.46		.46	Cost includes material and labor to
	Install	SF Flr.	.75	.68		1.43	install 3/4" particle board subfloor.
	Demolish and Install	SF Flr.	.75	1.14		1.89	
	Minimum Charge	Job		211		211	
Plank Board							
	Demolish	S.F.		1.45		1.45	Cost includes material and labor to
	Install	SF Flr.	1.98	.94		2.92	install 1" x 6" standard grade plank
	Demolish and Install	SF Flr.	1.98	2.39		4.37	flooring.
	Reinstall	SF Flr.		.94		.94	
	Minimum Charge	Job		211		211	
Felt Underlay							
	Demolish	S.F.		.08		.08	Includes material and labor to install
	Install	S.F.	.06	.11		.17	#15 felt building paper.
	Demolish and Install	S.F.	.06	.19		.25	
	Minimum Charge	Job		211		211	
Prep (for flooring)							
	Install	S.F.		.35		.35	Preparation of subflooring for
	Minimum Charge	Job		186		186	installation of new finished flooring. Cost reflects average time to prep area.

Wall Framing System		Unit	Material	Labor	Equip.	Total	Specification
2" x 4"							
8' High							
	Demolish	L.F.		4.59		4.59	Cost includes material and labor to
	Install	L.F.	4.04	4.97		9.01	install 2" x 4" x 8' high wall system,
	Demolish and Install	L.F.	4.04	9.56		13.60	including studs, treated bottom plate,
	Reinstall	L.F.		3.98		3.98	double top plate and one row of fire
	Minimum Charge	Job		211		211	blocking.
9' High							
	Demolish	L.F.		5.15		5.15	Cost includes material and labor to
	Install	L.F.	4.33	4.97		9.30	install 2" x 4" x 9' wall system
	Demolish and Install	L.F.	4.33	10.12		14.45	including studs, treated bottom plate,
	Reinstall	L.F.		3.98		3.98	double top plate and one row of fire
	Minimum Charge	Job		211		211	blocking.
10' High							
	Demolish	L.F.		5.75		5.75	Cost includes material and labor to
	Install	L.F.	4.63	4.97		9.60	install 2" x 4" x 10' wall system
	Demolish and Install	L.F.	4.63	10.72		15.35	including studs, treated bottom plate,
	Reinstall	L.F.		3.98		3.98	double top plate and one row of fire
	Minimum Charge	Job		211		211	blocking.

Wall Framing System	Unit	Material	Labor	Equip.	Total	Specification
12' High						
Demolish	L.F.		6.90		6.90	Cost includes material and labor to
Install	L.F.	5.25	6.05		11.30	install 2" x 4" x 12' wall system
Demolish and Install	L.F.	5.25	12.95		18.20	including studs, treated bottom plate,
Reinstall	L.F.		4.83		4.83	double top plate and one row of fire
Minimum Charge	Job		211		211	blocking.
2" x 6"						
8' High						
Demolish	L.F.		4.30		4.30	Cost includes material and labor to
Install	L.F.	6.25	9.20		15.45	install 2" x 6" x 8' wall system
Demolish and Install	L.F.	6.25	13.50		19.75	including studs, treated bottom plate,
Reinstall	L.F.		7.35		7.35	double top plate and one row of fire
Minimum Charge	Job		211		211	blocking.
9' High						
Demolish	L.F.		4.85		4.85	Cost includes material and labor to
Install	L.F.	6.70	9.20		15.90	install 2" x 6" x 9' wall system
Demolish and Install	L.F.	6.70	14.05		20.75	including studs, treated bottom plate,
Reinstall	L.F.		7.35		7.35	double top plate and one row of fire
Minimum Charge	Job		211		211	blocking.
10' High						
Demolish	L.F.		5.30		5.30	Cost includes material and labor to
Install	L.F.	7.20	9.20		16.40	install 2" x 6" x 10' wall system
Demolish and Install	L.F.	7.20	14.50		21.70	including studs, treated bottom plate,
Reinstall	L.F.		7.35		7.35	double top plate and one row of fire
Minimum Charge	Job		211		211	blocking.
12' High						
Demolish	L.F.		6.25		6.25	Cost includes material and labor to
Install	L.F.	8.10	11.30		19.40	install 2" x 6" x 12' wall system
Demolish and Install	L.F.	8.10	17.55		25.65	including studs, treated bottom plate,
Reinstall	L.F.		9.04		9.04	double top plate and one row of fire
Minimum Charge	Job		211		211	blocking.
Fireblock						
2" x 4", 16" O.C. System						
Demolish	L.F.		.57		.57	Cost includes material and labor to
Install	L.F.	.40	1.41		1.81	install 2" x 4" fireblocks in wood frame
Demolish and Install	L.F.	.40	1.98		2.38	walls per L.F. of wall to be blocked.
Reinstall	L.F.		1.13		1.13	
Minimum Charge	Job		211		211	
2" x 4", 24" O.C. System						
Demolish	L.F.		.57		.57	Cost includes material and labor to
Install	L.F.	.40	1.41		1.81	install 2" x 4" fireblocks in wood frame
Demolish and Install	L.F.	.40	1.98		2.38	walls per L.F. of wall to be blocked.
Reinstall	L.F.		1.13		1.13	
Minimum Charge	Job		211		211	
2" x 6", 16" O.C. System						
Demolish	L.F.		.86		.86	Cost includes material and labor to
Install	L.F.	.62	1.41		2.03	install 2" x 6" fireblocks in wood frame
Demolish and Install	L.F.	.62	2.27		2.89	walls per L.F. of wall to be blocked.
Reinstall	L.F.		1.13		1.13	
Minimum Charge	Job		211		211	
2" x 6", 24" O.C. System						
Demolish	L.F.		.86		.86	Cost includes material and labor to
Install	L.F.	.62	1.41		2.03	install 2" x 6" fireblocks in wood frame
Demolish and Install	L.F.	.62	2.27		2.89	walls per L.F. of wall to be blocked.
Reinstall	L.F.		1.13		1.13	
Minimum Charge	Job		211		211	

Rough Frame / Structure

Wall Framing System		Unit	Material	Labor	Equip.	Total	Specification
Bracing							
1″ x 3″							
	Demolish	L.F.		.33		.33	Cost includes material and labor to
	Install	L.F.	.37	2.64		3.01	install 1″ x 3″ let-in bracing.
	Demolish and Install	L.F.	.37	2.97		3.34	
	Minimum Charge	Job		211		211	
1″ x 4″							
	Demolish	L.F.		.33		.33	Cost includes material and labor to
	Install	L.F.	.45	2.64		3.09	install 1″ x 4″ let-in bracing.
	Demolish and Install	L.F.	.45	2.97		3.42	
	Minimum Charge	Job		211		211	
1″ x 6″							
	Demolish	L.F.		.33		.33	Cost includes material and labor to
	Install	L.F.	.91	2.82		3.73	install 1″ x 6″ let-in bracing.
	Demolish and Install	L.F.	.91	3.15		4.06	
	Minimum Charge	Job		211		211	
2″ x 3″							
	Demolish	L.F.		.43		.43	Cost includes material and labor to
	Install	L.F.	.36	2.82		3.18	install 2″ x 3″ let-in bracing.
	Demolish and Install	L.F.	.36	3.25		3.61	
	Reinstall	L.F.		2.25		2.25	
	Minimum Charge	Job		211		211	
2″ x 4″							
	Demolish	L.F.		.43		.43	Cost includes material and labor to
	Install	L.F.	.40	2.82		3.22	install 2″ x 4″ let-in bracing.
	Demolish and Install	L.F.	.40	3.25		3.65	
	Reinstall	L.F.		2.25		2.25	
	Minimum Charge	Job		211		211	
2″ x 6″							
	Demolish	L.F.		.43		.43	Cost includes material and labor to
	Install	L.F.	.62	3.02		3.64	install 2″ x 6″ let-in bracing.
	Demolish and Install	L.F.	.62	3.45		4.07	
	Reinstall	L.F.		2.42		2.42	
	Minimum Charge	Job		211		211	
2″ x 8″							
	Demolish	L.F.		.43		.43	Cost includes material and labor to
	Install	L.F.	.87	3.02		3.89	install 2″ x 8″ let-in bracing.
	Demolish and Install	L.F.	.87	3.45		4.32	
	Reinstall	L.F.		2.42		2.42	
	Minimum Charge	Job		211		211	
Earthquake Strapping							
	Install	Ea.	2.99	2.64		5.63	Includes labor and material to replace
	Minimum Charge	Job		211		211	an earthquake strap.
Hurricane Clips							
	Install	Ea.	1.12	2.92		4.04	Includes labor and material to install a
	Minimum Charge	Job		211		211	hurricane clip.
Stud							
2″ x 4″							
	Demolish	L.F.		.34		.34	Cost includes material and labor to
	Install	L.F.	.40	.77		1.17	install 2″ x 4″ wall stud per L.F. of stud.
	Demolish and Install	L.F.	.40	1.11		1.51	
	Reinstall	L.F.		.61		.61	
	Minimum Charge	Job		211		211	

Rough Frame / Structure

Wall Framing System		Unit	Material	Labor	Equip.	Total	Specification
2" x 6"							
	Demolish	L.F.		.43		.43	Cost includes material and labor to
	Install	L.F.	.62	.85		1.47	install 2" x 6" stud per L.F. of stud.
	Demolish and Install	L.F.	.62	1.28		1.90	
	Reinstall	L.F.		.68		.68	
	Minimum Charge	Job		211		211	
Plates							
2" x 4"							
	Demolish	L.F.		.31		.31	Cost includes material and labor to
	Install	L.F.	.40	1.06		1.46	install 2" x 4" plate per L.F. of plate.
	Demolish and Install	L.F.	.40	1.37		1.77	
	Reinstall	L.F.		.85		.85	
	Minimum Charge	Job		211		211	
2" x 6"							
	Demolish	L.F.		.33		.33	Cost includes material and labor to
	Install	L.F.	.62	1.13		1.75	install 2" x 6" plate per L.F. of plate.
	Demolish and Install	L.F.	.62	1.46		2.08	
	Reinstall	L.F.		.90		.90	
	Minimum Charge	Job		211		211	
Headers							
2" x 6"							
	Demolish	L.F.		.31		.31	Includes material and labor to install
	Install	L.F.	.62	2.35		2.97	header over wall openings and
	Demolish and Install	L.F.	.62	2.66		3.28	around floor, ceiling and roof
	Clean	L.F.	.01	.39		.40	openings or flush beam.
	Paint	L.F.	.11	1.44		1.55	
	Minimum Charge	Job		211		211	
2" x 8"							
	Demolish	L.F.		.33		.33	Includes material and labor to install
	Install	L.F.	.87	2.49		3.36	header over wall openings and
	Demolish and Install	L.F.	.87	2.82		3.69	around floor, ceiling and roof
	Clean	L.F.	.02	.39		.41	openings or flush beam.
	Paint	L.F.	.14	1.95		2.09	
	Minimum Charge	Job		211		211	
2" x 10"							
	Demolish	L.F.		.33		.33	Includes material and labor to install
	Install	L.F.	1.19	2.64		3.83	header over wall openings and
	Demolish and Install	L.F.	1.19	2.97		4.16	around floor, ceiling and roof
	Clean	L.F.	.02	.39		.41	openings or flush beam.
	Paint	L.F.	.18	2.41		2.59	
	Minimum Charge	Job		211		211	
2" x 12"							
	Demolish	L.F.		.34		.34	Includes material and labor to install
	Install	L.F.	1.45	2.82		4.27	header over wall openings and
	Demolish and Install	L.F.	1.45	3.16		4.61	around floor, ceiling and roof
	Clean	L.F.	.03	.40		.43	openings or flush beam.
	Paint	L.F.	.22	2.89		3.11	
	Minimum Charge	Job		211		211	
4" x 8"							
	Demolish	L.F.		.61		.61	Includes material and labor to install
	Install	L.F.	3.74	3.25		6.99	header over wall openings and
	Demolish and Install	L.F.	3.74	3.86		7.60	around floor, ceiling and roof
	Clean	L.F.	.02	.39		.41	openings or flush beam.
	Paint	L.F.	.14	1.95		2.09	
	Minimum Charge	Job		211		211	

Rough Frame / Structure

Wall Framing System		Unit	Material	Labor	Equip.	Total	Specification
4" x 10"							
	Demolish	L.F.		.66		.66	Includes material and labor to install
	Install	L.F.	4.68	3.52		8.20	header over wall openings and
	Demolish and Install	L.F.	4.68	4.18		8.86	around floor, ceiling and roof
	Clean	L.F.	.02	.39		.41	openings or flush beam.
	Paint	L.F.	.18	2.41		2.59	
	Minimum Charge	Job		211		211	
4" x 12"							
	Demolish	L.F.		.69		.69	Includes material and labor to install
	Install	L.F.	5.60	4.45		10.05	header over wall openings and
	Demolish and Install	L.F.	5.60	5.14		10.74	around floor, ceiling and roof
	Clean	L.F.	.03	.40		.43	openings or flush beam.
	Paint	L.F.	.22	2.89		3.11	
	Minimum Charge	Job		211		211	
6" x 8"							
	Demolish	L.F.		.61		.61	Includes material and labor to install
	Install	L.F.	5.40	2.35		7.75	header over wall openings and
	Demolish and Install	L.F.	5.40	2.96		8.36	around floor, ceiling and roof
	Clean	L.F.	.02	.39		.41	openings or flush beam.
	Paint	L.F.	.14	1.95		2.09	
	Minimum Charge	Job		211		211	
6" x 10"							
	Demolish	L.F.		.66		.66	Includes material and labor to install
	Install	L.F.	7.55	5.10		12.65	header over wall openings and
	Demolish and Install	L.F.	7.55	5.76		13.31	around floor, ceiling and roof
	Clean	L.F.	.02	.39		.41	openings or flush beam.
	Paint	L.F.	.18	2.41		2.59	
	Minimum Charge	Job		211		211	
6" x 12"							
	Demolish	L.F.		.69		.69	Includes material and labor to install
	Install	L.F.	11.15	6.05		17.20	header over wall openings and
	Demolish and Install	L.F.	11.15	6.74		17.89	around floor, ceiling and roof
	Clean	L.F.	.03	.40		.43	openings or flush beam.
	Paint	L.F.	.22	1		1.22	
	Minimum Charge	Job		211		211	

Rough-in Opening		Unit	Material	Labor	Equip.	Total	Specification
Door w / 2" x 4" Lumber							
3' Wide							
	Demolish	Ea.		10.75		10.75	Includes material and labor to install
	Install	Ea.	16.70	13.20		29.90	header, double studs each side,
	Demolish and Install	Ea.	16.70	23.95		40.65	cripples, blocking and nails, up to 3'
	Reinstall	Ea.		10.57		10.57	opening in 2" x 4" stud wall 8' high.
	Minimum Charge	Job		211		211	
4' Wide							
	Demolish	Ea.		10.75		10.75	Includes material and labor to install
	Install	Ea.	18	13.20		31.20	header, double studs each side,
	Demolish and Install	Ea.	18	23.95		41.95	cripples, blocking and nails, up to 4'
	Reinstall	Ea.		10.57		10.57	opening in 2" x 4" stud wall 8' high.
	Minimum Charge	Job		211		211	
5' Wide							
	Demolish	Ea.		10.75		10.75	Includes material and labor to install
	Install	Ea.	22	13.20		35.20	header, double studs each side,
	Demolish and Install	Ea.	22	23.95		45.95	cripples, blocking and nails, up to 5'
	Reinstall	Ea.		10.57		10.57	opening in 2" x 4" stud wall 8' high.
	Minimum Charge	Job		211		211	

Rough Frame / Structure

Rough-in Opening		Unit	Material	Labor	Equip.	Total	Specification
6' Wide							
	Demolish	Ea.		10.75		10.75	Includes material and labor to install
	Install	Ea.	23.50	13.20		36.70	header, double studs each side,
	Demolish and Install	Ea.	23.50	23.95		47.45	cripples, blocking and nails, up to 6'
	Reinstall	Ea.		10.57		10.57	opening in 2" x 4" stud wall 8' high.
	Minimum Charge	Job		211		211	
8' Wide							
	Demolish	Ea.		11.45		11.45	Includes material and labor to install
	Install	Ea.	32.50	14.10		46.60	header, double studs each side,
	Demolish and Install	Ea.	32.50	25.55		58.05	cripples, blocking and nails, up to 8'
	Reinstall	Ea.		11.27		11.27	opening in 2" x 4" stud wall 8' high.
	Minimum Charge	Job		211		211	
10' Wide							
	Demolish	Ea.		11.45		11.45	Includes material and labor to install
	Install	Ea.	42.50	14.10		56.60	header, double studs each side,
	Demolish and Install	Ea.	42.50	25.55		68.05	cripples, blocking and nails, up to 10'
	Reinstall	Ea.		11.27		11.27	opening in 2" x 4" stud wall 8' high.
	Minimum Charge	Job		211		211	
12' Wide							
	Demolish	Ea.		11.45		11.45	Includes material and labor to install
	Install	Ea.	59.50	14.10		73.60	header, double studs each side,
	Demolish and Install	Ea.	59.50	25.55		85.05	cripples, blocking and nails, up to 12'
	Reinstall	Ea.		11.27		11.27	opening in 2" x 4" stud wall 8' high.
	Minimum Charge	Job		211		211	
Door w / 2" x 6" Lumber							
3' Wide							
	Demolish	Ea.		10.75		10.75	Includes material and labor to install
	Install	Ea.	23.50	13.20		36.70	header, double studs each side,
	Demolish and Install	Ea.	23.50	23.95		47.45	cripples, blocking and nails, up to 3'
	Reinstall	Ea.		10.57		10.57	opening in 2" x 6" stud wall 8' high.
	Minimum Charge	Job		211		211	
4' Wide							
	Demolish	Ea.		10.75		10.75	Includes material and labor to install
	Install	Ea.	25	13.20		38.20	header, double studs each side,
	Demolish and Install	Ea.	25	23.95		48.95	cripples, blocking and nails, up to 4'
	Reinstall	Ea.		10.57		10.57	opening in 2" x 6" stud wall 8' high.
	Minimum Charge	Job		211		211	
5' Wide							
	Demolish	Ea.		10.75		10.75	Includes material and labor to install
	Install	Ea.	29	13.20		42.20	header, double studs each side,
	Demolish and Install	Ea.	29	23.95		52.95	cripples, blocking and nails up to 5'
	Reinstall	Ea.		10.57		10.57	opening in 2" x 6" stud wall 8' high.
	Minimum Charge	Job		211		211	
6' Wide							
	Demolish	Ea.		10.75		10.75	Includes material and labor to install
	Install	Ea.	31	13.20		44.20	header, double studs each side,
	Demolish and Install	Ea.	31	23.95		54.95	cripples, blocking and nails up to 6'
	Reinstall	Ea.		10.57		10.57	opening in 2" x 6" stud wall 8' high.
	Minimum Charge	Job		211		211	
8' Wide							
	Demolish	Ea.		11.45		11.45	Includes material and labor to install
	Install	Ea.	39.50	14.10		53.60	header, double studs each side,
	Demolish and Install	Ea.	39.50	25.55		65.05	cripples, blocking and nails up to 8'
	Reinstall	Ea.		11.27		11.27	opening in 2" x 6" stud wall 8' high.
	Minimum Charge	Job		211		211	
10' Wide							
	Demolish	Ea.		11.45		11.45	Includes material and labor to install
	Install	Ea.	49.50	14.10		63.60	header, double studs each side,
	Demolish and Install	Ea.	49.50	25.55		75.05	cripples, blocking and nails up to 10'
	Reinstall	Ea.		11.27		11.27	opening in 2" x 6" stud wall 8' high.
	Minimum Charge	Job		211		211	

Rough Frame / Structure

Rough-in Opening	Unit	Material	Labor	Equip.	Total	Specification
12' Wide						Includes material and labor to install
Demolish	Ea.		11.45		11.45	header, double studs each side,
Install	Ea.	66	14.10		80.10	cripples, blocking and nails, up to 12'
Demolish and Install	Ea.	66	25.55		91.55	opening in 2" x 6" stud wall 8' high.
Reinstall	Ea.		11.27		11.27	
Minimum Charge	Job		211		211	

Rough-in Opening	Unit	Material	Labor	Equip.	Total	Specification
Window w / 2" x 4" Lumber						
2' Wide						Includes material and labor to install
Demolish	Ea.		14.35		14.35	header, double studs each side of
Install	Ea.	17.70	17.60		35.30	opening, cripples, blocking, nails and
Demolish and Install	Ea.	17.70	31.95		49.65	sub-sills for opening up to 2' in a 2" x
Reinstall	Ea.		14.09		14.09	4" stud wall 8' high.
Minimum Charge	Job		211		211	
3' Wide						Includes material and labor to install
Demolish	Ea.		14.35		14.35	header, double studs each side of
Install	Ea.	20.50	17.60		38.10	opening, cripples, blocking, nails and
Demolish and Install	Ea.	20.50	31.95		52.45	sub-sills for opening up to 3' in a 2" x
Reinstall	Ea.		14.09		14.09	4" stud wall 8' high.
Minimum Charge	Job		211		211	
4' Wide						Includes material and labor to install
Demolish	Ea.		14.35		14.35	header, double studs each side of
Install	Ea.	22.50	17.60		40.10	opening, cripples, blocking, nails and
Demolish and Install	Ea.	22.50	31.95		54.45	sub-sills for opening up to 4' in a 2" x
Reinstall	Ea.		14.09		14.09	4" stud wall 8' high.
Minimum Charge	Job		211		211	
5' Wide						Includes material and labor to install
Demolish	Ea.		14.35		14.35	header, double studs each side of
Install	Ea.	26.50	17.60		44.10	opening, cripples, blocking, nails and
Demolish and Install	Ea.	26.50	31.95		58.45	sub-sills for opening up to 5' in a 2" x
Reinstall	Ea.		14.09		14.09	4" stud wall 8' high.
Minimum Charge	Job		211		211	
6' Wide						Includes material and labor to install
Demolish	Ea.		14.35		14.35	header, double studs each side of
Install	Ea.	29.50	17.60		47.10	opening, cripples, blocking, nails and
Demolish and Install	Ea.	29.50	31.95		61.45	sub-sills for opening up to 6' in a 2" x
Reinstall	Ea.		14.09		14.09	4" stud wall 8' high.
Minimum Charge	Job		211		211	
7' Wide						Includes material and labor to install
Demolish	Ea.		14.35		14.35	header, double studs each side of
Install	Ea.	37.50	17.60		55.10	opening, cripples, blocking, nails and
Demolish and Install	Ea.	37.50	31.95		69.45	sub-sills for opening up to 7' in a 2" x
Reinstall	Ea.		14.09		14.09	4" stud wall 8' high.
Minimum Charge	Job		211		211	
8' Wide						Includes material and labor to install
Demolish	Ea.		15.65		15.65	header, double studs each side of
Install	Ea.	40.50	19.20		59.70	opening, cripples, blocking, nails and
Demolish and Install	Ea.	40.50	34.85		75.35	sub-sills for opening up to 8' in a 2" x
Reinstall	Ea.		15.37		15.37	4" stud wall 8' high.
Minimum Charge	Job		211		211	

Rough Frame / Structure

Rough-in Opening		Unit	Material	Labor	Equip.	Total	Specification
10' Wide							Includes material and labor to install
	Demolish	Ea.		15.65		15.65	header, double studs each side of
	Install	Ea.	52.50	19.20		71.70	opening, cripples, blocking, nails and
	Demolish and Install	Ea.	52.50	34.85		87.35	sub-sills for opening up to 10' in a 2"
	Reinstall	Ea.		15.37		15.37	x 4" stud wall 8' high.
	Minimum Charge	Job		211		211	
12' Wide							Includes material and labor to install
	Demolish	Ea.		15.65		15.65	header, double studs each side of
	Install	Ea.	71	19.20		90.20	opening, cripples, blocking, nails and
	Demolish and Install	Ea.	71	34.85		105.85	sub-sills for opening up to 12' in a 2"
	Reinstall	Ea.		15.37		15.37	x 4" stud wall 8' high.
	Minimum Charge	Job		211		211	

Window w / 2" x 6" Lumber

		Unit	Material	Labor	Equip.	Total	Specification
2' Wide							Includes material and labor to install
	Demolish	Ea.		14.35		14.35	header, double studs each side of
	Install	Ea.	26.50	17.60		44.10	opening, cripples, blocking, nails and
	Demolish and Install	Ea.	26.50	31.95		58.45	sub-sills for opening up to 2' in a 2" x
	Reinstall	Ea.		14.09		14.09	6" stud wall 8' high.
	Minimum Charge	Job		211		211	
3' Wide							Includes material and labor to install
	Demolish	Ea.		14.35		14.35	header, double studs each side of
	Install	Ea.	29.50	17.60		47.10	opening, cripples, blocking, nails and
	Demolish and Install	Ea.	29.50	31.95		61.45	sub-sills for opening up to 3' in a 2" x
	Reinstall	Ea.		14.09		14.09	6" stud wall 8' high.
	Minimum Charge	Job		211		211	
4' Wide							Includes material and labor to install
	Demolish	Ea.		14.35		14.35	header, double studs each side of
	Install	Ea.	32	17.60		49.60	opening, cripples, blocking, nails and
	Demolish and Install	Ea.	32	31.95		63.95	sub-sills for opening up to 4' in a 2" x
	Reinstall	Ea.		14.09		14.09	6" stud wall 8' high.
	Minimum Charge	Job		211		211	
5' Wide							Includes material and labor to install
	Demolish	Ea.		14.35		14.35	header, double studs each side of
	Install	Ea.	36.50	17.60		54.10	opening, cripples, blocking, nails and
	Demolish and Install	Ea.	36.50	31.95		68.45	sub-sills for opening up to 5' in a 2" x
	Reinstall	Ea.		14.09		14.09	6" stud wall 8' high.
	Minimum Charge	Job		211		211	
6' Wide							Includes material and labor to install
	Demolish	Ea.		14.35		14.35	header, double studs each side of
	Install	Ea.	40	17.60		57.60	opening, cripples, blocking, nails and
	Demolish and Install	Ea.	40	31.95		71.95	sub-sills for opening up to 6' in a 2" x
	Reinstall	Ea.		14.09		14.09	6" stud wall 8' high.
	Minimum Charge	Job		211		211	
7' Wide							Includes material and labor to install
	Demolish	Ea.		14.35		14.35	header, double studs each side of
	Install	Ea.	48	17.60		65.60	opening, cripples, blocking, nails and
	Demolish and Install	Ea.	48	31.95		79.95	sub-sills for opening up to 7' in a 2" x
	Reinstall	Ea.		14.09		14.09	6" stud wall 8' high.
	Minimum Charge	Job		211		211	

Rough Frame / Structure

Rough-in Opening

	Unit	Material	Labor	Equip.	Total	Specification
8' Wide						
Demolish	Ea.		15.65		15.65	Includes material and labor to install
Install	Ea.	52.50	19.20		71.70	header, double studs each side of
Demolish and Install	Ea.	52.50	34.85		87.35	opening, cripples, blocking, nails and
Reinstall	Ea.		15.37		15.37	sub-sills for opening up to 8' in a 2" x
Minimum Charge	Job		211		211	6" stud wall 8' high.
10' Wide						
Demolish	Ea.		15.65		15.65	Includes material and labor to install
Install	Ea.	64.50	19.20		83.70	header, double studs each side of
Demolish and Install	Ea.	64.50	34.85		99.35	opening, cripples, blocking, nails and
Reinstall	Ea.		15.37		15.37	sub-sills for opening up to 10' in a 2"
Minimum Charge	Job		211		211	x 6" stud wall 8' high.
12' Wide						
Demolish	Ea.		15.65		15.65	Includes material and labor to install
Install	Ea.	84.50	19.20		103.70	header, double studs each side of
Demolish and Install	Ea.	84.50	34.85		119.35	opening, cripples, blocking, nails and
Reinstall	Ea.		15.37		15.37	sub-sills for opening up to 12' in a 2"
Minimum Charge	Job		211		211	x 6" stud wall 8' high.

Glue-Laminated Beams

	Unit	Material	Labor	Equip.	Total	Specification
3-1/2" x 6"						
Demolish	L.F.		2.57		2.57	Includes material and labor to install
Install	L.F.	5.65	3.19	1.18	10.02	glue-laminated wood beam.
Demolish and Install	L.F.	5.65	5.76	1.18	12.59	
Reinstall	L.F.		2.55	.95	3.50	
Clean	L.F.	.11	.39		.50	
Paint	L.F.	.35	1.06		1.41	
Minimum Charge	Job		211		211	
3-1/2" x 9"						
Demolish	L.F.		2.57		2.57	Includes material and labor to install
Install	L.F.	7.60	3.19	1.18	11.97	glue-laminated wood beam.
Demolish and Install	L.F.	7.60	5.76	1.18	14.54	
Reinstall	L.F.		2.55	.95	3.50	
Clean	L.F.	.14	.40		.54	
Paint	L.F.	.45	1.38		1.83	
Minimum Charge	Job		211		211	
3-1/2" x 12"						
Demolish	L.F.		3.86		3.86	Includes material and labor to install
Install	L.F.	10.10	3.19	1.18	14.47	glue-laminated wood beam.
Demolish and Install	L.F.	10.10	7.05	1.18	18.33	
Reinstall	L.F.		2.55	.95	3.50	
Clean	L.F.	.18	.41		.59	
Paint	L.F.	.56	1.72		2.28	
Minimum Charge	Job		211		211	
3-1/2" x 15"						
Demolish	L.F.		3.86		3.86	Includes material and labor to install
Install	L.F.	12.65	3.30	1.23	17.18	glue-laminated wood beam.
Demolish and Install	L.F.	12.65	7.16	1.23	21.04	
Reinstall	L.F.		2.64	.98	3.62	
Clean	L.F.	.21	.43		.64	
Paint	L.F.	.66	2		2.66	
Minimum Charge	Job		211		211	

Rough Frame / Structure

Glue-Laminated Beams	Unit	Material	Labor	Equip.	Total	Specification
3-1/2" x 18"						
Demolish	L.F.		4.83		4.83	Includes material and labor to install
Install	L.F.	17.05	3.30	1.23	21.58	glue-laminated wood beam.
Demolish and Install	L.F.	17.05	8.13	1.23	26.41	
Reinstall	L.F.		2.64	.98	3.62	
Clean	L.F.	.24	.44		.68	
Paint	L.F.	.78	2.33		3.11	
Minimum Charge	Job		211		211	
5-1/8" x 6"						
Demolish	L.F.		3.86		3.86	Includes material and labor to install
Install	L.F.	8.30	3.19	1.18	12.67	glue-laminated wood beam.
Demolish and Install	L.F.	8.30	7.05	1.18	16.53	
Reinstall	L.F.		2.55	.95	3.50	
Clean	L.F.	.12	.39		.51	
Paint	L.F.	.37	1.16		1.53	
Minimum Charge	Job		211		211	
5-1/8" x 9"						
Demolish	L.F.		4.83		4.83	Includes material and labor to install
Install	L.F.	12.50	3.19	1.18	16.87	glue-laminated wood beam.
Demolish and Install	L.F.	12.50	8.02	1.18	21.70	
Reinstall	L.F.		2.55	.95	3.50	
Clean	L.F.	.15	.40		.55	
Paint	L.F.	.48	1.49		1.97	
Minimum Charge	Job		211		211	
5-1/8" x 12"						
Demolish	L.F.		5.75		5.75	Includes material and labor to install
Install	L.F.	16.60	3.19	1.18	20.97	glue-laminated wood beam.
Demolish and Install	L.F.	16.60	8.94	1.18	26.72	
Reinstall	L.F.		2.55	.95	3.50	
Clean	L.F.	.19	.41		.60	
Paint	L.F.	.61	1.80		2.41	
Minimum Charge	Job		211		211	
5-1/8" x 18"						
Demolish	L.F.		9.65		9.65	Includes material and labor to install
Install	L.F.	25	3.30	1.23	29.53	glue-laminated wood beam.
Demolish and Install	L.F.	25	12.95	1.23	39.18	
Reinstall	L.F.		2.64	.98	3.62	
Clean	L.F.	.25	.44		.69	
Paint	L.F.	.69	2.44		3.13	
Minimum Charge	Job		211		211	
6-3/4" x 12"						
Demolish	L.F.		5.75		5.75	Includes material and labor to install
Install	L.F.	22	3.30	1.23	26.53	glue-laminated wood beam.
Demolish and Install	L.F.	22	9.05	1.23	32.28	
Reinstall	L.F.		2.64	.98	3.62	
Clean	L.F.	.21	.43		.64	
Paint	L.F.	.66	2		2.66	
Minimum Charge	Job		211		211	
6-3/4" x 15"						
Demolish	L.F.		9.65		9.65	Includes material and labor to install
Install	L.F.	27.50	3.30	1.23	32.03	glue-laminated wood beam.
Demolish and Install	L.F.	27.50	12.95	1.23	41.68	
Reinstall	L.F.		2.64	.98	3.62	
Clean	L.F.	.24	.44		.68	
Paint	L.F.	.78	2.33		3.11	
Minimum Charge	Job		211		211	

Rough Frame / Structure

Glue-Laminated Beams

Glue-Laminated Beams		Unit	Material	Labor	Equip.	Total	Specification
6-3/4" x 18"							
	Demolish	L.F.		9.65		9.65	Includes material and labor to install
	Install	L.F.	33	3.42	1.27	37.69	glue-laminated wood beam.
	Demolish and Install	L.F.	33	13.07	1.27	47.34	
	Reinstall	L.F.		2.73	1.02	3.75	
	Clean	L.F.	.28	.46		.74	
	Paint	L.F.	.88	2.65		3.53	
	Minimum Charge	Job		211		211	
Hardware							
5-1/4" Glue-Lam Seat							
	Install	Ea.	211	2.35		213.35	Includes labor and material to install
	Minimum Charge	Job		211		211	beam hangers.
6-3/4" Glue-Lam Seat							
	Install	Ea.	217	2.35		219.35	Includes labor and material to install
	Minimum Charge	Job		211		211	beam hangers.
Earthquake Strapping							
	Install	Ea.	2.99	2.64		5.63	Includes labor and material to replace
	Minimum Charge	Job		211		211	an earthquake strap.
Hurricane Clips							
	Install	Ea.	1.12	2.92		4.04	Includes labor and material to install a
	Minimum Charge	Job		211		211	hurricane clip.

Metal Stud Framing

Metal Stud Framing		Unit	Material	Labor	Equip.	Total	Specification
16" O.C. System							
4", 16 Ga.							
	Demolish	S.F.		1.26		1.26	Includes material and labor to install
	Install	S.F.	1.32	1.30		2.62	load bearing cold rolled metal stud
	Demolish and Install	S.F.	1.32	2.56		3.88	walls, to 10' high, including studs, top
	Reinstall	S.F.		1.04		1.04	and bottom track and screws.
	Minimum Charge	Job		211		211	
6", 16 Ga.							
	Demolish	S.F.		1.26		1.26	Includes material and labor to install
	Install	S.F.	1.65	1.32		2.97	load bearing cold rolled metal stud
	Demolish and Install	S.F.	1.65	2.58		4.23	walls, to 10' high, including studs, top
	Reinstall	S.F.		1.06		1.06	and bottom track and screws.
	Minimum Charge	Job		211		211	
4", 25 Ga.							
	Demolish	S.F.		.88		.88	Includes material and labor to install
	Install	S.F.	.44	.89		1.33	cold rolled metal stud walls, to 10'
	Demolish and Install	S.F.	.44	1.77		2.21	high, including studs, top and bottom
	Reinstall	S.F.		.71		.71	track and screws.
	Minimum Charge	Job		211		211	
6", 25 Ga.							
	Demolish	S.F.		.88		.88	Includes material and labor to install
	Install	S.F.	.54	.90		1.44	cold rolled metal stud walls, to 10'
	Demolish and Install	S.F.	.54	1.78		2.32	high, including studs, top and bottom
	Reinstall	S.F.		.72		.72	track and screws.
	Minimum Charge	Job		211		211	
24" O.C. System							
4", 25 Ga.							
	Demolish	S.F.		.75		.75	Includes material and labor to install
	Install	S.F.	.32	.57		.89	cold rolled metal stud walls, to 10'
	Demolish and Install	S.F.	.32	1.32		1.64	high, including studs, top and bottom
	Reinstall	S.F.		.46		.46	track and screws.
	Minimum Charge	Job		211		211	

Rough Frame / Structure

Metal Stud Framing

	Unit	Material	Labor	Equip.	Total	Specification
6", 25 Ga.						
Demolish	S.F.		.75		.75	Includes material and labor to install
Install	S.F.	.40	.58		.98	cold rolled metal stud walls, to 10'
Demolish and Install	S.F.	.40	1.33		1.73	high, including studs, top and bottom
Reinstall	S.F.		.47		.47	track and screws.
Minimum Charge	Job		211		211	
4", 16 Ga.						
Demolish	L.F.		8.50		8.50	Includes material and labor to install
Install	S.F.	.96	.94		1.90	load bearing cold rolled metal stud
Demolish and Install	S.F.	.96	9.44		10.40	walls, to 10' high, including studs, top
Reinstall	S.F.		.75		.75	and bottom track and screws.
Minimum Charge	Job		211		211	
6", 16 Ga.						
Demolish	L.F.		8.50		8.50	Includes material and labor to install
Install	S.F.	1.20	.96		2.16	load bearing cold rolled metal stud
Demolish and Install	S.F.	1.20	9.46		10.66	walls, to 10' high, including studs, top
Reinstall	S.F.		.77		.77	and bottom track and screws.
Minimum Charge	Job		211		211	

Metal Joist

	Unit	Material	Labor	Equip.	Total	Specification
Bar Joist						
18K9						
Demolish	L.F.		1.02	.40	1.42	Includes material, labor and
Install	L.F.	9	2.30	1	12.30	equipment to install open web joist.
Demolish and Install	L.F.	9	3.32	1.40	13.72	
Reinstall	L.F.		1.84	.80	2.64	
Minimum Charge	Job		1900	710	2610	
16K6						
Demolish	L.F.		1.02	.40	1.42	Includes material, labor and
Install	L.F.	5.10	1.55	.12	6.77	equipment to install open web joist.
Demolish and Install	L.F.	5.10	2.57	.52	8.19	
Reinstall	L.F.		1.24	.09	1.33	
Minimum Charge	Job		1900	710	2610	

Exterior Sheathing

	Unit	Material	Labor	Equip.	Total	Specification
CDX Plywood						
5/16"						
Demolish	S.F.		.59		.59	Cost includes material and labor to
Install	S.F.	.61	.53		1.14	install 5/16" CDX plywood sheathing.
Demolish and Install	S.F.	.61	1.12		1.73	
Minimum Charge	Job		211		211	
3/8"						
Demolish	S.F.		.60		.60	Cost includes material and labor to
Install	S.F.	.56	.70		1.26	install 3/8" CDX plywood sheathing.
Demolish and Install	S.F.	.56	1.30		1.86	
Minimum Charge	Job		211		211	
1/2"						
Demolish	S.F.		.61		.61	Cost includes material and labor to
Install	S.F.	.64	.75		1.39	install 1/2" CDX plywood sheathing.
Demolish and Install	S.F.	.64	1.36		2	
Minimum Charge	Job		211		211	

Rough Frame / Structure

Exterior Sheathing		Unit	Material	Labor	Equip.	Total	Specification
5/8"							
	Demolish	S.F.		.63		.63	Cost includes material and labor to
	Install	S.F.	.80	.81		1.61	install 5/8" CDX plywood sheathing.
	Demolish and Install	S.F.	.80	1.44		2.24	
	Minimum Charge	Job		211		211	
3/4"							
	Demolish	S.F.		.64		.64	Cost includes material and labor to
	Install	S.F.	.85	.87		1.72	install 3/4" CDX plywood sheathing.
	Demolish and Install	S.F.	.85	1.51		2.36	
	Minimum Charge	Job		211		211	
OSB							
1/2"							
	Demolish	S.F.		.61		.61	Cost includes material and labor to
	Install	S.F.	.41	.64		1.05	install 4' x 8' x 1/2" OSB sheathing.
	Demolish and Install	S.F.	.41	1.25		1.66	
	Minimum Charge	Job		211		211	
5/8"							
	Demolish	S.F.		.63		.63	Cost includes material and labor to
	Install	S.F.	.41	.51		.92	install 4' x 8' x 5/8" OSB sheathing.
	Demolish and Install	S.F.	.41	1.14		1.55	
	Minimum Charge	Job		211		211	
Vapor Barrier							
Black Paper							
	Install	S.F.	.06	.11		.17	Includes material and labor to install
	Minimum Charge	Job		211		211	#15 felt building paper.

Plywood Sheathing		Unit	Material	Labor	Equip.	Total	Specification
Finish Plywood							
5/16"							
	Demolish	S.F.		.59		.59	Includes material and labor to install
	Install	S.F.	.89	.26		1.15	exterior 5/16" AC plywood on walls.
	Demolish and Install	S.F.	.89	.85		1.74	
	Minimum Charge	Job		211		211	
3/8"							
	Demolish	S.F.		.60		.60	Includes material and labor to install
	Install	S.F.	.89	.35		1.24	exterior 3/8" AC plywood on walls.
	Demolish and Install	S.F.	.89	.95		1.84	
	Minimum Charge	Job		211		211	
1/2"							
	Demolish	S.F.		.61		.61	Includes material and labor to install
	Install	S.F.	1.03	.38		1.41	exterior 1/2" AC plywood on walls.
	Demolish and Install	S.F.	1.03	.99		2.02	
	Minimum Charge	Job		211		211	
5/8"							
	Demolish	S.F.		.63		.63	Includes material and labor to install
	Install	S.F.	1.20	.38		1.58	exterior 5/8" AC plywood on walls.
	Demolish and Install	S.F.	1.20	1.01		2.21	
	Reinstall	S.F.		.38		.38	
	Minimum Charge	Job		211		211	
3/4"							
	Demolish	S.F.		.64		.64	Includes material and labor to install
	Install	S.F.	1.36	.38		1.74	exterior 3/4" AC plywood on walls.
	Demolish and Install	S.F.	1.36	1.02		2.38	
	Reinstall	S.F.		.38		.38	
	Minimum Charge	Job		211		211	

Rough Frame / Structure

Stairs

	Unit	Material	Labor	Equip.	Total	Specification
Job-built						
Treads and Risers						Includes material and labor to install
Demolish	Ea.		43		43	three 2" x 12" stringers, treads and
Install	Ea.	605	425		1030	risers of 3/4" CDX plywood, installed
Demolish and Install	Ea.	605	468		1073	in a straight or "L" shaped run
Clean	Flight	3.27	26.50		29.77	including carpet.
Paint	Ea.	.11	72		72.11	
Minimum Charge	Job		211		211	
Landing						Includes material and labor to install
Demolish	S.F.		.34		.34	landing framing, 3/4" CDX plywood
Install	S.F.	72	2.94		74.94	surface and carpet.
Demolish and Install	S.F.	72	3.28		75.28	
Clean	S.F.		.39		.39	
Paint	S.F.	.44	.89		1.33	
Minimum Charge	Job		211		211	

Ceiling Framing

	Unit	Material	Labor	Equip.	Total	Specification
16" O.C. System						
2" x 6" Joists						Cost includes material and labor to
Demolish	S.F.		.53		.53	install 2" x 6" joists including end and
Install	S.F.	.47	.70		1.17	header joist installed 16" O.C. Does
Demolish and Install	S.F.	.47	1.23		1.70	not include beams, ledger strips,
Reinstall	S.F.		.56		.56	blocking or bridging.
Minimum Charge	Job		211		211	
2" x 8" Joists						Cost includes material and labor to
Demolish	S.F.		.55		.55	install 2" x 8" joists including end and
Install	S.F.	.66	.86		1.52	header joist installed 16" O.C. Does
Demolish and Install	S.F.	.66	1.41		2.07	not include beams, ledger strips,
Reinstall	S.F.		.68		.68	blocking or bridging.
Minimum Charge	Job		211		211	
2" x 10" Joists						Cost includes material and labor to
Demolish	S.F.		.57		.57	install 2" x 10" joists including end
Install	S.F.	.90	1.01		1.91	and header joist installed 16" O.C.
Demolish and Install	S.F.	.90	1.58		2.48	Does not include beams, ledger strips,
Reinstall	S.F.		.81		.81	blocking or bridging.
Minimum Charge	Job		211		211	
2" x 12" Joists						Cost includes material and labor to
Demolish	S.F.		.59		.59	install 2" x 12" joists including end
Install	S.F.	1.11	1.24		2.35	and header joist installed 16" O.C.
Demolish and Install	S.F.	1.11	1.83		2.94	Does not include beams, ledger strips,
Reinstall	S.F.		.99		.99	blocking or bridging.
Minimum Charge	Job		211		211	
Block / Bridge						Includes material and labor to install
Demolish	Ea.		1.08		1.08	set of cross bridging or per block of
Install	Ea.	1.60	2.38		3.98	solid bridging for 2" x 10" joists cut to
Demolish and Install	Ea.	1.60	3.46		5.06	size on site.
Minimum Charge	Job		211		211	

Rough Frame / Structure

Ceiling Framing		Unit	Material	Labor	Equip.	Total	Specification
24″ O.C. System							
2″ x 6″ Joists							
	Demolish	S.F.		.35		.35	Cost includes material and labor to
	Install	S.F.	.32	.46		.78	install 2″ x 6″ joists including end and
	Demolish and Install	S.F.	.32	.81		1.13	header joist installed 24″ O.C. Does
	Reinstall	S.F.		.37		.37	not include beams, ledger strips,
	Minimum Charge	Job		211		211	blocking or bridging.
2″ x 8″ Joists							
	Demolish	S.F.		.37		.37	Cost includes material and labor to
	Install	S.F.	.44	.57		1.01	install 2″ x 8″ joists including end and
	Demolish and Install	S.F.	.44	.94		1.38	header joist installed 24″ O.C. Does
	Reinstall	S.F.		.46		.46	not include beams, ledger strips,
	Minimum Charge	Job		211		211	blocking or bridging.
2″ x 10″ Joists							
	Demolish	S.F.		.38		.38	Cost includes material and labor to
	Install	S.F.	.61	.67		1.28	install 2″ x 10″ joists including end
	Demolish and Install	S.F.	.61	1.05		1.66	and header joist installed 24″ O.C.
	Reinstall	S.F.		.54		.54	Does not include beams, ledger strips,
	Minimum Charge	Job		211		211	blocking or bridging.
2″ x 12″ Joists							
	Demolish	S.F.		.39		.39	Cost includes material and labor to
	Install	S.F.	.75	.82		1.57	install 2″ x 12″ joists including end
	Demolish and Install	S.F.	.75	1.21		1.96	and header joist installed 24″ O.C.
	Reinstall	S.F.		.66		.66	Does not include beams, ledger strips,
	Minimum Charge	Job		211		211	blocking or bridging.
Block / Bridge							
	Demolish	Ea.		1.08		1.08	Includes material and labor to install
	Install	Ea.	1.60	2.38		3.98	set of cross bridging or per block of
	Demolish and Install	Ea.	1.60	3.46		5.06	solid bridging for 2″ x 10″ joists cut to
	Minimum Charge	Job		211		211	size on site.
Ledger Strips							
1″ x 2″							
	Demolish	L.F.		.29		.29	Cost includes material and labor to
	Install	L.F.	.28	1.23		1.51	install 1″ x 2″ ledger strip nailed to the
	Demolish and Install	L.F.	.28	1.52		1.80	face of studs, beams or joist.
	Minimum Charge	Job		211		211	
1″ x 3″							
	Demolish	L.F.		.29		.29	Includes material and labor to install
	Install	L.F.	.45	1.41		1.86	up to 1″ x 4″ ledger strip nailed to the
	Demolish and Install	L.F.	.45	1.70		2.15	face of studs, beams or joist.
	Minimum Charge	Job		211		211	
1″ x 4″							
	Demolish	L.F.		.29		.29	Includes material and labor to install
	Install	L.F.	.45	1.41		1.86	up to 1″ x 4″ ledger strip nailed to the
	Demolish and Install	L.F.	.45	1.70		2.15	face of studs, beams or joist.
	Minimum Charge	Job		211		211	
2″ x 2″							
	Demolish	L.F.		.31		.31	Cost includes material and labor to
	Install	L.F.	.36	1.28		1.64	install 2″ x 2″ ledger strip nailed to the
	Demolish and Install	L.F.	.36	1.59		1.95	face of studs, beams or joist.
	Minimum Charge	Job		211		211	
2″ x 4″							
	Demolish	L.F.		.34		.34	Cost includes material and labor to
	Install	L.F.	.40	1.69		2.09	install 2″ x 4″ ledger strip nailed to the
	Demolish and Install	L.F.	.40	2.03		2.43	face of studs, beams or joist.
	Reinstall	L.F.		1.35		1.35	
	Minimum Charge	Job		211		211	

Rough Frame / Structure

Ceiling Framing

	Unit	Material	Labor	Equip.	Total	Specification
Ledger Boards						
2" x 4"						
Demolish	L.F.		.34		.34	Cost includes material and labor to
Install	L.F.	4.63	2.35		6.98	install 2" x 4" ledger board fastened to
Demolish and Install	L.F.	4.63	2.69		7.32	a wall, joists or studs.
Reinstall	L.F.		1.88		1.88	
Minimum Charge	Job		211		211	
4" x 6"						
Demolish	L.F.		.57		.57	Cost includes material and labor to
Install	L.F.	7.90	3.02		10.92	install 4" x 6" ledger board fastened to
Demolish and Install	L.F.	7.90	3.59		11.49	a wall, joists or studs.
Reinstall	L.F.		2.42		2.42	
Minimum Charge	Job		211		211	
4" x 8"						
Demolish	L.F.		.76		.76	Cost includes material and labor to
Install	L.F.	9	3.52		12.52	install 4" x 8" ledger board fastened to
Demolish and Install	L.F.	9	4.28		13.28	a wall, joists or studs.
Reinstall	L.F.		2.82		2.82	
Minimum Charge	Job		211		211	
Earthquake Strapping						
Install	Ea.	2.99	2.64		5.63	Includes labor and material to replace
Minimum Charge	Job		211		211	an earthquake strap.
Hurricane Clips						
Install	Ea.	1.12	2.92		4.04	Includes labor and material to install a
Minimum Charge	Job		211		211	hurricane clip.

Roof Framing

	Unit	Material	Labor	Equip.	Total	Specification
16" O.C. System						
2" x 4" Rafters						
Demolish	S.F.		.78		.78	Cost includes material and labor to
Install	S.F.	.30	.76		1.06	install 2" x 4", 16" O.C. rafter framing
Demolish and Install	S.F.	.30	1.54		1.84	for flat, shed or gable roofs, 25' span,
Reinstall	S.F.		.60		.60	up to 5/12 slope, per S.F. of roof.
Minimum Charge	Job		211		211	
2" x 6" Rafters						
Demolish	S.F.		.82		.82	Cost includes material and labor to
Install	S.F.	.46	.79		1.25	install 2" x 6", 16" O.C. rafter framing
Demolish and Install	S.F.	.46	1.61		2.07	for flat, shed or gable roofs, 25' span,
Reinstall	S.F.		.63		.63	up to 5/12 slope, per S.F. of roof.
Minimum Charge	Job		211		211	
2" x 8" Rafters						
Demolish	S.F.		.84		.84	Cost includes material and labor to
Install	S.F.	.65	.85		1.50	install 2" x 8", 16" O.C. rafter framing
Demolish and Install	S.F.	.65	1.69		2.34	for flat, shed or gable roofs, 25' span,
Reinstall	S.F.		.68		.68	up to 5/12 slope, per S.F. of roof.
Minimum Charge	Job		211		211	
2" x 10" Rafters						
Demolish	S.F.		.84		.84	Cost includes material and labor to
Install	S.F.	.89	1.28		2.17	install 2" x 10", 16" O.C. rafter
Demolish and Install	S.F.	.89	2.12		3.01	framing for flat, shed or gable roofs,
Reinstall	S.F.		1.02		1.02	25' span, up to 5/12 slope, per S.F.
Minimum Charge	Job		211		211	of roof.

Rough Frame / Structure

Roof Framing		Unit	Material	Labor	Equip.	Total	Specification
2" x 12" Rafters							
	Demolish	S.F.		.85		.85	Cost includes material and labor to
	Install	S.F.	1.09	1.39		2.48	install 2" x 12", 16" O.C. rafter
	Demolish and Install	S.F.	1.09	2.24		3.33	framing for flat, shed or gable roofs,
	Reinstall	S.F.		1.11		1.11	25' span, up to 5/12 slope, per S.F.
	Minimum Charge	Job		211		211	of roof.
24" O.C. System							
2" x 4" Rafters							
	Demolish	S.F.		.59		.59	Cost includes material and labor to
	Install	S.F.	.20	.57		.77	install 2" x 4", 24" O.C. rafter framing
	Demolish and Install	S.F.	.20	1.16		1.36	for flat, shed or gable roofs, 25' span,
	Reinstall	S.F.		.45		.45	up to 5/12 slope, per S.F. of roof.
	Minimum Charge	Job		211		211	
2" x 6" Rafters							
	Demolish	S.F.		.62		.62	Cost includes material and labor to
	Install	S.F.	.31	.60		.91	install 2" x 6", 24" O.C. rafter framing
	Demolish and Install	S.F.	.31	1.22		1.53	for flat, shed or gable roofs, 25' span,
	Reinstall	S.F.		.48		.48	up to 5/12 slope, per S.F. of roof.
	Minimum Charge	Job		211		211	
2" x 8" Rafters							
	Demolish	S.F.		.63		.63	Cost includes material and labor to
	Install	S.F.	.44	.64		1.08	install 2" x 8", 24" O.C. rafter framing
	Demolish and Install	S.F.	.44	1.27		1.71	for flat, shed or gable roofs, 25' span,
	Reinstall	S.F.		.51		.51	up to 5/12 slope, per S.F. of roof.
	Minimum Charge	Job		211		211	
2" x 10" Rafters							
	Demolish	S.F.		.63		.63	Cost includes material and labor to
	Install	S.F.	.59	.96		1.55	install 2" x 10", 24" O.C. rafter
	Demolish and Install	S.F.	.59	1.59		2.18	framing for flat, shed or gable roofs,
	Reinstall	S.F.		.77		.77	25' span, up to 5/12 slope, per S.F.
	Minimum Charge	Job		211		211	of roof.
2" x 12" Rafters							
	Demolish	S.F.		.64		.64	Cost includes material and labor to
	Install	S.F.	.73	1.05		1.78	install 2" x 12", 24" O.C. rafter
	Demolish and Install	S.F.	.73	1.69		2.42	framing for flat, shed or gable roofs,
	Reinstall	S.F.		.84		.84	25' span, up to 5/12 slope, per S.F.
	Minimum Charge	Job		211		211	of roof.
Rafter							
2" x 4"							
	Demolish	L.F.		.80		.80	Cost includes material and labor to
	Install	L.F.	.40	1.06		1.46	install 2" x 4" rafters for flat, shed or
	Demolish and Install	L.F.	.40	1.86		2.26	gable roofs, up to 5/12 slope, 25'
	Reinstall	L.F.		.85		.85	span, per L.F.
	Minimum Charge	Job		211		211	
2" x 6"							
	Demolish	L.F.		.81		.81	Cost includes material and labor to
	Install	L.F.	.62	1.06		1.68	install 2" x 6" rafters for flat, shed or
	Demolish and Install	L.F.	.62	1.87		2.49	gable roofs, up to 5/12 slope, 25'
	Reinstall	L.F.		.85		.85	span, per L.F.
	Minimum Charge	Job		211		211	
2" x 8"							
	Demolish	L.F.		.82		.82	Cost includes material and labor to
	Install	L.F.	.87	1.13		2	install 2" x 8" rafters for flat, shed or
	Demolish and Install	L.F.	.87	1.95		2.82	gable roofs, up to 5/12 slope, 25'
	Reinstall	L.F.		.90		.90	span, per L.F.
	Minimum Charge	Job		211		211	

Rough Frame / Structure

Roof Framing	Unit	Material	Labor	Equip.	Total	Specification
2" x 10"						
Demolish	L.F.		.83		.83	Cost includes material and labor to
Install	L.F.	1.19	1.71		2.90	install 2" x 10" rafters for flat, shed or
Demolish and Install	L.F.	1.19	2.54		3.73	gable roofs, up to 5/12 slope, 25'
Reinstall	L.F.		1.37		1.37	span, per L.F.
Minimum Charge	Job		211		211	
2" x 12"						
Demolish	L.F.		.85		.85	Cost includes material and labor to
Install	L.F.	1.45	1.86		3.31	install 2" x 12" rafters for flat, shed or
Demolish and Install	L.F.	1.45	2.71		4.16	gable roofs, up to 5/12 slope, 25'
Reinstall	L.F.		1.49		1.49	span, per L.F.
Minimum Charge	Job		211		211	
2"x 4" Valley / Jack						
Demolish	L.F.		.80		.80	Includes material and labor to install
Install	L.F.	.62	1.78		2.40	up to 2" x 6" valley/jack rafters for
Demolish and Install	L.F.	.62	2.58		3.20	flat, shed or gable roofs, up to 5/12
Reinstall	L.F.		1.42		1.42	slope, 25' span, per L.F.
Minimum Charge	Job		211		211	
2"x 6" Valley / Jack						
Demolish	L.F.		.81		.81	Includes material and labor to install
Install	L.F.	.62	1.78		2.40	up to 2" x 6" valley/jack rafters for
Demolish and Install	L.F.	.62	2.59		3.21	flat, shed or gable roofs, up to 5/12
Reinstall	L.F.		1.42		1.42	slope, 25' span, per L.F.
Minimum Charge	Job		211		211	
Ridgeboard						
2" x 4"						
Demolish	L.F.		.76		.76	Cost includes material and labor to
Install	L.F.	.40	1.54		1.94	install 2" x 4" ridgeboard for flat, shed
Demolish and Install	L.F.	.40	2.30		2.70	or gable roofs, up to 5/12 slope, 25'
Reinstall	L.F.		1.23		1.23	span, per L.F.
Minimum Charge	Job		211		211	
2" x 6"						
Demolish	L.F.		.79		.79	Cost includes material and labor to
Install	L.F.	.62	1.69		2.31	install 2" x 6" ridgeboard for flat, shed
Demolish and Install	L.F.	.62	2.48		3.10	or gable roofs, up to 5/12 slope, 25'
Reinstall	L.F.		1.35		1.35	span, per L.F.
Minimum Charge	Job		211		211	
2" x 8"						
Demolish	L.F.		.81		.81	Cost includes material and labor to
Install	L.F.	.87	1.88		2.75	install 2" x 8" ridgeboard for flat, shed
Demolish and Install	L.F.	.87	2.69		3.56	or gable roofs, up to 5/12 slope, 25'
Reinstall	L.F.		1.50		1.50	span, per L.F.
Minimum Charge	Job		211		211	
2" x 10"						
Demolish	L.F.		.83		.83	Cost includes material and labor to
Install	L.F.	1.19	2.11		3.30	install 2" x 10" ridgeboard for flat,
Demolish and Install	L.F.	1.19	2.94		4.13	shed or gable roofs, up to 5/12
Reinstall	L.F.		1.69		1.69	slope, 25' span, per L.F.
Minimum Charge	Job		211		211	
2" x 12"						
Demolish	L.F.		.86		.86	Cost includes material and labor to
Install	L.F.	1.45	1.21		2.66	install 2" x 12" ridgeboard for flat,
Demolish and Install	L.F.	1.45	2.07		3.52	shed or gable roofs, up to 5/12
Reinstall	L.F.		.97		.97	slope, 25' span, per L.F.
Minimum Charge	Job		211		211	

Rough Frame / Structure

Roof Framing		Unit	Material	Labor	Equip.	Total	Specification
Collar Beam							
1″ x 6″							
	Demolish	L.F.		.61		.61	Cost includes material and labor to
	Install	L.F.	.91	.53		1.44	install 1″ x 6″ collar beam or tie for
	Demolish and Install	L.F.	.91	1.14		2.05	roof framing.
	Reinstall	L.F.		.42		.42	
	Minimum Charge	Job		211		211	
2″ x 6″							
	Demolish	L.F.		.71		.71	Cost includes material and labor to
	Install	L.F.	.62	.53		1.15	install 2″ x 6″ collar beam or tie for
	Demolish and Install	L.F.	.62	1.24		1.86	roof framing.
	Reinstall	L.F.		.42		.42	
	Minimum Charge	Job		211		211	
Purlins							
2″ x 6″							
	Demolish	L.F.		.69		.69	Cost includes material and labor to
	Install	L.F.	.62	.47		1.09	install 2″ x 6″ purlins below roof
	Demolish and Install	L.F.	.62	1.16		1.78	rafters.
	Reinstall	L.F.		.38		.38	
	Minimum Charge	Job		211		211	
2″ x 8″							
	Demolish	L.F.		.70		.70	Cost includes material and labor to
	Install	L.F.	.87	.48		1.35	install 2″ x 8″ purlins below roof
	Demolish and Install	L.F.	.87	1.18		2.05	rafters.
	Reinstall	L.F.		.38		.38	
	Minimum Charge	Job		211		211	
2″ x 10″							
	Demolish	L.F.		.70		.70	Cost includes material and labor to
	Install	L.F.	1.19	.48		1.67	install 2″ x 10″ purlins below roof
	Demolish and Install	L.F.	1.19	1.18		2.37	rafters.
	Reinstall	L.F.		.38		.38	
	Minimum Charge	Job		211		211	
2″ x 12″							
	Demolish	L.F.		.71		.71	Cost includes material and labor to
	Install	L.F.	1.45	.49		1.94	install 2″ x 12″ purlins below roof
	Demolish and Install	L.F.	1.45	1.20		2.65	rafters.
	Reinstall	L.F.		.39		.39	
	Minimum Charge	Job		211		211	
4″ x 6″							
	Demolish	L.F.		.71		.71	Cost includes material and labor to
	Install	L.F.	2.64	.49		3.13	install 4″ x 6″ purlins below roof
	Demolish and Install	L.F.	2.64	1.20		3.84	rafters.
	Reinstall	L.F.		.39		.39	
	Minimum Charge	Job		211		211	
4″ x 8″							
	Demolish	L.F.		.72		.72	Cost includes material and labor to
	Install	L.F.	3.74	.50		4.24	install 4″ x 8″ purlins below roof
	Demolish and Install	L.F.	3.74	1.22		4.96	rafters.
	Reinstall	L.F.		.40		.40	
	Minimum Charge	Job		211		211	
Ledger Board							
2″ x 6″							
	Demolish	L.F.		.34		.34	Cost includes material and labor to
	Install	L.F.	.62	1.90		2.52	install 2″ x 6″ ledger board nailed to
	Demolish and Install	L.F.	.62	2.24		2.86	the face of a rafter.
	Reinstall	L.F.		1.52		1.52	
	Minimum Charge	Job		211		211	

Rough Frame / Structure

Roof Framing

Roof Framing	Unit	Material	Labor	Equip.	Total	Specification
4" x 6"						
Demolish	L.F.		.57		.57	Cost includes material and labor to
Install	L.F.	2.64	2.80		5.44	install 4" x 6" ledger board nailed to
Demolish and Install	L.F.	2.64	3.37		6.01	the face of a rafter.
Reinstall	L.F.		2.24		2.24	
Minimum Charge	Job		211		211	
4" x 8"						
Demolish	L.F.		.76		.76	Cost includes material and labor to
Install	L.F.	3.74	3.23		6.97	install 4" x 8" ledger board nailed to
Demolish and Install	L.F.	3.74	3.99		7.73	the face of a rafter.
Reinstall	L.F.		2.58		2.58	
Minimum Charge	Job		211		211	
Outriggers						
2" x 4"						
Demolish	L.F.		.24		.24	Cost includes material and labor to
Install	L.F.	.40	1.69		2.09	install 2" x 4" outrigger rafters for flat,
Demolish and Install	L.F.	.40	1.93		2.33	shed or gabled roofs, up to 5/12
Reinstall	L.F.		1.35		1.35	slope, per L.F.
Minimum Charge	Job		211		211	
2" x 6"						
Demolish	L.F.		.36		.36	Cost includes material and labor to
Install	L.F.	.62	1.69		2.31	install 2" x 6" outrigger rafters for flat,
Demolish and Install	L.F.	.62	2.05		2.67	shed or gabled roofs, up to 5/12
Reinstall	L.F.		1.35		1.35	slope, per L.F.
Minimum Charge	Job		211		211	
Lookout Rafter						
2" x 4"						
Demolish	L.F.		.80		.80	Cost includes material and labor to
Install	L.F.	.40	.70		1.10	install 2" x 4" lookout rafters for flat,
Demolish and Install	L.F.	.40	1.50		1.90	shed or gabled roofs, up to 5/12
Reinstall	L.F.		.56		.56	slope, per L.F.
Minimum Charge	Job		211		211	
2" x 6"						
Demolish	L.F.		.81		.81	Cost includes material and labor to
Install	L.F.	.62	.72		1.34	install 2" x 6" lookout rafters for flat,
Demolish and Install	L.F.	.62	1.53		2.15	shed or gabled roofs, up to 5/12
Reinstall	L.F.		.57		.57	slope, per L.F.
Minimum Charge	Job		211		211	
Fly Rafter						
2" x 4"						
Demolish	L.F.		.80		.80	Cost includes material and labor to
Install	L.F.	.40	.70		1.10	install 2" x 4" fly rafters for flat, shed
Demolish and Install	L.F.	.40	1.50		1.90	or gabled roofs, up to 5/12 slope,
Reinstall	L.F.		.56		.56	per L.F.
Minimum Charge	Job		211		211	
2" x 6"						
Demolish	L.F.		.81		.81	Cost includes material and labor to
Install	L.F.	.62	.72		1.34	install 2" x 6" fly rafters for flat, shed
Demolish and Install	L.F.	.62	1.53		2.15	or gabled roofs, up to 5/12 slope,
Reinstall	L.F.		.57		.57	per L.F.
Minimum Charge	Job		211		211	

Rough Frame / Structure

Residential Trusses

	Unit	Material	Labor	Equip.	Total	Specification
W or Fink Truss						
24' Span						
Demolish	Ea.		32	11.95	43.95	Includes material, labor and
Install	Ea.	84.50	32	11.85	128.35	equipment to install wood gang-nailed
Demolish and Install	Ea.	84.50	64	23.80	172.30	residential truss, up to 24' span.
Reinstall	Ea.		31.82	11.85	43.67	
Minimum Charge	Job		211		211	
28' Span						
Demolish	Ea.		34.50	12.75	47.25	Includes material, labor and
Install	Ea.	102	36	13.40	151.40	equipment to install wood gang-nailed
Demolish and Install	Ea.	102	70.50	26.15	198.65	residential truss, up to 28' span.
Reinstall	Ea.		36.02	13.41	49.43	
Minimum Charge	Job		211		211	
32' Span						
Demolish	Ea.		38	14.10	52.10	Includes material, labor and
Install	Ea.	120	38	14.20	172.20	equipment to install wood gang-nailed
Demolish and Install	Ea.	120	76	28.30	224.30	residential truss, up to 32' span.
Reinstall	Ea.		38.18	14.22	52.40	
Minimum Charge	Job		211		211	
36' Span						
Demolish	Ea.		41	15.20	56.20	Includes material, labor and
Install	Ea.	146	41.50	15.45	202.95	equipment to install wood gang-nailed
Demolish and Install	Ea.	146	82.50	30.65	259.15	residential truss, up to 36' span.
Reinstall	Ea.		41.50	15.45	56.95	
Minimum Charge	Job		211		211	
Gable End						
24' Span						
Demolish	Ea.		32	11.95	43.95	Includes material, labor and
Install	Ea.	118	34	12.70	164.70	equipment to install wood gang-nailed
Demolish and Install	Ea.	118	66	24.65	208.65	residential gable end truss, up to 24'
Reinstall	Ea.		34.09	12.69	46.78	span.
Clean	Ea.	1.36	41.50		42.86	
Paint	Ea.	4.40	48		52.40	
Minimum Charge	Job		211		211	
28' Span						
Demolish	Ea.		34.50	12.75	47.25	Includes material, labor and
Install	Ea.	145	40.50	15.10	200.60	equipment to install wood gang-nailed
Demolish and Install	Ea.	145	75	27.85	247.85	residential gable end truss, up to 30'
Reinstall	Ea.		40.62	15.12	55.74	span.
Clean	Ea.	1.42	56.50		57.92	
Paint	Ea.	4.59	65.50		70.09	
Minimum Charge	Job		211		211	
32' Span						
Demolish	Ea.		38	14.10	52.10	Includes material, labor and
Install	Ea.	160	41.50	15.45	216.95	equipment to install wood gang-nailed
Demolish and Install	Ea.	160	79.50	29.55	269.05	residential gable end truss, up to 32'
Reinstall	Ea.		41.50	15.45	56.95	span.
Clean	Ea.	1.49	73.50		74.99	
Paint	Ea.	4.80	86		90.80	
Minimum Charge	Job		211		211	
36' Span						
Demolish	Ea.		41	15.20	56.20	Includes material, labor and
Install	Ea.	206	47.50	17.75	271.25	equipment to install wood gang-nailed
Demolish and Install	Ea.	206	88.50	32.95	327.45	residential gable end truss, up to 36'
Reinstall	Ea.		47.73	17.77	65.50	span.
Clean	Ea.	1.55	93		94.55	
Paint	Ea.	5.05	108		113.05	
Minimum Charge	Job		211		211	

Rough Frame / Structure

Residential Trusses

	Unit	Material	Labor	Equip.	Total	Specification
Truss Hardware						
5-1/4" Glue-Lam Seat						
Install	Ea.	211	2.35		213.35	Includes labor and material to install
Minimum Charge	Job		211		211	beam hangers.
6-3/4" Glue-Lam Seat						
Install	Ea.	217	2.35		219.35	Includes labor and material to install
Minimum Charge	Job		211		211	beam hangers.
2" x 4" Joist Hanger						
Install	Ea.	.74	2.42		3.16	Includes labor and material to install
Minimum Charge	Job		211		211	joist and beam hangers.
2" x 12" Joist Hanger						
Install	Ea.	1.41	2.56		3.97	Cost includes labor and material to
Minimum Charge	Job		211		211	install 2" x 12" joist hanger.

Commercial Trusses

	Unit	Material	Labor	Equip.	Total	Specification
Engineered Lumber, Truss / Joist						
9-1/2"						
Demolish	S.F.		.57		.57	Includes material and labor to install
Install	S.F.	1.65	.42		2.07	engineered lumber truss / joists per
Demolish and Install	S.F.	1.65	.99		2.64	S.F. of floor area based on joists 16"
Reinstall	S.F.		.34		.34	O.C. Beams, supports and bridging
Minimum Charge	Job		211		211	are not included.
11-7/8"						
Demolish	S.F.		.59		.59	Includes material and labor to install
Install	S.F.	1.86	.43		2.29	engineered lumber truss / joists per
Demolish and Install	S.F.	1.86	1.02		2.88	S.F. of floor area based on joists 16"
Reinstall	S.F.		.35		.35	O.C. Beams, supports and bridging
Minimum Charge	Job		211		211	are not included.
14"						
Demolish	S.F.		.61		.61	Includes material and labor to install
Install	S.F.	2.49	.46		2.95	engineered lumber truss / joists per
Demolish and Install	S.F.	2.49	1.07		3.56	S.F. of floor area based on joists 16"
Reinstall	S.F.		.37		.37	O.C. Beams, supports and bridging
Minimum Charge	Job		211		211	are not included.
16"						
Demolish	S.F.		.63		.63	Includes material and labor to install
Install	S.F.	3.09	.49		3.58	engineered lumber truss / joists per
Demolish and Install	S.F.	3.09	1.12		4.21	S.F. of floor area based on joists 16"
Reinstall	S.F.		.39		.39	O.C. Beams, supports and bridging
Minimum Charge	Job		211		211	are not included.
Bowstring Truss						
100' Clear Span						
Demolish	SF Flr.		.58	.49	1.07	Includes material, labor and
Install	SF Flr.	7.50	.48	.31	8.29	equipment to install bow string truss
Demolish and Install	SF Flr.	7.50	1.06	.80	9.36	system up to 100' clear span per S.F.
Minimum Charge	Job		211		211	of floor area.
120' Clear Span						
Demolish	SF Flr.		.65	.55	1.20	Includes material, labor and
Install	SF Flr.	8	.53	.34	8.87	equipment to install bow string truss
Demolish and Install	SF Flr.	8	1.18	.89	10.07	system up to 120' clear span per S.F.
Minimum Charge	Job		211		211	of floor area.

Roof Sheathing		Unit	Material	Labor	Equip.	Total	Specification
Plywood							
3/8"							
	Demolish	S.F.		.45		.45	Cost includes material and labor to
	Install	S.F.	.56	.55		1.11	install 3/8" CDX plywood roof
	Demolish and Install	S.F.	.56	1		1.56	sheathing.
	Reinstall	S.F.		.55		.55	
	Minimum Charge	Job		211		211	
1/2"							
	Demolish	S.F.		.45		.45	Cost includes material and labor to
	Install	S.F.	.64	.60		1.24	install 1/2" CDX plywood roof
	Demolish and Install	S.F.	.64	1.05		1.69	sheathing.
	Minimum Charge	Job		211		211	
5/8"							
	Demolish	S.F.		.45		.45	Cost includes material and labor to
	Install	S.F.	.80	.65		1.45	install 5/8" CDX plywood roof
	Demolish and Install	S.F.	.80	1.10		1.90	sheathing.
	Minimum Charge	Job		211		211	
3/4"							
	Demolish	S.F.		.45		.45	Cost includes material and labor to
	Install	S.F.	.85	.70		1.55	install 3/4" CDX plywood roof
	Demolish and Install	S.F.	.85	1.15		2	sheathing.
	Minimum Charge	Job		211		211	
OSB							
1/2"							
	Demolish	S.F.		.45		.45	Cost includes material and labor to
	Install	S.F.	.41	.64		1.05	install 4' x 8' x 1/2" OSB sheathing.
	Demolish and Install	S.F.	.41	1.09		1.50	
	Minimum Charge	Job		211		211	
5/8"							
	Demolish	S.F.		.45		.45	Cost includes material and labor to
	Install	S.F.	.41	.51		.92	install 4' x 8' x 5/8" OSB sheathing.
	Demolish and Install	S.F.	.41	.96		1.37	
	Minimum Charge	Job		211		211	
Plank							
	Demolish	S.F.		.45		.45	Cost includes material and labor to
	Install	S.F.	1.99	1.17		3.16	install 1" x 6" or 1" x 8" utility T&G
	Demolish and Install	S.F.	1.99	1.62		3.61	board sheathing laid diagonal.
	Reinstall	S.F.		1.17		1.17	
	Minimum Charge	Job		211		211	
Skip-type							
1" x 4"							
	Demolish	S.F.		.69		.69	Includes material and labor to install
	Install	S.F.	.56	.35		.91	sheathing board material 1" x 4", 7"
	Demolish and Install	S.F.	.56	1.04		1.60	O.C.
	Minimum Charge	Job		211		211	
1" x 6"							
	Demolish	S.F.		.69		.69	Includes material and labor to install
	Install	S.F.	.91	.29		1.20	sheathing board material 1" x 6", 7"
	Demolish and Install	S.F.	.91	.98		1.89	or 9" O.C.
	Minimum Charge	Job		211		211	

Rough Frame / Structure

Wood Deck		Unit	Material	Labor	Equip.	Total	Specification
Treated							
	Demolish	S.F.		1.43		1.43	Includes material and labor to install
	Install	S.F.	4.07	10.55		14.62	unfinished deck using 2" x 4" treated
	Demolish and Install	S.F.	4.07	11.98		16.05	decking, with 2" x 6" double beams
	Reinstall	S.F.		8.46		8.46	24" O.C., 4" x 4" posts 5' O.C. set in
	Clean	S.F.	.07	.32		.39	concrete.
	Paint	S.F.	.13	.28		.41	
	Minimum Charge	Job		211		211	
Redwood							
	Demolish	S.F.		1.43		1.43	Includes material and labor to install
	Install	S.F.	10.65	10.55		21.20	unfinished deck using 2" x 4" select
	Demolish and Install	S.F.	10.65	11.98		22.63	redwood decking, with 2" x 6" double
	Reinstall	S.F.		8.46		8.46	beams 24" O.C., 4" x 4" posts 5'
	Clean	S.F.	.07	.32		.39	O.C. set in concrete.
	Paint	S.F.	.13	.28		.41	
	Minimum Charge	Job		211		211	
Cedar							
	Demolish	S.F.		1.43		1.43	Includes material and labor to install
	Install	S.F.	10.20	10.55		20.75	unfinished deck using 2" x 4" select
	Demolish and Install	S.F.	10.20	11.98		22.18	cedar decking, with 2" x 6" double
	Reinstall	S.F.		8.46		8.46	beams 24" O.C., 4" x 4" posts 5'
	Clean	S.F.	.07	.32		.39	O.C. set in concrete.
	Paint	S.F.	.13	.28		.41	
	Minimum Charge	Job		211		211	
Stairs							
	Demolish	Ea.		410		410	Includes material and labor to install
	Install	Ea.	128	735		863	wood steps. 5 - 7 risers included.
	Demolish and Install	Ea.	128	1145		1273	
	Reinstall	Ea.		586.88		586.88	
	Clean	Flight	3.27	26.50		29.77	
	Paint	Flight	21	48		69	
	Minimum Charge	Job		211		211	
Railing							
	Demolish	L.F.		5.30		5.30	Includes material and labor to install
	Install	L.F.	28	19.20		47.20	porch rail with balusters.
	Demolish and Install	L.F.	28	24.50		52.50	
	Clean	L.F.	.65	1.94		2.59	
	Paint	L.F.	.83	3.01		3.84	
	Minimum Charge	Job		211		211	
Bench Seating							
	Demolish	L.F.		5.75		5.75	Includes material and labor to install
	Install	L.F.	1.86	21		22.86	pressure treated bench seating.
	Demolish and Install	L.F.	1.86	26.75		28.61	
	Reinstall	L.F.		16.91		16.91	
	Clean	L.F.	.13	1.55		1.68	
	Paint	L.F.	.63	2.89		3.52	
	Minimum Charge	Job		211		211	
Privacy Wall							
	Demolish	SF Flr.		1.38		1.38	Includes material, labor and
	Install	SF Flr.	1.47	.72	.23	2.42	equipment to install privacy wall.
	Demolish and Install	SF Flr.	1.47	2.10	.23	3.80	
	Reinstall	SF Flr.		.58	.18	.76	
	Clean	SF Flr.	.07	.32		.39	
	Paint	S.F.	.13	.28		.41	
	Minimum Charge	Job		211		211	

Porch	Unit	Material	Labor	Equip.	Total	Specification
Treated						
Demolish	S.F.		2.87		2.87	Includes material and labor to install
Install	S.F.	5.15	21		26.15	floor and roof frame and decking
Demolish and Install	S.F.	5.15	23.87		29.02	using 2″ x 4″ treated decking, with 2″
Reinstall	S.F.		16.91		16.91	x 6″ double beams 24″ O.C., 4″ x 4″
Clean	S.F.	.13	.64		.77	posts 5′ O.C. set in concrete.
Paint	S.F.	.21	.72		.93	
Minimum Charge	Job		211		211	
Redwood						
Demolish	S.F.		2.87		2.87	Includes material and labor to install
Install	S.F.	12.05	21		33.05	floor and roof frame and decking
Demolish and Install	S.F.	12.05	23.87		35.92	using 2″ x 4″ redwood decking, with
Reinstall	S.F.		16.91		16.91	2″ x 6″ double beams 24″ O.C., 4″ x
Clean	S.F.	.13	.64		.77	4″ posts 5′ O.C. set in concrete.
Paint	S.F.	.21	.72		.93	
Minimum Charge	Job		211		211	
Cedar						
Demolish	S.F.		2.87		2.87	Includes material and labor to install
Install	S.F.	11.50	21		32.50	floor and roof frame and cedar
Demolish and Install	S.F.	11.50	23.87		35.37	decking using 2″ x 4″ select cedar,
Reinstall	S.F.		16.91		16.91	with 2″ x 6″ double beams 24″ O.C.,
Clean	S.F.	.13	.64		.77	4″ x 4″ posts 5′ O.C. set in concrete.
Paint	S.F.	.21	.72		.93	
Minimum Charge	Job		211		211	
Stairs						
Demolish	Ea.		410		410	Includes material and labor to install
Install	Ea.	128	735		863	wood steps. 5 - 7 risers included.
Demolish and Install	Ea.	128	1145		1273	
Reinstall	Ea.		586.88		586.88	
Clean	Flight	3.27	26.50		29.77	
Paint	Flight	21	48		69	
Minimum Charge	Job		211		211	
Railing						
Demolish	L.F.		5.30		5.30	Includes material and labor to install
Install	L.F.	28	19.20		47.20	porch rail with balusters.
Demolish and Install	L.F.	28	24.50		52.50	
Clean	L.F.	.65	2.42		3.07	
Paint	L.F.	.83	3.01		3.84	
Minimum Charge	Job		211		211	
Bench Seating						
Demolish	L.F.		5.75		5.75	Includes material and labor to install
Install	L.F.	1.86	21		22.86	pressure treated bench seating.
Demolish and Install	L.F.	1.86	26.75		28.61	
Reinstall	L.F.		16.91		16.91	
Clean	L.F.	.13	1.55		1.68	
Paint	L.F.	.63	2.89		3.52	
Minimum Charge	Job		211		211	
Privacy Wall						
Demolish	SF Flr.		1.38		1.38	Includes material, labor and
Install	SF Flr.	1.47	.72	.23	2.42	equipment to install privacy wall.
Demolish and Install	SF Flr.	1.47	2.10	.23	3.80	
Reinstall	SF Flr.		.58	.18	.76	
Clean	SF Flr.	.07	.32		.39	
Paint	S.F.	.13	.28		.41	
Minimum Charge	Job		211		211	

Rough Frame / Structure

Porch	Unit	Material	Labor	Equip.	Total	Specification
Wood Board Ceiling						
Demolish	S.F.		.57		.57	Includes material and labor to install
Install	S.F.	3.18	1.88		5.06	wood board on ceiling.
Demolish and Install	S.F.	3.18	2.45		5.63	
Clean	S.F.	.02	.18		.20	
Paint	S.F.	.19	.45		.64	
Minimum Charge	Job		211		211	
Re-screen						
Install	S.F.	1.77	1.69		3.46	Includes labor and material to
Minimum Charge	Job		211		211	re-screen wood frame.

Exterior Trim

Exterior Trim	Unit	Material	Labor	Equip.	Total	Specification
Stock Lumber						
1″ x 2″						
Demolish	L.F.		.29		.29	Cost includes material and labor to
Install	L.F.	.29	1.28		1.57	install 1″ x 2″ pine trim.
Demolish and Install	L.F.	.29	1.57		1.86	
Clean	L.F.	.01	.21		.22	
Paint	L.F.	.02	.56		.58	
Minimum Charge	Job		211		211	
1″ x 3″						
Demolish	L.F.		.29		.29	Cost includes material and labor to
Install	L.F.	.26	1.46		1.72	install 1″ x 3″ pine trim.
Demolish and Install	L.F.	.26	1.75		2.01	
Clean	L.F.	.01	.21		.22	
Paint	L.F.	.02	.56		.58	
Minimum Charge	Job		211		211	
1″ x 4″						
Demolish	L.F.		.29		.29	Cost includes material and labor to
Install	L.F.	.48	1.69		2.17	install 1″ x 4″ pine trim.
Demolish and Install	L.F.	.48	1.98		2.46	
Clean	L.F.	.01	.21		.22	
Paint	L.F.	.02	.56		.58	
Minimum Charge	Job		211		211	
1″ x 6″						
Demolish	L.F.		.29		.29	Cost includes material and labor to
Install	L.F.	.94	1.69		2.63	install 1″ x 6″ pine trim.
Demolish and Install	L.F.	.94	1.98		2.92	
Clean	L.F.	.01	.21		.22	
Paint	L.F.	.08	.56		.64	
Minimum Charge	Job		211		211	
1″ x 8″						
Demolish	L.F.		.31		.31	Cost includes material and labor to
Install	L.F.	1.28	2.11		3.39	install 1″ x 8″ pine trim.
Demolish and Install	L.F.	1.28	2.42		3.70	
Reinstall	L.F.		1.69		1.69	
Clean	L.F.	.02	.22		.24	
Paint	L.F.	.08	.56		.64	
Minimum Charge	Job		211		211	

Exterior Trim

Exterior Trim	Unit	Material	Labor	Equip.	Total	Specification
1″ x 10″						
Demolish	L.F.		.48		.48	Cost includes material and labor to
Install	L.F.	1.65	2.35		4	install 1″ x 10″ pine trim.
Demolish and Install	L.F.	1.65	2.83		4.48	
Minimum Charge	Job		211		211	
1″ x 12″						
Demolish	L.F.		.48		.48	Cost includes material and labor to
Install	L.F.	2.07	2.35		4.42	install 1″ x 12″ pine trim.
Demolish and Install	L.F.	2.07	2.83		4.90	
Minimum Charge	Job		211		211	

Fascia	Unit	Material	Labor	Equip.	Total	Specification
Aluminum						
Demolish	L.F.		.60		.60	Includes material and labor to install
Install	S.F.	2.06	2.01		4.07	aluminum fascia.
Demolish and Install	S.F.	2.06	2.61		4.67	
Reinstall	S.F.		1.61		1.61	
Clean	L.F.		.39		.39	
Minimum Charge	Job		211		211	
Vinyl						
Demolish	L.F.		.60		.60	Cost includes material and labor to
Install	L.F.	4.79	2.42		7.21	install 6″ vinyl fascia with 12″ vinyl
Demolish and Install	L.F.	4.79	3.02		7.81	soffit and J-channel.
Reinstall	L.F.		1.93		1.93	
Clean	L.F.		.39		.39	
Minimum Charge	Job		211		211	
Plywood						
Demolish	L.F.		.59		.59	Cost includes material and labor to
Install	L.F.	1.41	.94		2.35	install 12″ wide plywood fascia.
Demolish and Install	L.F.	1.41	1.53		2.94	
Reinstall	L.F.		.75		.75	
Clean	L.F.	.03	.26		.29	
Paint	L.F.	.18	.90		1.08	
Minimum Charge	Job		211		211	
Cedar						
Demolish	L.F.		.59		.59	Cost includes material and labor to
Install	L.F.	3	.94		3.94	install 12″ wide cedar fascia.
Demolish and Install	L.F.	3	1.53		4.53	
Reinstall	L.F.		.75		.75	
Clean	L.F.	.03	.26		.29	
Paint	L.F.	.18	.90		1.08	
Minimum Charge	Job		211		211	
Redwood						
1″ x 6″						
Demolish	L.F.		.46		.46	Cost includes material and labor to
Install	L.F.	1.95	1.69		3.64	install 1″ x 6″ redwood facia board.
Demolish and Install	L.F.	1.95	2.15		4.10	
Reinstall	L.F.		1.35		1.35	
Minimum Charge	Job		211		211	
1″ x 8″						
Demolish	L.F.		.52		.52	Cost includes material and labor to
Install	L.F.	2.64	1.84		4.48	install 1″ x 8″ redwood facia board.
Demolish and Install	L.F.	2.64	2.36		5	
Reinstall	L.F.		1.47		1.47	
Minimum Charge	Job		211		211	

Exterior Trim

Fascia		Unit	Material	Labor	Equip.	Total	Specification
2″ x 6″							
	Demolish	L.F.		.46		.46	Cost includes material and labor to
	Install	L.F.	.62	3.38		4	install 2″ x 6″ hem-fir exterior trim
	Demolish and Install	L.F.	.62	3.84		4.46	board.
	Reinstall	L.F.		2.71		2.71	
	Minimum Charge	Job		211		211	
2″ x 8″							
	Demolish	L.F.		.52		.52	Cost includes material and labor to
	Install	L.F.	.87	3.76		4.63	install 2″ x 8″ hem-fir exterior trim
	Demolish and Install	L.F.	.87	4.28		5.15	board.
	Reinstall	L.F.		3.01		3.01	
	Minimum Charge	Job		211		211	

Hem-fir Std & Better

		Unit	Material	Labor	Equip.	Total	Specification
2″ x 6″							
	Demolish	L.F.		.46		.46	Cost includes material and labor to
	Install	L.F.	.62	3.38		4	install 2″ x 6″ hem-fir exterior trim
	Demolish and Install	L.F.	.62	3.84		4.46	board.
	Reinstall	L.F.		2.71		2.71	
	Clean	L.F.	.02	.26		.28	
	Paint	L.F.	.18	.90		1.08	
	Minimum Charge	Job		211		211	
2″ x 8″							
	Demolish	L.F.		.52		.52	Cost includes material and labor to
	Install	L.F.	.87	3.76		4.63	install 2″ x 8″ hem-fir exterior trim
	Demolish and Install	L.F.	.87	4.28		5.15	board.
	Reinstall	L.F.		3.01		3.01	
	Clean	L.F.	.02	.26		.28	
	Paint	L.F.	.18	.90		1.08	
	Minimum Charge	Job		211		211	
2″ x 10″							
	Demolish	L.F.		.59		.59	Cost includes material and labor to
	Install	L.F.	1.19	4.70		5.89	install 2″ x 10″ hem-fir exterior trim
	Demolish and Install	L.F.	1.19	5.29		6.48	board.
	Reinstall	L.F.		3.76		3.76	
	Clean	L.F.	.03	.28		.31	
	Paint	L.F.	.18	.90		1.08	
	Minimum Charge	Job		211		211	

Barge Rafter

		Unit	Material	Labor	Equip.	Total	Specification
2″ x 6″							
	Demolish	L.F.		.92		.92	Cost includes material and labor to
	Install	L.F.	.62	.47		1.09	install 2″ x 6″ barge rafters.
	Demolish and Install	L.F.	.62	1.39		2.01	
	Reinstall	L.F.		.38		.38	
	Clean	L.F.	.02	.26		.28	
	Paint	L.F.	.18	.90		1.08	
	Minimum Charge	Job		211		211	
2″ x 8″							
	Demolish	L.F.		.93		.93	Cost includes material and labor to
	Install	L.F.	.87	.50		1.37	install 2″ x 8″ barge rafters.
	Demolish and Install	L.F.	.87	1.43		2.30	
	Reinstall	L.F.		.40		.40	
	Clean	L.F.	.02	.26		.28	
	Paint	L.F.	.18	.90		1.08	
	Minimum Charge	Job		211		211	

Exterior Trim

Fascia

Fascia	Unit	Material	Labor	Equip.	Total	Specification
2" x 10"						
Demolish	L.F.		.95		.95	Cost includes material and labor to
Install	L.F.	1.19	.51		1.70	install 2" x 10" barge rafters.
Demolish and Install	L.F.	1.19	1.46		2.65	
Reinstall	L.F.		.41		.41	
Clean	L.F.	.03	.28		.31	
Paint	L.F.	.18	.90		1.08	
Minimum Charge	Job		211		211	

Soffit

Soffit	Unit	Material	Labor	Equip.	Total	Specification
Aluminum						
12" w / Fascia						
Demolish	L.F.		.60		.60	Cost includes material and labor to
Install	L.F.	4.84	2.42		7.26	install 12" aluminum soffit with 6"
Demolish and Install	L.F.	4.84	3.02		7.86	fascia and J-channel.
Reinstall	L.F.		1.93		1.93	
Clean	L.F.		.39		.39	
Paint	L.F.	.22	.90		1.12	
Minimum Charge	Job		211		211	
18" w / Fascia						
Demolish	L.F.		.60		.60	Cost includes material and labor to
Install	L.F.	5.90	2.82		8.72	install 18" aluminum soffit with 6"
Demolish and Install	L.F.	5.90	3.42		9.32	fascia and J-channel.
Reinstall	L.F.		2.25		2.25	
Clean	L.F.		.39		.39	
Paint	L.F.	.22	.90		1.12	
Minimum Charge	Job		211		211	
24" w / Fascia						
Demolish	L.F.		.60		.60	Cost includes material and labor to
Install	L.F.	6.90	3.13		10.03	install 24" aluminum soffit with 6"
Demolish and Install	L.F.	6.90	3.73		10.63	fascia and J-channel.
Reinstall	L.F.		2.51		2.51	
Clean	L.F.		.56		.56	
Paint	L.F.	.22	.90		1.12	
Minimum Charge	Job		211		211	
Vinyl						
Demolish	L.F.		.60		.60	Cost includes material and labor to
Install	L.F.	4.79	2.42		7.21	install 6" vinyl fascia with 12" vinyl
Demolish and Install	L.F.	4.79	3.02		7.81	soffit and J-channel.
Reinstall	L.F.		1.93		1.93	
Clean	L.F.		.39		.39	
Paint	L.F.	.22	.90		1.12	
Minimum Charge	Job		211		211	
Plywood						
Demolish	L.F.		.66		.66	Cost includes material and labor to
Install	L.F.	1.36	1.21		2.57	install 12" wide plywood soffit.
Demolish and Install	L.F.	1.36	1.87		3.23	
Reinstall	L.F.		.97		.97	
Clean	L.F.	.03	.26		.29	
Paint	L.F.	.18	.90		1.08	
Minimum Charge	Job		211		211	

Exterior Trim

Soffit	Unit	Material	Labor	Equip.	Total	Specification
Redwood						
Demolish	L.F.		.66		.66	Cost includes material and labor to
Install	L.F.	5.15	1.21		6.36	install 12″ wide redwood soffit.
Demolish and Install	L.F.	5.15	1.87		7.02	
Reinstall	L.F.		.97		.97	
Clean	L.F.	.03	.26		.29	
Paint	L.F.	.18	.90		1.08	
Minimum Charge	Job		211		211	
Cedar						
Demolish	L.F.		.66		.66	Cost includes material and labor to
Install	L.F.	2.96	1.21		4.17	install 12″ wide cedar soffit.
Demolish and Install	L.F.	2.96	1.87		4.83	
Reinstall	L.F.		.97		.97	
Clean	L.F.	.03	.26		.29	
Paint	L.F.	.18	.90		1.08	
Minimum Charge	Job		211		211	
Douglas Fir						
Demolish	L.F.		.66		.66	Cost includes material and labor to
Install	L.F.	2.02	1.21		3.23	install 12″ wide Douglas fir soffit.
Demolish and Install	L.F.	2.02	1.87		3.89	
Reinstall	L.F.		.97		.97	
Clean	L.F.	.03	.26		.29	
Paint	L.F.	.18	.90		1.08	
Minimum Charge	Job		211		211	

Roofing

Roofing and Flashing

Tile Roofing

Metal Flashing	Unit	Material	Labor	Equip.	Total	Specification
Galvanized						
6″						
Demolish	L.F.		.23		.23	Includes material and labor to install
Install	L.F.	.72	.96		1.68	flat galvanized steel flashing, 6″ wide.
Demolish and Install	L.F.	.72	1.19		1.91	
Paint	L.F.	.12	.45		.57	
Minimum Charge	Job		192		192	
14″						
Demolish	L.F.		.23		.23	Includes material and labor to install
Install	L.F.	1.67	.96		2.63	flat galvanized steel flashing, 14″
Demolish and Install	L.F.	1.67	1.19		2.86	wide.
Paint	L.F.	.12	.45		.57	
Minimum Charge	Job		192		192	
20″						
Demolish	L.F.		.23		.23	Includes material and labor to install
Install	L.F.	2.39	.96		3.35	flat galvanized steel flashing, 20″
Demolish and Install	L.F.	2.39	1.19		3.58	wide.
Paint	L.F.	.12	.45		.57	
Minimum Charge	Job		192		192	
Aluminum						
6″						
Demolish	L.F.		.23		.23	Includes material and labor to install
Install	L.F.	.40	.96		1.36	flat aluminum flashing, 6″ wide.
Demolish and Install	L.F.	.40	1.19		1.59	
Paint	L.F.	.12	.45		.57	
Minimum Charge	Job		192		192	
14″						
Demolish	L.F.		.23		.23	Includes material and labor to install
Install	L.F.	.67	.96		1.63	flat aluminum flashing, 14″ wide.
Demolish and Install	L.F.	.67	1.19		1.86	
Paint	L.F.	.12	.45		.57	
Minimum Charge	Job		192		192	
20″						
Demolish	L.F.		.23		.23	Includes material and labor to install
Install	L.F.	.98	.96		1.94	flat aluminum flashing, 20″ wide.
Demolish and Install	L.F.	.98	1.19		2.17	
Paint	L.F.	.12	.45		.57	
Minimum Charge	Job		192		192	

Roofing

Metal Flashing		Unit	Material	Labor	Equip.	Total	Specification
Copper							
6"							
	Demolish	L.F.		.23		.23	Includes material and labor to install
	Install	L.F.	4.17	1.28		5.45	flat copper flashing, 6" wide.
	Demolish and Install	L.F.	4.17	1.51		5.68	
	Minimum Charge	Job		234		234	
14"							
	Demolish	L.F.		.23		.23	Includes material and labor to install
	Install	L.F.	9.75	1.28		11.03	flat copper flashing, 14" wide.
	Demolish and Install	L.F.	9.75	1.51		11.26	
	Minimum Charge	Job		234		234	
18"							
	Demolish	L.F.		.23		.23	Includes material and labor to install
	Install	L.F.	13.90	1.28		15.18	flat copper flashing, 18" wide.
	Demolish and Install	L.F.	13.90	1.51		15.41	
	Minimum Charge	Job		234		234	
Aluminum Valley							
	Demolish	L.F.		.23		.23	Includes material and labor to install
	Install	L.F.	3.77	2.49		6.26	aluminum valley flashing, .024" thick.
	Demolish and Install	L.F.	3.77	2.72		6.49	
	Paint	L.F.	.12	.45		.57	
	Minimum Charge	Job		211		211	
Gravel Stop							
	Demolish	L.F.		.36		.36	Includes material and labor to install
	Install	L.F.	6.60	3.23		9.83	aluminum gravel stop, .050" thick, 4"
	Demolish and Install	L.F.	6.60	3.59		10.19	high, mill finish.
	Reinstall	L.F.		2.59		2.59	
	Paint	L.F.	.19	.45		.64	
	Minimum Charge	Job		211		211	
Drip Edge							
	Demolish	L.F.		.34		.34	Includes material and labor to install
	Install	L.F.	.44	1.06		1.50	aluminum drip edge, .016" thick, 5"
	Demolish and Install	L.F.	.44	1.40		1.84	wide, mill finish.
	Reinstall	L.F.		.85		.85	
	Paint	L.F.	.19	.45		.64	
	Minimum Charge	Job		211		211	
End Wall							
	Demolish	L.F.		1.06		1.06	Includes material and labor to install
	Install	L.F.	2.19	2.49		4.68	aluminum end wall flashing, .024"
	Demolish and Install	L.F.	2.19	3.55		5.74	thick.
	Minimum Charge	Job		211		211	
Side Wall							
	Demolish	L.F.		.21		.21	Includes material and labor to install
	Install	L.F.	2.19	2.49		4.68	aluminum side wall flashing, .024"
	Demolish and Install	L.F.	2.19	2.70		4.89	thick.
	Minimum Charge	Job		211		211	
Coping and Wall Cap							
	Demolish	L.F.		1.43		1.43	Includes material and labor to install
	Install	L.F.	2.35	2.65		5	aluminum flashing, mill finish, 0.40
	Demolish and Install	L.F.	2.35	4.08		6.43	thick, 12" wide.
	Clean	L.F.	.08	.97		1.05	
	Minimum Charge	Job		192		192	

Roofing

Roof Shingles		Unit	Material	Labor	Equip.	Total	Specification
Underlayment							
15#							
	Demolish	Sq.		11.45		11.45	Includes material and labor to install
	Install	Sq.	5.70	6		11.70	#15 felt underlayment.
	Demolish and Install	Sq.	5.70	17.45		23.15	
	Minimum Charge	Job		192		192	
30#							
	Demolish	Sq.		11.45		11.45	Includes material and labor to install
	Install	Sq.	11.40	6.65		18.05	#30 felt underlayment.
	Demolish and Install	Sq.	11.40	18.10		29.50	
	Minimum Charge	Job		192		192	
Self Adhering							
	Demolish	Sq.		28.50		28.50	Includes material and labor to install
	Install	Sq.	86	17.50		103.50	self adhering ice barrier roofing
	Demolish and Install	Sq.	86	46		132	underlayment.
	Minimum Charge	Job		192		192	

Rolled Roofing		Unit	Material	Labor	Equip.	Total	Specification
90 lb							
	Demolish	Sq.		28.50		28.50	Cost includes material and labor to
	Install	Sq.	40	32		72	install 90 lb mineral surface rolled
	Demolish and Install	Sq.	40	60.50		100.50	roofing.
	Minimum Charge	Job		192		192	

Composition Shingles		Unit	Material	Labor	Equip.	Total	Specification
Fiberglass (3-tab)							
225 lb, 20 Year							
	Demolish	Sq.		31		31	Cost includes material and labor to
	Install	Sq.	83	75		158	install 225 lb, 20 year fiberglass
	Demolish and Install	Sq.	83	106		189	shingles.
	Minimum Charge	Job		192		192	
300 lb, 25 Year							
	Demolish	Sq.		31		31	Cost includes material and labor to
	Install	Sq.	83	75		158	install 300 lb, 25 year fiberglass
	Demolish and Install	Sq.	83	106		189	shingles.
	Minimum Charge	Job		192		192	
Emergency Repairs							
	Install	Ea.	154	192		346	Includes labor and material for
	Minimum Charge	Job		192		192	emergency repairs.
Architectural							
Laminated, 25 Year							
	Demolish	Sq.		31		31	Cost includes material and labor to
	Install	Sq.	129	75		204	install 25 year laminated roof
	Demolish and Install	Sq.	129	106		235	shingles.
	Minimum Charge	Job		192		192	
Laminated, 30 Year							
	Demolish	Sq.		31		31	Cost includes material and labor to
	Install	Sq.	103	66.50		169.50	install 30 year laminated roof
	Demolish and Intall	S.F.	103	97.50		200.50	shingles.
	Minimum Charge	Job		192		192	

Roofing

Composition Shingles	Unit	Material	Labor	Equip.	Total	Specification
Laminated, 40 Year						
Demolish	Sq.		31		31	Cost includes material and labor to
Install	Sq.	143	75		218	install 40 year laminated roof
Demolish and Install	Sq.	143	106		249	shingles.
Minimum Charge	Job		192		192	
Laminated Shake, 40 Year						
Demolish	Sq.		31		31	Cost includes material and labor to
Install	Sq.	150	86		244	install 40 year laminated roof shakes.
Demolish and Install	Sq.	158	117		275	
Minimum Charge	Job		192		192	
Emergency Repairs						
Install	Ea.	154	192		346	Includes labor and material for
Minimum Charge	Job		192		192	emergency repairs.
Roof Jack						
Demolish	Ea.		6.25		6.25	Includes material and labor to install
Install	Ea.	26.50	42.50		69	residential roof jack, w/bird screen,
Demolish and Install	Ea.	26.50	48.75		75.25	backdraft damper, 3" & 4" dia. round
Paint	Ea.	.07	4.51		4.58	duct.
Minimum Charge	Job		230		230	
Ridge Vent						
Demolish	L.F.		1.11		1.11	Includes material and labor to install a
Install	L.F.	3.28	3.02		6.30	mill finish aluminum ridge vent strip.
Demolish and Install	L.F.	3.28	4.13		7.41	
Minimum Charge	Job		211		211	
Shingle Molding						
1" x 2"						
Demolish	L.F.		.22		.22	Cost includes material and labor to
Install	L.F.	.29	1.28		1.57	install 1" x 2" pine trim.
Demolish and Install	L.F.	.29	1.50		1.79	
Paint	L.F.	.02	.56		.58	
Minimum Charge	Job		211		211	
1" x 3"						
Demolish	L.F.		.22		.22	Cost includes material and labor to
Install	L.F.	.26	1.46		1.72	install 1" x 3" pine trim.
Demolish and Install	L.F.	.26	1.68		1.94	
Paint	L.F.	.02	.56		.58	
Minimum Charge	Job		211		211	
1" x 4"						
Demolish	L.F.		.22		.22	Cost includes material and labor to
Install	L.F.	.48	1.69		2.17	install 1" x 4" pine trim.
Demolish and Install	L.F.	.48	1.91		2.39	
Paint	L.F.	.02	.56		.58	
Minimum Charge	Job		211		211	
Add for Steep Pitch						
Install	Sq.		24		24	Includes labor and material added for
Add for Additional Story						2nd story or steep roofs.
Install	Sq.		24		24	Includes labor and material added for 2nd story or steep roofs.

Roofing

Wood Shingles		Unit	Material	Labor	Equip.	Total	Specification
Cedar Wood Shingles							
16" Red Label (#2)							
	Demolish	Sq.		76.50		76.50	Includes material and labor to install
	Install	Sq.	149	169		318	wood shingles/shakes.
	Demolish and Install	Sq.	149	245.50		394.50	
	Minimum Charge	Job		211		211	
16" Blue Label (#1)							
	Demolish	Sq.		76.50		76.50	Includes material and labor to install
	Install	Sq.	272	130		402	wood shingles/shakes.
	Demolish and Install	Sq.	272	206.50		478.50	
	Minimum Charge	Job		211		211	
18" Red Label (#2)							
	Demolish	Sq.		76.50		76.50	Includes material and labor to install
	Install	Sq.	315	108		423	wood shingles/shakes.
	Demolish and Install	Sq.	315	184.50		499.50	
	Minimum Charge	Job		211		211	
18" Blue Label (#1)							
	Demolish	Sq.		76.50		76.50	Includes material and labor to install
	Install	Sq.	289	118		407	wood shingles/shakes.
	Demolish and Install	Sq.	289	194.50		483.50	
	Minimum Charge	Job		211		211	
Replace Shingle							
	Install	Ea.	71.50	10.55		82.05	Includes labor and material to replace
	Minimum Charge	Job		211		211	individual wood shingle/shake. Cost is per shingle.
Cedar Shake							
18" Medium Handsplit							
	Demolish	Sq.		76.50		76.50	Includes material and labor to install
	Install	Sq.	222	163		385	wood shingles/shakes.
	Demolish and Install	Sq.	222	239.50		461.50	
	Minimum Charge	Job		211		211	
18" Heavy Handsplit							
	Demolish	Sq.		76.50		76.50	Includes material and labor to install
	Install	Sq.	222	181		403	wood shingles/shakes.
	Demolish and Install	Sq.	222	257.50		479.50	
	Minimum Charge	Job		211		211	
24" Heavy Handsplit							
	Demolish	Sq.		76.50		76.50	Includes material and labor to install
	Install	Sq.	253	145		398	wood shingles/shakes.
	Demolish and Install	Sq.	253	221.50		474.50	
	Minimum Charge	Job		211		211	
24" Medium Handsplit							
	Demolish	Sq.		76.50		76.50	Includes material and labor to install
	Install	Sq.	253	130		383	wood shingles/shakes.
	Demolish and Install	Sq.	253	206.50		459.50	
	Minimum Charge	Job		211		211	
24" Straight Split							
	Demolish	Sq.		76.50		76.50	Includes material and labor to install
	Install	Sq.	288	192		480	wood shingles/shakes.
	Demolish and Install	Sq.	288	268.50		556.50	
	Minimum Charge	Job		211		211	
24" Tapersplit							
	Demolish	Sq.		76.50		76.50	Includes material and labor to install
	Install	Sq.	340	192		532	wood shingles/shakes.
	Demolish and Install	Sq.	340	268.50		608.50	
	Minimum Charge	Job		211		211	

Roofing

Wood Shingles

Wood Shingles		Unit	Material	Labor	Equip.	Total	Specification
18" Straight Split							
	Demolish	Sq.		76.50		76.50	Includes material and labor to install
	Install	Sq.	288	192		480	wood shingles/shakes.
	Demolish and Install	Sq.	288	268.50		556.50	
	Minimum Charge	Job		211		211	
Replace Shingle							
	Install	Ea.	71.50	10.55		82.05	Includes labor and material to replace
	Minimum Charge	Job		211		211	individual wood shingle/shake. Cost is per shingle.

Slate Tile Roofing

Slate Tile Roofing		Unit	Material	Labor	Equip.	Total	Specification
Unfading Green							
	Demolish	Sq.		69		69	Includes material and labor to install
	Install	Sq.	530	221		751	clear Vermont slate tile.
	Demolish and Install	Sq.	530	290		820	
	Reinstall	Sq.		177.19		177.19	
	Minimum Charge	Job		194		194	
Unfading Purple							
	Demolish	Sq.		69		69	Includes material and labor to install
	Install	Sq.	470	221		691	clear Vermont slate tile.
	Demolish and Install	Sq.	470	290		760	
	Reinstall	Sq.		177.19		177.19	
	Minimum Charge	Job		194		194	
Replace Slate Tiles							
	Install	Ea.	6.50	20.50		27	Includes labor and material to replace
	Minimum Charge	Job		194		194	individual slate tiles. Cost is per tile.
Variegated Purple							
	Demolish	Sq.		69		69	Includes material and labor to install
	Install	Sq.	455	221		676	clear Vermont slate tile.
	Demolish and Install	Sq.	455	290		745	
	Reinstall	Sq.		177.19		177.19	
	Minimum Charge	Job		194		194	
Unfading Grey / Black							
	Demolish	Sq.		69		69	Includes material and labor to install
	Install	Sq.	505	221		726	clear Vermont slate tile.
	Demolish and Install	Sq.	505	290		795	
	Reinstall	Sq.		177.19		177.19	
	Minimum Charge	Job		194		194	
Unfading Red							
	Demolish	Sq.		69		69	Includes material and labor to install
	Install	Sq.	1300	221		1521	clear Vermont slate tile.
	Demolish and Install	Sq.	1300	290		1590	
	Reinstall	Sq.		177.19		177.19	
	Minimum Charge	Job		194		194	
Weathering Black							
	Demolish	Sq.		69		69	Includes material and labor to install
	Install	Sq.	540	221		761	clear Pennsylvania slate tile.
	Demolish and Install	Sq.	540	290		830	
	Reinstall	Sq.		177.19		177.19	
	Minimum Charge	Job		194		194	

Roofing

Slate Tile Roofing

Slate Tile Roofing	Unit	Material	Labor	Equip.	Total	Specification
Weathering Green						
Demolish	Sq.		69		69	Includes material and labor to install
Install	Sq.	385	221		606	clear Vermont slate tile.
Demolish and Install	Sq.	385	290		675	
Reinstall	Sq.		177.19		177.19	
Minimum Charge	Job		194		194	

Flat Clay Tile Roofing

Flat Clay Tile Roofing	Unit	Material	Labor	Equip.	Total	Specification
Glazed						
Demolish	Sq.		76.50		76.50	Includes material and labor to install
Install	Sq.	390	194		584	tile roofing. Cost based on terra cotta
Demolish and Install	Sq.	390	270.50		660.50	red flat clay tile and includes ridge
Reinstall	Sq.		155.04		155.04	and rake tiles.
Minimum Charge	Job		194		194	
Terra Cotta Red						
Demolish	Sq.		76.50		76.50	Includes material and labor to install
Install	Sq.	390	194		584	tile roofing. Cost based on terra cotta
Demolish and Install	Sq.	390	270.50		660.50	red flat clay tile and includes ridge
Reinstall	Sq.		155.04		155.04	and rake tiles.
Minimum Charge	Job		194		194	

Mission Tile Roofing

Mission Tile Roofing	Unit	Material	Labor	Equip.	Total	Specification
Glazed Red						
Demolish	Sq.		76.50		76.50	Includes material and labor to install
Install	Sq.	915	211		1126	tile roofing. Cost based on glazed red
Demolish and Install	Sq.	915	287.50		1202.50	tile and includes birdstop, ridge and
Reinstall	Sq.		169.13		169.13	rake tiles.
Minimum Charge	Job		194		194	
Unglazed Red						
Demolish	Sq.		76.50		76.50	Includes material and labor to install
Install	Sq.	915	211		1126	tile roofing. Cost based on unglazed
Demolish and Install	Sq.	915	287.50		1202.50	red tile and includes birdstop, ridge
Reinstall	Sq.		169.13		169.13	and rake tiles.
Minimum Charge	Job		194		194	
Glazed Blue						
Demolish	Sq.		76.50		76.50	Includes material and labor to install
Install	Sq.	915	211		1126	tile roofing. Cost based on glazed
Demolish and Install	Sq.	915	287.50		1202.50	blue tile and includes birdstop, ridge
Reinstall	Sq.		169.13		169.13	and rake tiles.
Minimum Charge	Job		194		194	
Glazed White						
Demolish	Sq.		76.50		76.50	Includes material and labor to install
Install	Sq.	915	211		1126	tile roofing. Cost based on glazed
Demolish and Install	Sq.	915	287.50		1202.50	white tile and includes birdstop, ridge
Reinstall	Sq.		169.13		169.13	and rake tiles.
Minimum Charge	Job		194		194	
Unglazed White						
Demolish	Sq.		76.50		76.50	Includes material and labor to install
Install	Sq.	915	211		1126	tile roofing. Cost based on unglazed
Demolish and Install	Sq.	915	287.50		1202.50	white tile and includes birdstop, ridge
Reinstall	Sq.		169.13		169.13	and rake tiles.
Minimum Charge	Job		194		194	

Roofing

Mission Tile Roofing

Mission Tile Roofing	Unit	Material	Labor	Equip.	Total	Specification
Color Blend						
Demolish	Sq.		76.50		76.50	Includes material and labor to install
Install	Sq.	915	211		1126	tile roofing. Cost based on color blend
Demolish and Install	Sq.	915	287.50		1202.50	tile and includes birdstop, ridge and
Reinstall	Sq.		169.13		169.13	rake tiles.
Minimum Charge	Job		194		194	
Accessories / Extras						
Add for Cap Furring Strips						
Install	Sq.	98	242		340	Includes labor and material to install vertically placed wood furring strips under cap tiles.
Add for Tile Adhesive						
Install	Sq.	23	14.15		37.15	Includes material and labor for adhesive material placed between tiles.
Add for Wire Attachment						
Install	Sq.	15.15	169		184.15	Includes labor and material to install single wire anchors for attaching tile.
Add for Braided Runners						
Install	Sq.	38	162		200	Includes labor and material to install braided wire runners for secure attachment in high wind areas.
Add for Stainless Steel Nails						
Install	Sq.	25.50			25.50	Includes material only for stainless steel nails.
Add for Brass Nails						
Install	Sq.	44			44	Includes material only for brass nails.
Add for Hurricane Clips						
Install	Sq.	22.50	228		250.50	Includes material and labor for Hurricane or wind clips for high wind areas.
Add for Copper Nails						
Install	Sq.	44			44	Includes material only for copper nails.
Add for Vertical Furring Strips						
Install	Sq.	19.30	45		64.30	Includes labor and material to install vertically placed wood furring strips under tiles.
Add for Horizontal Furring Strips						
Install	Sq.	12.45	20		32.45	Includes labor and material to install horizontally placed wood furring strips under tiles.

Spanish Tile Roofing

Spanish Tile Roofing	Unit	Material	Labor	Equip.	Total	Specification
Glazed Red						
Demolish	Sq.		76.50		76.50	Includes material and labor to install
Install	Sq.	755	162		917	tile roofing. Cost based on glazed red
Demolish and Install	Sq.	755	238.50		993.50	tile and includes birdstop, ridge and
Reinstall	Sq.		129.20		129.20	rake tiles.
Minimum Charge	Job		194		194	

Roofing

Spanish Tile Roofing		Unit	Material	Labor	Equip.	Total	Specification
Glazed White							
	Demolish	Sq.		76.50		76.50	Includes material and labor to install
	Install	Sq.	755	215		970	tile roofing. Cost based on glazed
	Demolish and Install	Sq.	755	291.50		1046.50	white tile and includes birdstop, ridge
	Reinstall	Sq.		172.27		172.27	and rake tiles.
	Minimum Charge	Job		194		194	
Glazed Blue							
	Demolish	Sq.		76.50		76.50	Includes material and labor to install
	Install	Sq.	755	162		917	tile roofing. Cost based on glazed
	Demolish and Install	Sq.	755	238.50		993.50	blue tile and includes birdstop, ridge
	Reinstall	Sq.		129.20		129.20	and rake tiles.
	Minimum Charge	Job		194		194	
Unglazed Red							
	Demolish	Sq.		76.50		76.50	Includes material and labor to install
	Install	Sq.	590	162		752	tile roofing. Cost based on unglazed
	Demolish and Install	Sq.	590	238.50		828.50	red tile and includes birdstop, ridge
	Reinstall	Sq.		129.20		129.20	and rake tiles.
	Minimum Charge	Job		194		194	
Unglazed White							
	Demolish	Sq.		76.50		76.50	Includes material and labor to install
	Install	Sq.	660	162		822	tile roofing. Cost based on unglazed
	Demolish and Install	Sq.	660	238.50		898.50	white tile and includes birdstop, ridge
	Reinstall	Sq.		129.20		129.20	and rake tiles.
	Minimum Charge	Job		194		194	
Color Blend							
	Demolish	Sq.		76.50		76.50	Includes material and labor to install
	Install	Sq.	660	162		822	tile roofing. Cost based on color blend
	Demolish and Install	Sq.	660	238.50		898.50	tile and includes birdstop, ridge and
	Reinstall	Sq.		129.20		129.20	rake tiles.
	Minimum Charge	Job		194		194	
Accessories / Extras							
Add for Vertical Furring Strips							
	Install	Sq.	19.30	45		64.30	Includes labor and material to install vertically placed wood furring strips under tiles.
Add for Wire Attachment							
	Install	Sq.	7.75	129		136.75	Includes labor and material to install single wire ties.
Add for Braided Runners							
	Install	Sq.	29	111		140	Includes labor and material to install braided wire runners for secure attachment in high wind areas.
Add for Stainless Steel Nails							
	Install	Sq.	19			19	Includes material only for stainless steel nails.
Add for Copper Nails							
	Install	Sq.	33			33	Includes material only for copper nails.
Add for Brass Nails							
	Install	Sq.	32.50			32.50	Includes material only for brass nails.
Add for Hurricane Clips							
	Install	Sq.	11.60	134		145.60	Includes material and labor to install hurricane or wind clips for high wind areas.

Roofing

Spanish Tile Roofing		Unit	Material	Labor	Equip.	Total	Specification
Add for Mortar Set Tiles	Install	Sq.	48	194		242	Includes material and labor to install setting tiles by using mortar material placed between tiles.
Add for Tile Adhesive	Install	Sq.	11.75	12.60		24.35	Includes material and labor for adhesive placed between tiles.
Add for Cap Furring Strips	Install	Sq.	98	242		340	Includes labor and material to install vertically placed wood furring strips under cap tiles.
Add for Horizontal Furring Strips	Install	Sq.	12.45	20		32.45	Includes labor and material to install horizontally placed wood furring strips under tiles.

Concrete Tile Roofing		Unit	Material	Labor	Equip.	Total	Specification
Corrugated Red / Brown	Demolish	Sq.		76.50		76.50	Includes material and labor to install tile roofing. Cost includes birdstop, booster, ridge and rake tiles.
	Install	Sq.	107	287		394	
	Demolish and Install	Sq.	107	363.50		470.50	
	Reinstall	Sq.		229.69		229.69	
	Minimum Charge	Job		194		194	
Corrugated Gray	Demolish	Sq.		76.50		76.50	Includes material and labor to install tile roofing. Cost includes birdstop, booster, ridge and rake tiles.
	Install	Sq.	107	287		394	
	Demolish and Install	Sq.	107	363.50		470.50	
	Reinstall	Sq.		229.69		229.69	
	Minimum Charge	Job		194		194	
Corrugated Bright Red	Demolish	Sq.		76.50		76.50	Includes material and labor to install tile roofing. Cost includes birdstop, booster, ridge and rake tiles.
	Install	Sq.	111	287		398	
	Demolish and Install	Sq.	111	363.50		474.50	
	Reinstall	Sq.		229.69		229.69	
	Minimum Charge	Job		194		194	
Corrugated Black	Demolish	Sq.		76.50		76.50	Includes material and labor to install tile roofing. Cost includes birdstop, booster, ridge and rake tiles.
	Install	Sq.	111	287		398	
	Demolish and Install	Sq.	111	363.50		474.50	
	Reinstall	Sq.		229.69		229.69	
	Minimum Charge	Job		194		194	
Corrugated Green	Demolish	Sq.		76.50		76.50	Includes material and labor to install tile roofing. Cost includes birdstop, booster, ridge and rake tiles.
	Install	Sq.	111	287		398	
	Demolish and Install	Sq.	111	363.50		474.50	
	Reinstall	Sq.		229.69		229.69	
	Minimum Charge	Job		194		194	
Corrugated Blue	Demolish	Sq.		76.50		76.50	Includes material and labor to install tile roofing. Cost includes birdstop, booster, ridge and rake tiles.
	Install	Sq.	111	287		398	
	Demolish and Install	Sq.	111	363.50		474.50	
	Reinstall	Sq.		229.69		229.69	
	Minimum Charge	Job		194		194	

Roofing

Concrete Tile Roofing	Unit	Material	Labor	Equip.	Total	Specification
Flat Natural Gray						
Demolish	Sq.		76.50		76.50	Includes material and labor to install
Install	Sq.	101	287		388	tile roofing. Cost includes birdstop,
Demolish and Install	Sq.	101	363.50		464.50	booster, ridge and rake tiles.
Reinstall	Sq.		229.69		229.69	
Minimum Charge	Job		194		194	
Flat Red / Brown						
Demolish	Sq.		76.50		76.50	Includes material and labor to install
Install	Sq.	101	287		388	tile roofing. Cost includes birdstop,
Demolish and Install	Sq.	101	363.50		464.50	booster, ridge and rake tiles.
Reinstall	Sq.		229.69		229.69	
Minimum Charge	Job		194		194	
Flat Bright Red						
Demolish	Sq.		76.50		76.50	Includes material and labor to install
Install	Sq.	101	287		388	tile roofing. Cost includes birdstop,
Demolish and Install	Sq.	101	363.50		464.50	booster, ridge and rake tiles.
Reinstall	Sq.		229.69		229.69	
Minimum Charge	Job		194		194	
Flat Green						
Demolish	Sq.		76.50		76.50	Includes material and labor to install
Install	Sq.	101	287		388	tile roofing. Cost includes birdstop,
Demolish and Install	Sq.	101	363.50		464.50	booster, ridge and rake tiles.
Minimum Charge	Job		194		194	
Flat Blue						
Demolish	Sq.		76.50		76.50	Includes material and labor to install
Install	Sq.	101	287		388	tile roofing. Cost includes birdstop,
Demolish and Install	Sq.	101	363.50		464.50	booster, ridge and rake tiles.
Reinstall	Sq.		229.69		229.69	
Minimum Charge	Job		194		194	
Flat Black						
Demolish	Sq.		76.50		76.50	Includes material and labor to install
Install	Sq.	101	287		388	tile roofing. Cost includes birdstop,
Demolish and Install	Sq.	101	363.50		464.50	booster, ridge and rake tiles.
Reinstall	Sq.		229.69		229.69	
Minimum Charge	Job		194		194	

Aluminum Sheet	Unit	Material	Labor	Equip.	Total	Specification
Corrugated						
.019″ Thick Natural						
Demolish	S.F.		.81		.81	Includes material and labor to install
Install	S.F.	1.43	1.95		3.38	aluminum sheet metal roofing.
Demolish and Install	S.F.	1.43	2.76		4.19	
Reinstall	S.F.		1.56		1.56	
Minimum Charge	Job		470		470	
.016″ Thick Natural						
Demolish	S.F.		.81		.81	Includes material and labor to install
Install	S.F.	1.09	1.95		3.04	aluminum sheet metal roofing.
Demolish and Install	S.F.	1.09	2.76		3.85	
Reinstall	S.F.		1.56		1.56	
Minimum Charge	Job		470		470	
.016″ Colored Finish						
Demolish	S.F.		.81		.81	Includes material and labor to install
Install	S.F.	1.58	1.95		3.53	aluminum sheet metal roofing.
Demolish and Install	S.F.	1.58	2.76		4.34	
Reinstall	S.F.		1.56		1.56	
Minimum Charge	Job		470		470	

Roofing

Aluminum Sheet		Unit	Material	Labor	Equip.	Total	Specification
.019″ Colored Finish							
	Demolish	S.F.		.81		.81	Includes material and labor to install
	Install	S.F.	1.76	1.95		3.71	aluminum sheet metal roofing.
	Demolish and Install	S.F.	1.76	2.76		4.52	
	Reinstall	S.F.		1.56		1.56	
	Minimum Charge	Job		470		470	
Ribbed							
.019″ Thick Natural							
	Demolish	S.F.		.81		.81	Includes material and labor to install
	Install	S.F.	1.43	1.95		3.38	aluminum sheet metal roofing.
	Demolish and Install	S.F.	1.43	2.76		4.19	
	Reinstall	S.F.		1.56		1.56	
	Minimum Charge	Job		470		470	
.016″ Thick Natural							
	Demolish	S.F.		.81		.81	Includes material and labor to install
	Install	S.F.	1.09	1.95		3.04	aluminum sheet metal roofing.
	Demolish and Install	S.F.	1.09	2.76		3.85	
	Reinstall	S.F.		1.56		1.56	
	Minimum Charge	Job		470		470	
.050″ Colored Finish							
	Demolish	S.F.		.81		.81	Includes material and labor to install
	Install	S.F.	4.90	1.95		6.85	aluminum sheet metal roofing.
	Demolish and Install	S.F.	4.90	2.76		7.66	
	Reinstall	S.F.		1.56		1.56	
	Minimum Charge	Job		470		470	
.032″ Thick Natural							
	Demolish	S.F.		.81		.81	Includes material and labor to install
	Install	S.F.	2.78	1.95		4.73	aluminum sheet metal roofing.
	Demolish and Install	S.F.	2.78	2.76		5.54	
	Reinstall	S.F.		1.56		1.56	
	Minimum Charge	Job		470		470	
.040″ Thick Natural							
	Demolish	S.F.		.81		.81	Includes material and labor to install
	Install	S.F.	3.38	1.95		5.33	aluminum sheet metal roofing.
	Demolish and Install	S.F.	3.38	2.76		6.14	
	Reinstall	S.F.		1.56		1.56	
	Minimum Charge	Job		470		470	
.050″ Thick Natural							
	Demolish	S.F.		.81		.81	Includes material and labor to install
	Install	S.F.	4.06	1.95		6.01	aluminum sheet metal roofing.
	Demolish and Install	S.F.	4.06	2.76		6.82	
	Reinstall	S.F.		1.56		1.56	
	Minimum Charge	Job		470		470	
.016″ Colored Finish							
	Demolish	S.F.		.81		.81	Includes material and labor to install
	Install	S.F.	1.58	1.95		3.53	aluminum sheet metal roofing.
	Demolish and Install	S.F.	1.58	2.76		4.34	
	Reinstall	S.F.		1.56		1.56	
	Minimum Charge	Job		470		470	
.019″ Colored Finish							
	Demolish	S.F.		.81		.81	Includes material and labor to install
	Install	S.F.	1.76	1.95		3.71	aluminum sheet metal roofing.
	Demolish and Install	S.F.	1.76	2.76		4.52	
	Reinstall	S.F.		1.56		1.56	
	Minimum Charge	Job		470		470	

Roofing

Aluminum Sheet

Aluminum Sheet	Unit	Material	Labor	Equip.	Total	Specification
.032″ Colored Finish						
Demolish	S.F.		.81		.81	Includes material and labor to install
Install	S.F.	3.41	1.95		5.36	aluminum sheet metal roofing.
Demolish and Install	S.F.	3.41	2.76		6.17	
Reinstall	S.F.		1.56		1.56	
Minimum Charge	Job		470		470	
.040″ Colored Finish						
Demolish	S.F.		.81		.81	Includes material and labor to install
Install	S.F.	4.07	1.95		6.02	aluminum sheet metal roofing.
Demolish and Install	S.F.	4.07	2.76		6.83	
Reinstall	S.F.		1.56		1.56	
Minimum Charge	Job		470		470	

Fiberglass Sheet

Fiberglass Sheet	Unit	Material	Labor	Equip.	Total	Specification
12 Ounce Corrugated						
Demolish	S.F.		1.46		1.46	Cost includes material and labor to
Install	S.F.	3.91	1.56		5.47	install 12 ounce corrugated fiberglass
Demolish and Install	S.F.	3.91	3.02		6.93	sheet roofing.
Reinstall	S.F.		1.25		1.25	
Minimum Charge	Job		192		192	
8 Ounce Corrugated						
Demolish	S.F.		1.46		1.46	Cost includes material and labor to
Install	S.F.	2.29	1.56		3.85	install 8 ounce corrugated fiberglass
Demolish and Install	S.F.	2.29	3.02		5.31	roofing.
Reinstall	S.F.		1.25		1.25	
Minimum Charge	Job		192		192	

Galvanized Steel

Galvanized Steel	Unit	Material	Labor	Equip.	Total	Specification
Corrugated						
24 Gauge						
Demolish	S.F.		.81		.81	Includes material and labor to install
Install	S.F.	3.43	1.64		5.07	galvanized steel sheet metal roofing.
Demolish and Install	S.F.	3.43	2.45		5.88	Cost based on 24 gauge corrugated
Reinstall	S.F.		1.31		1.31	or ribbed steel roofing.
Minimum Charge	Job		192		192	
26 Gauge						
Demolish	S.F.		.81		.81	Includes material and labor to install
Install	S.F.	2.18	1.56		3.74	galvanized steel sheet metal roofing.
Demolish and Install	S.F.	2.18	2.37		4.55	Cost based on 26 gauge corrugated
Reinstall	S.F.		1.25		1.25	or ribbed steel roofing.
Minimum Charge	Job		192		192	
28 Gauge						
Demolish	S.F.		.81		.81	Includes material and labor to install
Install	S.F.	2.11	1.48		3.59	galvanized steel sheet metal roofing.
Demolish and Install	S.F.	2.11	2.29		4.40	Cost based on 28 gauge corrugated
Reinstall	S.F.		1.19		1.19	or ribbed steel roofing.
Minimum Charge	Job		192		192	
30 Gauge						
Demolish	S.F.		.81		.81	Includes material and labor to install
Install	S.F.	1.98	1.42		3.40	galvanized steel sheet metal roofing.
Demolish and Install	S.F.	1.98	2.23		4.21	Cost based on 30 gauge corrugated
Reinstall	S.F.		1.13		1.13	or ribbed steel roofing.
Minimum Charge	Job		192		192	

Roofing

Galvanized Steel		Unit	Material	Labor	Equip.	Total	Specification
Ribbed							
24 Gauge							
	Demolish	S.F.		.81		.81	Includes material and labor to install
	Install	S.F.	3.43	1.64		5.07	galvanized steel sheet metal roofing.
	Demolish and Install	S.F.	3.43	2.45		5.88	Cost based on 24 gauge corrugated
	Reinstall	S.F.		1.31		1.31	or ribbed steel roofing.
	Minimum Charge	Job		192		192	
26 Gauge							
	Demolish	S.F.		.81		.81	Includes material and labor to install
	Install	S.F.	2.18	1.56		3.74	galvanized steel sheet metal roofing.
	Demolish and Install	S.F.	2.18	2.37		4.55	Cost based on 26 gauge corrugated
	Reinstall	S.F.		1.25		1.25	or ribbed steel roofing.
	Minimum Charge	Job		192		192	
28 Gauge							
	Demolish	S.F.		.81		.81	Includes material and labor to install
	Install	S.F.	2.11	1.48		3.59	galvanized steel sheet metal roofing.
	Demolish and Install	S.F.	2.11	2.29		4.40	Cost based on 28 gauge corrugated
	Reinstall	S.F.		1.19		1.19	or ribbed steel roofing.
	Minimum Charge	Job		192		192	
30 Gauge							
	Demolish	S.F.		.81		.81	Includes material and labor to install
	Install	S.F.	1.98	1.42		3.40	galvanized steel sheet metal roofing.
	Demolish and Install	S.F.	1.98	2.23		4.21	Cost based on 30 gauge corrugated
	Reinstall	S.F.		1.13		1.13	or ribbed steel roofing.
	Minimum Charge	Job		192		192	

Standing Seam		Unit	Material	Labor	Equip.	Total	Specification
Copper							
16 Ounce							
	Demolish	Sq.		81.50		81.50	Includes material and labor to install
	Install	Sq.	1050	360		1410	copper standing seam metal roofing.
	Demolish and Install	Sq.	1050	441.50		1491.50	Cost based on 16 ounce copper
	Reinstall	Sq.		288.49		288.49	roofing.
	Minimum Charge	Job		234		234	
18 Ounce							
	Demolish	Sq.		81.50		81.50	Includes material and labor to install
	Install	Sq.	1150	390		1540	copper standing seam metal roofing.
	Demolish and Install	Sq.	1150	471.50		1621.50	Cost based on 18 ounce copper
	Reinstall	Sq.		312.53		312.53	roofing.
	Minimum Charge	Job		234		234	
20 Ounce							
	Demolish	Sq.		81.50		81.50	Includes material and labor to install
	Install	Sq.	1350	425		1775	copper standing seam metal roofing.
	Demolish and Install	Sq.	1350	506.50		1856.50	Cost based on 20 ounce copper
	Reinstall	Sq.		340.95		340.95	roofing.
Stainless Steel							
26 Gauge							
	Demolish	Sq.		81.50		81.50	Cost includes material and labor to
	Install	Sq.	565	410		975	install 26 gauge stainless steel
	Demolish and Install	Sq.	565	491.50		1056.50	standing seam metal roofing.
	Reinstall	Sq.		326.12		326.12	
	Minimum Charge	Job		234		234	

Roofing

Standing Seam

	Unit	Material	Labor	Equip.	Total	Specification
28 Gauge						Includes material and labor to install
Demolish	Sq.		81.50		81.50	standing seam metal roofing. Cost
Install	Sq.	610	390		1000	based on 28 gauge stainless steel
Demolish and Install	Sq.	610	471.50		1081.50	roofing.
Reinstall	Sq.		312.53		312.53	
Minimum Charge	Job		234		234	

Elastomeric Roofing

	Unit	Material	Labor	Equip.	Total	Specification
Neoprene						
1/16"						Includes material and labor to install
Demolish	Sq.		28		28	neoprene membrane, fully adhered,
Install	Sq.	245	184	41	470	1/16" thick.
Demolish and Install	Sq.	245	212	41	498	
Minimum Charge	Job		880	108	988	
GRM Membrane						
1/16"						Includes material and labor to install
Demolish	Sq.		28		28	coal tar based elastomeric roofing
Install	Sq.	78	293	36	407	membrane, polyester fiber reinforced.
Demolish and Install	Sq.	78	321	36	435	
Minimum Charge	Job		880	108	988	
Cant Strips						
3"						Includes material and labor to install
Demolish	L.F.		.88		.88	cants 3" x 3", treated timber, cut
Install	L.F.	1.01	1.18		2.19	diagonally.
Demolish and Install	L.F.	1.01	2.06		3.07	
Minimum Charge	Job		96		96	
4"						Includes material and labor to install
Demolish	L.F.		.88		.88	cants 4" x 4", treated timber, cut
Install	L.F.	2.09	1.18		3.27	diagonally.
Demolish and Install	L.F.	2.09	2.06		4.15	
Minimum Charge	Job		96		96	
Polyvinyl Chloride (PVC)						
45 mil (loose laid)						Cost includes material and labor to
Demolish	Sq.		28		28	install 45 mil PVC elastomeric roofing,
Install	Sq.	103	34.50	4.25	141.75	loose-laid with stone.
Demolish and Install	Sq.	103	62.50	4.25	169.75	
Minimum Charge	Job		880	108	988	
48 mil (loose laid)						Cost includes material and labor to
Demolish	Sq.		28		28	install 48 mil PVC elastomeric roofing,
Install	Sq.	124	34.50	4.25	162.75	loose-laid with stone.
Demolish and Install	Sq.	124	62.50	4.25	190.75	
Minimum Charge	Job		880	108	988	
60 mil (loose laid)						Cost includes material and labor to
Demolish	Sq.		28		28	install 60 mil PVC elastomeric roofing,
Install	Sq.	127	34.50	4.25	165.75	loose-laid with stone.
Demolish and Install	Sq.	127	62.50	4.25	193.75	
Minimum Charge	Job		880	108	988	
45 mil (attached)						Cost includes material and labor to
Demolish	Sq.		45		45	install 45 mil PVC elastomeric roofing
Install	Sq.	93	67.50	8.35	168.85	fully attached to roof deck.
Demolish and Install	Sq.	93	112.50	8.35	213.85	
Minimum Charge	Job		880	108	988	

Elastomeric Roofing	Unit	Material	Labor	Equip.	Total	Specification
48 mil (attached)						
Demolish	Sq.		45		45	Cost includes material and labor to
Install	Sq.	155	67.50	8.35	230.85	install 48 mil PVC elastomeric roofing
Demolish and Install	Sq.	155	112.50	8.35	275.85	fully attached to roof deck.
Minimum Charge	Job		880	108	988	
60 mil (attached)						
Demolish	Sq.		45		45	Cost includes material and labor to
Install	Sq.	157	67.50	8.35	232.85	install 60 mil PVC elastomeric roofing
Demolish and Install	Sq.	157	112.50	8.35	277.85	fully attached to roof deck.
Minimum Charge	Job		880	108	988	
CSPE Type						
45 mil (loose laid w / stone)						
Demolish	Sq.		28		28	Cost includes material and labor to
Install	Sq.	223	34.50	4.25	261.75	install 45 mil CSPE (chlorosulfonated
Demolish and Install	Sq.	223	62.50	4.25	289.75	polyethylene-hypalon) roofing material
Minimum Charge	Job		880	108	988	loose laid on deck and ballasted with
						stone.
45 mil (attached at seams)						
Demolish	Sq.		33.50		33.50	Cost includes material and labor to
Install	Sq.	215	50	6.20	271.20	install 45 mil CSPE (chlorosulfonated
Demolish and Install	Sq.	215	83.50	6.20	304.70	polyethylene-hypalon) roofing material
Minimum Charge	Job		880	108	988	attached to deck at seams with batten
						strips.
45 mil (fully attached)						
Demolish	Sq.		45		45	Cost includes material and labor to
Install	Sq.	213	67.50	8.35	288.85	install 45 mil CSPE (chlorosulfonated
Demolish and Install	Sq.	213	112.50	8.35	333.85	polyethylene-hypalon) roofing material
Minimum Charge	Job		880	108	988	fully attached to deck.
EPDM Type						
45 mil (attached at seams)						
Demolish	Sq.		33.50		33.50	Cost includes material and labor to
Install	Sq.	84.50	50	6.20	140.70	install 45 mil EPDM (ethylene
Demolish and Install	Sq.	84.50	83.50	6.20	174.20	propylene diene monomer) roofing
Minimum Charge	Job		880	108	988	material attached to deck at seams.
45 mil (fully attached)						
Demolish	Sq.		45		45	Cost includes material and labor to
Install	Sq.	125	67.50	8.35	200.85	install 45 mil EPDM (ethylene
Demolish and Install	Sq.	125	112.50	8.35	245.85	propylene diene monomer) roofing
Minimum Charge	Job		880	108	988	material fully attached to deck.
45 mil (loose laid w / stone)						
Demolish	Sq.		28		28	Cost includes material and labor to
Install	Sq.	91.50	34.50	4.25	130.25	install 45 mil EPDM (ethylene
Demolish and Install	Sq.	91.50	62.50	4.25	158.25	propylene diene monomer) roofing
Minimum Charge	Job		880	108	988	material loose laid on deck and
						ballasted with stones (approx. 1/2 ton
						of stone per square).
60 mil (attached at seams)						
Demolish	Sq.		33.50		33.50	Cost includes material and labor to
Install	Sq.	101	50	6.20	157.20	install 60 mil EPDM (ethylene
Demolish and Install	Sq.	101	83.50	6.20	190.70	propylene diene monomer) roofing
Minimum Charge	Job		880	108	988	material attached to deck at seams.
60 mil (fully attached)						
Demolish	Sq.		45		45	Cost includes material and labor to
Install	Sq.	142	67.50	8.35	217.85	install 60 mil EPDM (ethylene
Demolish and Install	Sq.	142	112.50	8.35	262.85	propylene diene monomer) roofing
Minimum Charge	Job		880	108	988	material fully attached to deck.

Roofing

Elastomeric Roofing

	Unit	Material	Labor	Equip.	Total	Specification
60 mil (loose laid w / stone)						
Demolish	Sq.		28		28	Cost includes material and labor to
Install	Sq.	109	34.50	4.25	147.75	install 60 mil EPDM (ethylene
Demolish and Install	Sq.	109	62.50	4.25	175.75	propylene diene monomer) roofing
Minimum Charge	Job		880	108	988	material loose laid on deck and
						ballasted with stones (approx. 1/2 ton
						of stone per square).

Modified Bitumen Roofing

	Unit	Material	Labor	Equip.	Total	Specification
Fully Attached						
120 mil						
Demolish	Sq.		67.50		67.50	Cost includes material and labor to
Install	Sq.	76	8.55	1.91	86.46	install 120 mil modified bitumen
Demolish and Install	Sq.	76	76.05	1.91	153.96	roofing, fully attached to deck by
Minimum Charge	Job		96		96	torch.
150 mil						
Demolish	Sq.		67.50		67.50	Cost includes material and labor to
Install	Sq.	83.50	21	4.69	109.19	install 150 mil modified bitumen
Demolish and Install	Sq.	83.50	88.50	4.69	176.69	roofing, fully attached to deck by
Minimum Charge	Job		96		96	torch.
160 mil						
Demolish	Sq.		67.50		67.50	Cost includes material and labor to
Install	Sq.	95.50	22	4.89	122.39	install 160 mil modified bitumen
Demolish and Install	Sq.	95.50	89.50	4.89	189.89	roofing, fully attached to deck by
Minimum Charge	Job		96		96	torch.

Built-up Roofing

	Unit	Material	Labor	Equip.	Total	Specification
Emergency Repairs						
Install	Ea.	242	1800	400	2442	Minimum charge for work requiring
Minimum Charge	Job		1275	281	1556	hot asphalt.
Asphalt						
3-ply Asphalt						
Demolish	Sq.		70		70	Cost includes material and labor to
Install	Sq.	79	105	23.50	207.50	install 3-ply asphalt built-up roof
Demolish and Install	Sq.	79	175	23.50	277.50	including one 30 lb and two 15 lb
Minimum Charge	Job		1275	281	1556	layers of felt.
4-ply Asphalt						
Demolish	Sq.		83.50		83.50	Cost includes material and labor to
Install	Sq.	108	115	25.50	248.50	install 4-ply asphalt built-up roof
Demolish and Install	Sq.	108	198.50	25.50	332	including one 30 lb and three 15 lb
Minimum Charge	Job		1275	281	1556	layers of felt.
5-ply Asphalt						
Demolish	Sq.		109		109	Cost includes material and labor to
Install	Sq.	145	126	28	299	install 5-ply asphalt built-up roof
Demolish and Install	Sq.	145	235	28	408	including one 30 lb and four 15 lb
Minimum Charge	Job		1275	281	1556	layers of felt.

Roofing

Built-up Roofing

Built-up Roofing		Unit	Material	Labor	Equip.	Total	Specification
Components / Accessories							
Flood Coat	Install	Sq.	30.50	59		89.50	Includes labor and material to install flood coat.
R / R Gravel and Flood Coat	Install	Sq.	35	192		227	Includes labor and material to install flood coat and gravel.
Aluminum UV Coat	Install	Sq.	23	219	27	269	Includes labor and material to install aluminum-based UV coating.
Gravel Coat	Demolish	Sq.		35		35	Includes labor and material to install gravel coat.
	Install	Sq.	5.10	32		37.10	
	Demolish and Install	Sq.	5.10	67		72.10	
Mineral Surface Cap Sheet	Install	Sq.	102	6.65		108.65	Cost includes labor and material to install 50 # mineral surfaced cap sheet.
Fibrated Aluminum UV Coating	Install	Sq.	23	59		82	Includes labor and material to install fibered aluminum-based UV coating.
Impregnated Walk							
1"	Install	S.F.	3.09	2.05		5.14	Cost includes labor and material to install 1" layer of asphalt coating.
1/2"	Install	S.F.	2.21	1.62		3.83	Cost includes labor and material to install 1/2" layer of asphalt coating.
3/4"	Install	S.F.	2.76	1.71		4.47	Cost includes labor and material to install 3/4" layer of asphalt coating.

Exterior Doors and Windows

Doors	Roof Window	Double Hung Window

Exterior Type Door		Unit	Material	Labor	Equip.	Total	Specification
Entry							
Flush (1-3/4″)							
	Demolish	Ea.		39		39	Cost includes material and labor to
	Install	Ea.	252	26.50		278.50	install 3′ x 6′ 8″ pre-hung 1-3/4″
	Demolish and Install	Ea.	252	65.50		317.50	door, including casing and stop,
	Reinstall	Ea.		26.43		26.43	hinges, jamb, aluminum sill,
	Clean	Ea.	3.27	12.90		16.17	weatherstripped.
	Paint	Ea.	10.10	51.50		61.60	
	Minimum Charge	Job		211		211	
Raised Panel							
	Demolish	Ea.		39		39	Cost includes material and labor to
	Install	Ea.	485	23.50		508.50	install 3′ x 6′ 8″ pre-hung 1-3/4″
	Demolish and Install	Ea.	485	62.50		547.50	door, jamb, including casing and
	Reinstall	Ea.		23.49		23.49	stop, hinges, aluminum sill,
	Clean	Ea.	3.27	12.90		16.17	weatherstripping.
	Paint	Ea.	10.10	51.50		61.60	
	Minimum Charge	Job		211		211	
Entry w / Lights							
Good Quality							
	Demolish	Ea.		39		39	Cost includes material and labor to
	Install	Ea.	350	53		403	install 3′ x 6′ 8″ pre-hung 1-3/4″
	Demolish and Install	Ea.	350	92		442	door, solid wood lower panel, with up
	Reinstall	Ea.		52.85		52.85	to 9-light glass panels, casing, sill,
	Clean	Ea.	3.27	12.90		16.17	weatherstripped.
	Paint	Ea.	6.75	24		30.75	
	Minimum Charge	Job		211		211	
Better Quality							
	Demolish	Ea.		39		39	Cost includes material and labor to
	Install	Ea.	360	53		413	install 3′ x 6′ 8″ pre-hung 1-3/4″ Ash
	Demolish and Install	Ea.	360	92		452	door, solid wood, raised panel door
	Reinstall	Ea.		52.85		52.85	without glass panels.
	Clean	Ea.	3.27	12.90		16.17	
	Paint	Ea.	10.10	51.50		61.60	
	Minimum Charge	Job		211		211	
Premium Quality							
	Demolish	Ea.		39		39	Cost includes material and labor to
	Install	Ea.	965	25		990	install 3′ x 6′ 8″ pre-hung 1-3/4″ Oak
	Demolish and Install	Ea.	965	64		1029	door, solid wood, raised panel door
	Reinstall	Ea.		24.87		24.87	without glass panels.
	Clean	Ea.	3.27	12.90		16.17	
	Paint	Ea.	10.10	51.50		61.60	
	Minimum Charge	Job		211		211	

Exterior Doors and Windows

Exterior Type Door		Unit	Material	Labor	Equip.	Total	Specification
Custom Quality							
	Demolish	Ea.		39		39	Cost includes material and labor to
	Install	Ea.	510	56.50		566.50	install 3' x 6' 8" pre-hung 1-3/4"
	Demolish and Install	Ea.	510	95.50		605.50	hand-made, solid wood, raised panel
	Reinstall	Ea.		56.37		56.37	door with custom stained glass.
	Clean	Ea.	.22	12.90		13.12	
	Paint	Ea.	2.59	120		122.59	
	Minimum Charge	Job		211		211	
Miami Style							
	Demolish	Ea.		39		39	Cost includes material and labor to
	Install	Ea.	232	25		257	install 3' x 6' 8" pre-hung 1-3/4"
	Demolish and Install	Ea.	232	64		296	lauan door of solid wood.
	Reinstall	Ea.		24.87		24.87	
	Clean	Ea.	.22	12.90		13.12	
	Paint	Ea.	10.10	51.50		61.60	
	Minimum Charge	Job		211		211	
French Style							
1-Light							
	Demolish	Ea.		39		39	Cost includes material and labor to
	Install	Ea.	660	70.50		730.50	install 3' x 6' 8" pre-hung 1-3/4"
	Demolish and Install	Ea.	660	109.50		769.50	French style door with 1 light.
	Reinstall	Ea.		70.47		70.47	
	Clean	Ea.	2.05	19.35		21.40	
	Paint	Ea.	10.95	90		100.95	
	Minimum Charge	Job		211		211	
9-Light							
	Demolish	Ea.		39		39	Cost includes material and labor to
	Install	Ea.	725	30		755	install 3' x 6' 8" pre-hung 1-3/4"
	Demolish and Install	Ea.	725	69		794	French style door with 9 lights.
	Reinstall	Ea.		30.20		30.20	
	Clean	Ea.	.45	26		26.45	
	Paint	Ea.	10.95	90		100.95	
	Minimum Charge	Job		211		211	
15-Light							
	Demolish	Ea.		43		43	Cost includes material and labor to
	Install	Ea.	705	60.50		765.50	install 3' x 6' 8" pre-hung 1-3/4"
	Demolish and Install	Ea.	705	103.50		808.50	french style door with 15 lights.
	Reinstall	Ea.		60.40		60.40	
	Clean	Ea.	.45	26		26.45	
	Paint	Ea.	10.95	90		100.95	
	Minimum Charge	Job		211		211	
Jamb							
Wood							
	Demolish	Ea.		23.50		23.50	Includes material and labor to install
	Install	Ea.	157	11.45		168.45	flat pine jamb with square cut heads
	Demolish and Install	Ea.	157	34.95		191.95	and rabbeted sides for 6' 8" high and
	Clean	Ea.	.10	7.75		7.85	3-9/16" door including trim sets for
	Paint	Ea.	6.75	24		30.75	both sides.
	Minimum Charge	Job		211		211	
Metal 6' 8" High							
	Demolish	Ea.		23.50		23.50	Cost includes material and labor to
	Install	Ea.	168	26.50		194.50	install 18 gauge hollow metal door
	Demolish and Install	Ea.	168	50		218	frame to fit a 4-1/2" jamb 6' 8" high
	Reinstall	Ea.		26.43		26.43	and up to 3' 6" opening.
	Clean	Ea.	.68	6.45		7.13	
	Paint	Ea.	1.77	7.20		8.97	
	Minimum Charge	Job		211		211	

Exterior Doors and Windows

Exterior Type Door

Exterior Type Door		Unit	Material	Labor	Equip.	Total	Specification
Metal 7' High							
	Demolish	Ea.		23.50		23.50	Cost includes material and labor to
	Install	Ea.	175	28		203	install 18 gauge hollow metal door
	Demolish and Install	Ea.	175	51.50		226.50	frame to fit a 4-1/2" jamb 7' high and
	Reinstall	Ea.		28.19		28.19	up to 3' 6" opening.
	Clean	Ea.	.68	6.45		7.13	
	Paint	Ea.	1.77	7.20		8.97	
	Minimum Charge	Job		211		211	
Metal 8' High							
	Demolish	Ea.		23.50		23.50	Cost includes material and labor to
	Install	Ea.	216	35		251	install 18 gauge hollow metal door
	Demolish and Install	Ea.	216	58.50		274.50	frame to fit a 4-1/2" jamb 8' high and
	Reinstall	Ea.		35.23		35.23	up to 3' 6" opening.
	Clean	Ea.	.96	7.75		8.71	
	Paint	Ea.	1.98	8		9.98	
	Minimum Charge	Job		211		211	
Metal 9' High							
	Demolish	Ea.		23.50		23.50	Cost includes material and labor to
	Install	Ea.	278	42.50		320.50	install 18 gauge hollow metal door
	Demolish and Install	Ea.	278	66		344	frame to fit a 4-1/2" jamb 9' high and
	Reinstall	Ea.		42.28		42.28	up to 3' 6" opening.
	Clean	Ea.	.96	7.75		8.71	
	Paint	Ea.	1.98	8		9.98	
	Minimum Charge	Job		211		211	
Casing Trim							
Single Width							
	Demolish	Ea.		7.45		7.45	Cost includes material and labor to
	Install	Opng.	28.50	35		63.50	install 11/16" x 2-1/2" pine ranch
	Demolish and Install	Opng.	28.50	42.45		70.95	style casing for one side of a standard
	Reinstall	Opng.		28.19		28.19	door opening.
	Clean	Opng.	.46	3.10		3.56	
	Paint	Ea.	5.20	7.20		12.40	
	Minimum Charge	Job		211		211	
Double Width							
	Demolish	Ea.		8.85		8.85	Cost includes material and labor to
	Install	Opng.	33.50	42.50		76	install 11/16" x 2-1/2" pine ranch
	Demolish and Install	Opng.	33.50	51.35		84.85	style door casing for one side of a
	Reinstall	Opng.		33.82		33.82	double door opening.
	Clean	Opng.	.54	6.20		6.74	
	Paint	Ea.	5.20	8.20		13.40	
	Minimum Charge	Job		211		211	

Residential Metal Door

Residential Metal Door		Unit	Material	Labor	Equip.	Total	Specification
Pre-hung, Steel-clad							
2' 8" x 6' 8"							
	Demolish	Ea.		39		39	Cost includes material and labor to
	Install	Ea.	330	49.50		379.50	install 2' x 6'8" pre-hung 1-3/4" 24
	Demolish and Install	Ea.	330	88.50		418.50	gauge steel clad door with 4-1/2"
	Reinstall	Ea.		49.74		49.74	wood frame, trims, hinges, aluminum
	Clean	Ea.	3.27	12.90		16.17	sill, weatherstripped.
	Paint	Ea.	10.95	51.50		62.45	
	Minimum Charge	Job		211		211	

Exterior Doors and Windows

Residential Metal Door	Unit	Material	Labor	Equip.	Total	Specification
3' x 6' 8"						
Demolish	Ea.		39		39	Cost includes material and labor to
Install	Ea.	273	56.50		329.50	install 3' x 6' 8" pre-hung 1-3/4" 24
Demolish and Install	Ea.	273	95.50		368.50	gauge steel clad door with 4-1/2"
Reinstall	Ea.		56.37		56.37	wood frame, trims, hinges, aluminum
Clean	Ea.	3.27	12.90		16.17	sill, weatherstripped.
Paint	Ea.	10.95	51.50		62.45	
Minimum Charge	Job		211		211	

Metal Door	Unit	Material	Labor	Equip.	Total	Specification
2' 8" x 6' 8", 18 Gauge						
Demolish	Ea.		8.40		8.40	Cost includes material and labor to
Install	Ea.	550	26.50		576.50	install 18 gauge 2' 8" x 6' 8" steel
Demolish and Install	Ea.	550	34.90		584.90	door only, 1-3/4" thick, hinges.
Reinstall	Ea.		26.43		26.43	
Clean	Ea.	1.36	9.70		11.06	
Paint	Ea.	6.40	30		36.40	
Minimum Charge	Job		211		211	
3' x 6' 8", 18 Gauge						
Demolish	Ea.		8.40		8.40	Cost includes material and labor to
Install	Ea.	535	28		563	install 18 gauge 3' x 6' 8" steel door
Demolish and Install	Ea.	535	36.40		571.40	only, 1-3/4" thick, hinges.
Reinstall	Ea.		28.19		28.19	
Clean	Ea.	1.36	9.70		11.06	
Paint	Ea.	6.40	30		36.40	
Minimum Charge	Job		211		211	
2' 8" x 7', 18 Gauge						
Demolish	Ea.		8.40		8.40	Cost includes material and labor to
Install	Ea.	565	28		593	install 18 gauge 2' 8" x 7' steel door
Demolish and Install	Ea.	565	36.40		601.40	only, 1-3/4" thick, hinges.
Reinstall	Ea.		28.19		28.19	
Clean	Ea.	1.36	9.70		11.06	
Paint	Ea.	6.40	30		36.40	
Minimum Charge	Job		211		211	
3' x 7', 18 Gauge						
Demolish	Ea.		8.40		8.40	Cost includes material and labor to
Install	Ea.	665	30		695	install 18 gauge 3' x 7' steel door
Demolish and Install	Ea.	665	38.40		703.40	only, 1-3/4" thick, hinges.
Reinstall	Ea.		30.20		30.20	
Clean	Ea.	1.36	9.70		11.06	
Paint	Ea.	6.40	30		36.40	
Minimum Charge	Job		211		211	
3' 6" x 7', 18 Gauge						
Demolish	Ea.		8.40		8.40	Cost includes material and labor to
Install	Ea.	645	32.50		677.50	install 18 gauge 3' 6" x 7' steel door
Demolish and Install	Ea.	645	40.90		685.90	only, 1-3/4" thick, hinges.
Reinstall	Ea.		32.52		32.52	
Clean	Ea.	1.36	9.70		11.06	
Paint	Ea.	6.40	30		36.40	
Minimum Charge	Job		211		211	

Exterior Doors and Windows

Metal Door	Unit	Material	Labor	Equip.	Total	Specification
Casing Trim						
Single Width						
Demolish	Ea.		7.45		7.45	Cost includes material and labor to
Install	Opng.	28.50	35		63.50	install 11/16" x 2-1/2" pine ranch
Demolish and Install	Opng.	28.50	42.45		70.95	style casing for one side of a standard
Reinstall	Opng.		28.19		28.19	door opening.
Clean	Opng.	.46	3.10		3.56	
Paint	Ea.	5.20	7.20		12.40	
Minimum Charge	Job		211		211	
Double Width						
Demolish	Ea.		8.85		8.85	Cost includes material and labor to
Install	Opng.	33.50	42.50		76	install 11/16" x 2-1/2" pine ranch
Demolish and Install	Opng.	33.50	51.35		84.85	style door casing for one side of a
Reinstall	Opng.		33.82		33.82	double door opening.
Clean	Opng.	.54	6.20		6.74	
Paint	Ea.	5.20	8.20		13.40	
Minimum Charge	Job		211		211	

Solid Core Door	Unit	Material	Labor	Equip.	Total	Specification
Exterior Type						
Flush (1-3/4")						
Demolish	Ea.		39		39	Cost includes material and labor to
Install	Ea.	415	53		468	install 3' x 6' 8" pre-hung 1-3/4"
Demolish and Install	Ea.	415	92		507	door, including casing and stop,
Reinstall	Ea.		52.85		52.85	hinges, jamb, aluminum sill,
Clean	Ea.	3.27	12.90		16.17	weatherstripped.
Paint	Ea.	10.95	51.50		62.45	
Minimum Charge	Job		211		211	
Raised Panel						
Demolish	Ea.		39		39	Cost includes material and labor to
Install	Ea.	485	23.50		508.50	install 3' x 6' 8" pre-hung 1-3/4"
Demolish and Install	Ea.	485	62.50		547.50	door, jamb, including casing and
Reinstall	Ea.		23.49		23.49	stop, hinges, aluminum sill,
Clean	Ea.	3.27	12.90		16.17	weatherstripping.
Paint	Ea.	10.95	120		130.95	
Minimum Charge	Job		211		211	
Entry w / Lights						
Demolish	Ea.		39		39	Cost includes material and labor to
Install	Ea.	350	53		403	install 3' x 6' 8" pre-hung 1-3/4"
Demolish and Install	Ea.	350	92		442	door, solid wood lower panel, with up
Reinstall	Ea.		52.85		52.85	to 9-light glass panels, casing, sill,
Clean	Ea.	3.27	12.90		16.17	weatherstripped.
Paint	Ea.	2.59	120		122.59	
Minimum Charge	Job		211		211	
Custom Entry						
Demolish	Ea.		39		39	Cost includes material and labor to
Install	Ea.	1925	84.50		2009.50	install prehung, fir, single glazed with
Demolish and Install	Ea.	1925	123.50		2048.50	lead caming, 1-3/4" x 6' 8" x 3'
Reinstall	Ea.		84.56		84.56	
Clean	Ea.	3.27	12.90		16.17	
Paint	Ea.	6.40	30		36.40	
Minimum Charge	Job		211		211	

Exterior Doors and Windows

Solid Core Door		Unit	Material	Labor	Equip.	Total	Specification
Dutch Style							
	Demolish	Ea.		39		39	Includes material and labor to install
	Install	Ea.	810	70.50		880.50	dutch door, pine, 1-3/4" x 6' 8" x 2'
	Demolish and Install	Ea.	810	109.50		919.50	8" wide.
	Reinstall	Ea.		70.47		70.47	
	Clean	Ea.	3.27	12.90		16.17	
	Paint	Ea.	2.59	120		122.59	
	Minimum Charge	Job		211		211	
Miami Style							
	Demolish	Ea.		39		39	Cost includes material and labor to
	Install	Ea.	232	25		257	install 3' x 6' 8" pre-hung 1-3/4"
	Demolish and Install	Ea.	232	64		296	lauan door of solid wood.
	Reinstall	Ea.		24.87		24.87	
	Clean	Ea.	3.27	12.90		16.17	
	Paint	Ea.	10.10	51.50		61.60	
	Minimum Charge	Job		211		211	
French Style							
	Demolish	Ea.		43		43	Cost includes material and labor to
	Install	Ea.	705	60.50		765.50	install 3' x 6' 8" pre-hung 1-3/4"
	Demolish and Install	Ea.	705	103.50		808.50	french style door with 15 lights.
	Reinstall	Ea.		60.40		60.40	
	Clean	Ea.	.22	12.90		13.12	
	Paint	Ea.	7.80	180		187.80	
	Minimum Charge	Job		211		211	
Double							
	Demolish	Ea.		28.50		28.50	Includes material and labor to install a
	Install	Ea.	1475	106		1581	pre-hung double entry door including
	Demolish and Install	Ea.	1475	134.50		1609.50	frame and exterior trim.
	Clean	Ea.	.12	45.50		45.62	
	Paint	Ea.	22	180		202	
	Minimum Charge	Job		106		106	
Commercial Grade							
3' x 6' 8"							
	Demolish	Ea.		39		39	Includes material and labor to install
	Install	Ea.	315	25		340	flush solid core 1-3/4" hardwood
	Demolish and Install	Ea.	315	64		379	veneer 3' x 6' 8' door only.
	Reinstall	Ea.		24.87		24.87	
	Clean	Ea.	3.27	12.90		16.17	
	Paint	Ea.	10.65	30		40.65	
	Minimum Charge	Job		211		211	
3' 6" x 6' 8"							
	Demolish	Ea.		39		39	Includes material and labor to install
	Install	Ea.	390	25		415	flush solid core 1-3/4" hardwood
	Demolish and Install	Ea.	390	64		454	veneer 3' 6" x 6' 8' door only.
	Reinstall	Ea.		24.87		24.87	
	Clean	Ea.	3.27	12.90		16.17	
	Paint	Ea.	10.65	30		40.65	
	Minimum Charge	Job		211		211	
3' x 7'							
	Demolish	Ea.		39		39	Includes material and labor to install
	Install	Ea.	340	60.50		400.50	flush solid core 1-3/4" hardwood
	Demolish and Install	Ea.	340	99.50		439.50	veneer 3' x 7' door only.
	Reinstall	Ea.		60.40		60.40	
	Clean	Ea.	3.27	12.90		16.17	
	Paint	Ea.	10.65	30		40.65	
	Minimum Charge	Job		211		211	

Exterior Doors and Windows

Solid Core Door		Unit	Material	Labor	Equip.	Total	Specification
3' 6" x 7'							
	Demolish	Ea.		39		39	Includes material and labor to install
	Install	Ea.	455	30		485	flush solid core 1-3/4" hardwood
	Demolish and Install	Ea.	455	69		524	veneer 3' 6" x 7' door only.
	Reinstall	Ea.		30.20		30.20	
	Clean	Ea.	3.27	12.90		16.17	
	Paint	Ea.	10.65	30		40.65	
	Minimum Charge	Job		211		211	
3' x 8'							
	Demolish	Ea.		39		39	Includes material and labor to install
	Install	Ea.	400	53		453	flush solid core 1-3/4" hardwood
	Demolish and Install	Ea.	400	92		492	veneer 3' x 8' door only.
	Reinstall	Ea.		52.85		52.85	
	Clean	Ea.	3.92	19.35		23.27	
	Paint	Ea.	10.65	30		40.65	
	Minimum Charge	Job		211		211	
3' 6" x 8'							
	Demolish	Ea.		39		39	Includes material and labor to install
	Install	Ea.	490	53		543	flush solid core 1-3/4" hardwood
	Demolish and Install	Ea.	490	92		582	veneer 3' 6" x 8' door only.
	Reinstall	Ea.		52.85		52.85	
	Clean	Ea.	3.92	19.35		23.27	
	Paint	Ea.	10.65	30		40.65	
	Minimum Charge	Job		211		211	
3' x 9'							
	Demolish	Ea.		39		39	Includes material and labor to install
	Install	Ea.	550	60.50		610.50	flush solid core 1-3/4" hardwood
	Demolish and Install	Ea.	550	99.50		649.50	veneer 3' x 9' door only.
	Reinstall	Ea.		60.40		60.40	
	Clean	Ea.	3.92	19.35		23.27	
	Paint	Ea.	10.65	30		40.65	
	Minimum Charge	Job		211		211	
Stain							
	Install	Ea.	9.85	40		49.85	Includes labor and material to stain
	Minimum Charge	Job		180		180	exterior type door and trim on both
							sides.
Door Only (SC)							
2' 4" x 6' 8"							
	Demolish	Ea.		8.40		8.40	Cost includes material and labor to
	Install	Ea.	209	23.50		232.50	install 2' 4" x 6' 8" lauan solid core
	Demolish and Install	Ea.	209	31.90		240.90	door slab.
	Reinstall	Ea.		23.49		23.49	
	Clean	Ea.	1.36	9.70		11.06	
	Paint	Ea.	4.86	45		49.86	
	Minimum Charge	Job		211		211	
2' 8" x 6' 8"							
	Demolish	Ea.		8.40		8.40	Cost includes material and labor to
	Install	Ea.	217	23.50		240.50	install 2' 8" x 6' 8" lauan solid core
	Demolish and Install	Ea.	217	31.90		248.90	door slab.
	Reinstall	Ea.		23.49		23.49	
	Clean	Ea.	1.36	9.70		11.06	
	Paint	Ea.	4.86	45		49.86	
	Minimum Charge	Job		211		211	

Exterior Doors and Windows

Solid Core Door		Unit	Material	Labor	Equip.	Total	Specification
3' x 6' 8"							
	Demolish	Ea.		8.40		8.40	Cost includes material and labor to
	Install	Ea.	223	25		248	install 3' x 6' 8" lauan solid core door
	Demolish and Install	Ea.	223	33.40		256.40	slab.
	Reinstall	Ea.		24.87		24.87	
	Clean	Ea.	1.36	9.70		11.06	
	Paint	Ea.	4.86	45		49.86	
	Minimum Charge	Job		211		211	
3' 6" x 6' 8"							
	Demolish	Ea.		8.40		8.40	Cost includes material and labor to
	Install	Ea.	256	26.50		282.50	install 3' 6" x 6' 8" lauan solid core
	Demolish and Install	Ea.	256	34.90		290.90	door slab.
	Reinstall	Ea.		26.43		26.43	
	Clean	Ea.	1.36	9.70		11.06	
	Paint	Ea.	4.86	45		49.86	
	Minimum Charge	Job		211		211	
Panel							
	Demolish	Ea.		8.40		8.40	Cost includes material and labor to
	Install	Ea.	224	49.50		273.50	install 3' x 6' 8" six (6) panel interior
	Demolish and Install	Ea.	224	57.90		281.90	door.
	Reinstall	Ea.		49.74		49.74	
	Clean	Ea.	1.63	9.70		11.33	
	Paint	Ea.	4.86	45		49.86	
	Minimum Charge	Job		211		211	
Casing Trim							
Single Width							
	Demolish	Ea.		7.45		7.45	Cost includes material and labor to
	Install	Opng.	28.50	35		63.50	install 11/16" x 2-1/2" pine ranch
	Demolish and Install	Opng.	28.50	42.45		70.95	style casing for one side of a standard
	Reinstall	Opng.		28.19		28.19	door opening.
	Clean	Opng.	.46	3.10		3.56	
	Paint	Ea.	5.20	7.20		12.40	
	Minimum Charge	Job		211		211	
Double Width							
	Demolish	Ea.		8.85		8.85	Cost includes material and labor to
	Install	Opng.	33.50	42.50		76	install 11/16" x 2-1/2" pine ranch
	Demolish and Install	Opng.	33.50	51.35		84.85	style door casing for one side of a
	Reinstall	Opng.		33.82		33.82	double door opening.
	Clean	Opng.	.54	6.20		6.74	
	Paint	Ea.	5.20	8.20		13.40	
	Minimum Charge	Job		211		211	

Components		Unit	Material	Labor	Equip.	Total	Specification
Hardware							
Doorknob							
	Demolish	Ea.		16.80		16.80	Includes material and labor to install
	Install	Ea.	29.50	26.50		56	residential passage lockset, keyless
	Demolish and Install	Ea.	29.50	43.30		72.80	bored type.
	Reinstall	Ea.		21.14		21.14	
	Clean	Ea.	.33	9.70		10.03	
	Minimum Charge	Job		211		211	

Exterior Doors and Windows

Components		Unit	Material	Labor	Equip.	Total	Specification
Doorknob w / Lock							
	Demolish	Ea.		16.80		16.80	Includes material and labor to install
	Install	Ea.	45	26.50		71.50	privacy lockset.
	Demolish and Install	Ea.	45	43.30		88.30	
	Reinstall	Ea.		21.14		21.14	
	Clean	Ea.	.33	9.70		10.03	
	Minimum Charge	Job		211		211	
Lever Handle							
	Demolish	Ea.		16.80		16.80	Includes material and labor to install
	Install	Ea.	40	42.50		82.50	residential passage lockset, keyless
	Demolish and Install	Ea.	40	59.30		99.30	bored type with lever handle.
	Reinstall	Ea.		33.82		33.82	
	Clean	Ea.	.33	9.70		10.03	
	Minimum Charge	Job		211		211	
Deadbolt							
	Demolish	Ea.		14.70		14.70	Includes material and labor to install a
	Install	Ea.	70	30		100	deadbolt.
	Demolish and Install	Ea.	70	44.70		114.70	
	Reinstall	Ea.		24.16		24.16	
	Clean	Ea.	.33	9.70		10.03	
	Minimum Charge	Job		211		211	
Grip Handle Entry							
	Demolish	Ea.		13.40		13.40	Includes material and labor to install
	Install	Ea.	88.50	47		135.50	residential keyed entry lock set with
	Demolish and Install	Ea.	88.50	60.40		148.90	grip type handle.
	Reinstall	Ea.		37.58		37.58	
	Clean	Ea.	.33	9.70		10.03	
	Minimum Charge	Job		211		211	
Panic Device							
	Demolish	Ea.		34.50		34.50	Includes material and labor to install
	Install	Ea.	400	70.50		470.50	touch bar, low profile, exit only.
	Demolish and Install	Ea.	400	105		505	
	Reinstall	Ea.		56.37		56.37	
	Clean	Ea.	.17	17.20		17.37	
	Minimum Charge	Job		211		211	
Closer							
	Demolish	Ea.		8.80		8.80	Includes material and labor to install
	Install	Ea.	256	70.50		326.50	pneumatic heavy duty closer for
	Demolish and Install	Ea.	256	79.30		335.30	exterior type doors.
	Reinstall	Ea.		56.37		56.37	
	Clean	Ea.	.17	12.90		13.07	
	Minimum Charge	Job		211		211	
Peep-hole							
	Demolish	Ea.		7.65		7.65	Includes material and labor to install
	Install	Ea.	17.40	13.20		30.60	peep-hole for door.
	Demolish and Install	Ea.	17.40	20.85		38.25	
	Reinstall	Ea.		10.57		10.57	
	Clean	Ea.	.02	3.49		3.51	
	Minimum Charge	Job		211		211	
Exit Sign							
	Demolish	Ea.		23		23	Includes material and labor to install a
	Install	Ea.	118	57.50		175.50	wall mounted interior electric exit sign.
	Demolish and Install	Ea.	118	80.50		198.50	
	Reinstall	Ea.		45.92		45.92	
	Clean	Ea.	.02	7.10		7.12	
	Minimum Charge	Job		211		211	

Exterior Doors and Windows

Components		Unit	Material	Labor	Equip.	Total	Specification
Kickplate							
	Demolish	Ea.		13.05		13.05	Includes material and labor to install
	Install	Ea.	23.50	28		51.50	aluminum kickplate, 10″ x 28″.
	Demolish and Install	Ea.	23.50	41.05		64.55	
	Reinstall	Ea.		22.55		22.55	
	Clean	Ea.	.33	9.70		10.03	
	Minimum Charge	Job		211		211	
Threshold							
	Install	L.F.	3.92	4.23		8.15	Includes labor and material to install
	Minimum Charge	Job		106		106	oak threshold.
Weatherstripping							
	Install	L.F.	.52	1.06		1.58	Includes labor and material to install
	Minimum Charge	Job		106		106	rubber weatherstripping for door.
Metal Jamb							
6′ 8″ High							
	Demolish	Ea.		23.50		23.50	Cost includes material and labor to
	Install	Ea.	168	26.50		194.50	install 18 gauge hollow metal door
	Demolish and Install	Ea.	168	50		218	frame to fit a 4-1/2″ jamb 6′ 8″ high
	Reinstall	Ea.		26.43		26.43	and up to 3′ 6″ opening.
	Clean	Ea.	.68	6.45		7.13	
	Paint	Ea.	1.77	7.20		8.97	
	Minimum Charge	Job		211		211	
7′ High							
	Demolish	Ea.		23.50		23.50	Cost includes material and labor to
	Install	Ea.	175	28		203	install 18 gauge hollow metal door
	Demolish and Install	Ea.	175	51.50		226.50	frame to fit a 4-1/2″ jamb 7′ high and
	Reinstall	Ea.		28.19		28.19	up to 3′ 6″ opening.
	Clean	Ea.	.68	6.45		7.13	
	Paint	Ea.	1.77	7.20		8.97	
	Minimum Charge	Job		211		211	
8′ High							
	Demolish	Ea.		23.50		23.50	Cost includes material and labor to
	Install	Ea.	216	35		251	install 18 gauge hollow metal door
	Demolish and Install	Ea.	216	58.50		274.50	frame to fit a 4-1/2″ jamb 8′ high and
	Reinstall	Ea.		35.23		35.23	up to 3′ 6″ opening.
	Clean	Ea.	.96	7.75		8.71	
	Paint	Ea.	1.98	8		9.98	
	Minimum Charge	Job		211		211	
9′ High							
	Demolish	Ea.		23.50		23.50	Cost includes material and labor to
	Install	Ea.	278	42.50		320.50	install 18 gauge hollow metal door
	Demolish and Install	Ea.	278	66		344	frame to fit a 4-1/2″ jamb 9′ high and
	Reinstall	Ea.		42.28		42.28	up to 3′ 6″ opening.
	Clean	Ea.	.96	7.75		8.71	
	Paint	Ea.	1.98	8		9.98	
	Minimum Charge	Job		211		211	
Wood Jamb							
	Demolish	Ea.		29.50		29.50	Includes material and labor to install
	Install	Ea.	157	11.45		168.45	flat pine jamb with square cut heads
	Demolish and Install	Ea.	157	40.95		197.95	and rabbeted sides for 6′ 8″ high and
	Clean	Ea.	.10	7.75		7.85	3-9/16″ door including trim sets for
	Paint	Ea.	6.75	24		30.75	both sides.
	Minimum Charge	Job		106		106	
Stain							
	Install	Ea.	6.80	30		36.80	Includes labor and material to stain
	Minimum Charge	Job		180		180	single door and trim on both sides.

Exterior Doors and Windows

Components		Unit	Material	Labor	Equip.	Total	Specification
Shave & Refit							
	Install	Ea.		35		35	Includes labor to shave and rework
	Minimum Charge	Job		106		106	door to fit opening at the job site.

Sliding Patio Door		Unit	Material	Labor	Equip.	Total	Specification
Aluminum							
6' Single Pane							
	Demolish	Ea.		58.50		58.50	Cost includes material and labor to
	Install	Ea.	500	211		711	install 6' x 6' 8" sliding anodized
	Demolish and Install	Ea.	500	269.50		769.50	aluminum patio door with single pane
	Reinstall	Ea.		169.12		169.12	tempered safety glass, anodized
	Clean	Ea.	2.05	19.35		21.40	frame, screen, weatherstripped.
	Minimum Charge	Job		211		211	
8' Single Pane							
	Demolish	Ea.		58.50		58.50	Cost includes material and labor to
	Install	Ea.	635	282		917	install 8' x 6' 8" sliding aluminum
	Demolish and Install	Ea.	635	340.50		975.50	patio door with single pane tempered
	Reinstall	Ea.		225.49		225.49	safety glass, anodized frame, screen,
	Clean	Ea.	2.05	19.35		21.40	weatherstripped.
	Minimum Charge	Job		211		211	
10' Single Pane							
	Demolish	Ea.		58.50		58.50	Cost includes material and labor to
	Install	Ea.	710	425		1135	install 10' x 6' 8" sliding aluminum
	Demolish and Install	Ea.	710	483.50		1193.50	patio door with single pane tempered
	Reinstall	Ea.		338.24		338.24	safety glass, anodized frame, screen,
	Clean	Ea.	2.05	19.35		21.40	weatherstripped.
	Minimum Charge	Job		211		211	
6' Insulated							
	Demolish	Ea.		58.50		58.50	Cost includes material and labor to
	Install	Ea.	1000	211		1211	install 6' x 6' 8" sliding aluminum
	Demolish and Install	Ea.	1000	269.50		1269.50	patio door with double insulated
	Reinstall	Ea.		169.12		169.12	tempered glass, baked-on enamel
	Clean	Ea.	2.05	19.35		21.40	finish, screen, weatherstripped.
	Minimum Charge	Job		211		211	
8' Insulated							
	Demolish	Ea.		58.50		58.50	Cost includes material and labor to
	Install	Ea.	1175	282		1457	install 8' x 6' 8" sliding aluminum
	Demolish and Install	Ea.	1175	340.50		1515.50	patio door with insulated tempered
	Reinstall	Ea.		225.49		225.49	glass, baked-on enamel finish, screen,
	Clean	Ea.	2.05	19.35		21.40	weatherstripped.
	Minimum Charge	Job		211		211	
10' Insulated							
	Demolish	Ea.		58.50		58.50	Cost includes material and labor to
	Install	Ea.	1450	425		1875	install 10' x 6' 8" sliding aluminum
	Demolish and Install	Ea.	1450	483.50		1933.50	patio door with double insulated
	Reinstall	Ea.		338.24		338.24	tempered glass, baked-on enamel
	Clean	Ea.	2.05	19.35		21.40	finish, screen, weatherstripped.
	Minimum Charge	Job		211		211	

Exterior Doors and Windows

Sliding Patio Door	Unit	Material	Labor	Equip.	Total	Specification
Vinyl						
6' Single Pane						
Demolish	Ea.		58.50		58.50	Cost includes material and labor to
Install	Ea.	460	211		671	install 6' x 6' 8" sliding vinyl patio
Demolish and Install	Ea.	460	269.50		729.50	door with single pane tempered safety
Reinstall	Ea.		169.12		169.12	glass, anodized frame, screen,
Clean	Ea.	2.05	19.35		21.40	weatherstripped.
Minimum Charge	Job		211		211	
8' Single Pane						
Demolish	Ea.		58.50		58.50	Cost includes material and labor to
Install	Ea.	645	282		927	install 8' x 6' 8" sliding vinyl patio
Demolish and Install	Ea.	645	340.50		985.50	door with single pane tempered safety
Reinstall	Ea.		225.49		225.49	glass, anodized frame, screen,
Clean	Ea.	2.05	19.35		21.40	weatherstripped.
Minimum Charge	Job		211		211	
10' Single Pane						
Demolish	Ea.		58.50		58.50	Cost includes material and labor to
Install	Ea.	700	425		1125	install 10' x 6' 8" sliding vinyl patio
Demolish and Install	Ea.	700	483.50		1183.50	door with single pane tempered safety
Reinstall	Ea.		338.24		338.24	glass, anodized frame, screen,
Clean	Ea.	2.05	19.35		21.40	weatherstripped.
Minimum Charge	Job		211		211	
6' Insulated						
Demolish	Ea.		58.50		58.50	Cost includes material and labor to
Install	Ea.	1250	211		1461	install 6' x 6' 8" sliding vinyl patio
Demolish and Install	Ea.	1250	269.50		1519.50	door with double insulated tempered
Reinstall	Ea.		169.12		169.12	glass, anodized frame, screen,
Clean	Ea.	2.05	19.35		21.40	weatherstripped.
Minimum Charge	Job		211		211	
8' Insulated						
Demolish	Ea.		58.50		58.50	Cost includes material and labor to
Install	Ea.	1025	282		1307	install 8' x 6' 8" sliding vinyl patio
Demolish and Install	Ea.	1025	340.50		1365.50	door with double insulated tempered
Reinstall	Ea.		225.49		225.49	glass, anodized frame, screen,
Clean	Ea.	2.05	19.35		21.40	weatherstripped.
Minimum Charge	Job		211		211	
10' Insulated						
Demolish	Ea.		58.50		58.50	Cost includes material and labor to
Install	Ea.	1850	211		2061	install 10' x 6' 8" sliding vinyl patio
Demolish and Install	Ea.	1850	269.50		2119.50	door with double insulated tempered
Reinstall	Ea.		169.12		169.12	glass, anodized frame, screen,
Clean	Ea.	2.05	19.35		21.40	weatherstripped.
Minimum Charge	Job		211		211	
Wood Framed						
6' Premium Insulated						
Demolish	Ea.		58.50		58.50	Cost includes material and labor to
Install	Ea.	1575	211		1786	install 6' x 6' 8" sliding wood patio
Demolish and Install	Ea.	1575	269.50		1844.50	door with pine frame, double insulated
Reinstall	Ea.		169.12		169.12	glass, screen, lock and dead bolt,
Clean	Ea.	2.05	19.35		21.40	weatherstripped. Add for casing and
Minimum Charge	Job		211		211	finishing.

Exterior Doors and Windows

Sliding Patio Door		Unit	Material	Labor	Equip.	Total	Specification
8' Premium Insulated							Cost includes material and labor to
	Demolish	Ea.		58.50		58.50	install 8' x 6' 8" sliding wood patio
	Install	Ea.	2000	282		2282	door with pine frame, double insulated
	Demolish and Install	Ea.	2000	340.50		2340.50	glass, screen, lock and dead bolt,
	Reinstall	Ea.		225.49		225.49	weatherstripped. Add for casing and
	Clean	Ea.	2.05	19.35		21.40	finishing.
	Minimum Charge	Job		211		211	
9' Premium Insulated							Cost includes material and labor to
	Demolish	Ea.		58.50		58.50	install 9' x 6' 8" sliding wood patio
	Install	Ea.	2000	282		2282	door with pine frame, double insulated
	Demolish and Install	Ea.	2000	340.50		2340.50	glass, screen, lock and dead bolt,
	Reinstall	Ea.		225.49		225.49	weatherstripped. Add for casing and
	Clean	Ea.	2.05	19.35		21.40	finishing.
	Minimum Charge	Job		211		211	
Stain							Includes labor and material to stain
	Install	Ea.	9.85	40		49.85	exterior type door and trim on both
	Minimum Charge	Job		180		180	sides.
Casing Trim							Cost includes material and labor to
	Demolish	Ea.		8.85		8.85	install 11/16" x 2-1/2" pine ranch
	Install	Opng.	33.50	42.50		76	style door casing for one side of a
	Demolish and Install	Opng.	33.50	51.35		84.85	double door opening.
	Reinstall	Opng.		33.82		33.82	
	Clean	Opng.	.54	6.20		6.74	
	Paint	Ea.	5.20	8.20		13.40	
	Minimum Charge	Job		211		211	
Sliding Screen							Includes material and labor to install
	Demolish	Ea.		8.40		8.40	sliding screen door, 6' 8" x 3'.
	Install	Ea.	139	21		160	
	Demolish and Install	Ea.	139	29.40		168.40	
	Reinstall	Ea.		16.91		16.91	
	Clean	Ea.	.68	6.45		7.13	
	Minimum Charge	Job		211		211	

Screen Door		Unit	Material	Labor	Equip.	Total	Specification
Aluminum							Includes material and labor to install
	Demolish	Ea.		39		39	aluminum frame screen door with
	Install	Ea.	130	60.50		190.50	fiberglass screen cloth, closer, hinges,
	Demolish and Install	Ea.	130	99.50		229.50	latch.
	Reinstall	Ea.		48.32		48.32	
	Clean	Ea.	3.27	12.90		16.17	
	Paint	Ea.	4.99	14.45		19.44	
	Minimum Charge	Job		211		211	
Wood							Includes material and labor to install
	Demolish	Ea.		39		39	wood screen door with aluminum cloth
	Install	Ea.	340	70.50		410.50	screen. Frame and hardware not
	Demolish and Install	Ea.	340	109.50		449.50	included.
	Reinstall	Ea.		56.37		56.37	
	Clean	Ea.	3.27	12.90		16.17	
	Paint	Ea.	4.99	14.45		19.44	
	Minimum Charge	Job		211		211	
Shave & Refit							Includes labor to shave and rework
	Install	Ea.		35		35	door to fit opening at the job site.
	Minimum Charge	Job		211		211	

Exterior Doors and Windows

Storm Door		Unit	Material	Labor	Equip.	Total	Specification
Good Grade							
	Demolish	Ea.		39		39	Includes material and labor to install
	Install	Ea.	220	60.50		280.50	aluminum frame storm door with
	Demolish and Install	Ea.	220	99.50		319.50	fiberglass screen cloth, upper glass
	Reinstall	Ea.		48.32		48.32	panel, closer, hinges, latch.
	Clean	Ea.	3.27	12.90		16.17	
	Paint	Ea.	4.99	14.45		19.44	
	Minimum Charge	Job		211		211	
Better Grade							
	Demolish	Ea.		39		39	Includes material and labor to install
	Install	Ea.	271	60.50		331.50	aluminum frame storm door with
	Demolish and Install	Ea.	271	99.50		370.50	fiberglass screen cloth, upper and
	Reinstall	Ea.		48.32		48.32	lower glass panel, closer, latch.
	Clean	Ea.	3.27	12.90		16.17	
	Paint	Ea.	4.99	14.45		19.44	
	Minimum Charge	Job		211		211	
Premium Grade							
	Demolish	Ea.		39		39	Includes material and labor to install
	Install	Ea.	296	60.50		356.50	aluminum frame storm door with
	Demolish and Install	Ea.	296	99.50		395.50	fiberglass screen cloth, upper and
	Reinstall	Ea.		48.32		48.32	lower glass panel, closer, latch.
	Clean	Ea.	3.27	12.90		16.17	
	Paint	Ea.	4.99	14.45		19.44	
	Minimum Charge	Job		211		211	
Insulated							
	Demolish	Ea.		39		39	Includes material and labor to install
	Install	Ea.	272	30		302	aluminum frame storm door with
	Demolish and Install	Ea.	272	69		341	fiberglass screen cloth, upper and
	Reinstall	Ea.		24.16		24.16	lower glass panel, closer, latch.
	Clean	Ea.	3.27	12.90		16.17	
	Paint	Ea.	4.99	14.45		19.44	
	Minimum Charge	Job		211		211	

Garage Door		Unit	Material	Labor	Equip.	Total	Specification
9' One Piece Metal							
	Demolish	Ea.		99		99	Includes material and labor to install
	Install	Ea.	460	106		566	one piece steel garage door 9' x 7'
	Demolish and Install	Ea.	460	205		665	primed including hardware.
	Reinstall	Ea.		84.56		84.56	
	Clean	Ea.	4.08	38.50		42.58	
	Paint	Ea.	9.20	90		99.20	
	Minimum Charge	Job		211		211	
16' One Piece Metal							
	Demolish	Ea.		117		117	Includes material and labor to install
	Install	Ea.	755	141		896	one piece steel garage door 16' x 7'
	Demolish and Install	Ea.	755	258		1013	primed including hardware.
	Reinstall	Ea.		112.75		112.75	
	Clean	Ea.	8.15	51.50		59.65	
	Paint	Ea.	18.35	103		121.35	
	Minimum Charge	Job		211		211	

Exterior Doors and Windows

Garage Door		Unit	Material	Labor	Equip.	Total	Specification
9' Metal Sectional							
	Demolish	Ea.		99		99	Includes material and labor to install
	Install	Ea.	755	160		915	sectional metal garage door 9' x 7'
	Demolish and Install	Ea.	755	259		1014	primed including hardware.
	Reinstall	Ea.		128.12		128.12	
	Clean	Ea.	4.18	44		48.18	
	Paint	Ea.	18.35	103		121.35	
	Minimum Charge	Job		211		211	
16' Metal Sectional							
	Demolish	Ea.		117		117	Includes material and labor to install
	Install	Ea.	935	282		1217	sectional metal garage door 16' x 7'
	Demolish and Install	Ea.	935	399		1334	primed including hardware.
	Reinstall	Ea.		225.49		225.49	
	Clean	Ea.	8.15	51.50		59.65	
	Paint	Ea.	18.35	103		121.35	
	Minimum Charge	Job		211		211	
Metal / Steel Roll Up							
8' x 8' Steel							
	Demolish	Ea.		134		134	Cost includes material and labor to
	Install	Ea.	1200	645		1845	install 8' x 8' steel overhead roll-up,
	Demolish and Install	Ea.	1200	779		1979	chain hoist operated service door.
	Reinstall	Ea.		516.40		516.40	
	Clean	Ea.	4.18	44		48.18	
	Paint	Ea.	19.65	120		139.65	
	Minimum Charge	Job		1025		1025	
10' x 10' Steel							
	Demolish	Ea.		157		157	Cost includes material and labor to
	Install	Ea.	2000	740		2740	install 10' x 10' steel overhead roll-up,
	Demolish and Install	Ea.	2000	897		2897	chain hoist operated service door.
	Reinstall	Ea.		590.17		590.17	
	Clean	Ea.	6.55	62		68.55	
	Paint	Ea.	31	120		151	
	Minimum Charge	Job		1025		1025	
12' x 12' Steel							
	Demolish	Ea.		188		188	Cost includes material and labor to
	Install	Ea.	2075	860		2935	install 12' x 12' steel overhead roll-up,
	Demolish and Install	Ea.	2075	1048		3123	chain hoist operated service door.
	Reinstall	Ea.		688.53		688.53	
	Clean	Ea.	9.40	69		78.40	
	Paint	Ea.	44	144		188	
	Minimum Charge	Job		1025		1025	
14' x 14' Steel							
	Demolish	Ea.		235		235	Cost includes material and labor to
	Install	Ea.	3325	1300		4625	install 14' x 14' steel overhead roll-up,
	Demolish and Install	Ea.	3325	1535		4860	chain hoist operated service door.
	Reinstall	Ea.		1032.80		1032.80	
	Clean	Ea.	12.80	88.50		101.30	
	Paint	Ea.	51.50	180		231.50	
	Minimum Charge	Job		1025		1025	
18' x 18' Steel							
	Demolish	Ea.		261		261	Cost includes material and labor to
	Install	Opng.	2625	1725		4350	install 18' x 18' steel overhead roll-up,
	Demolish and Install	Opng.	2625	1986		4611	chain hoist operated service door.
	Reinstall	Opng.		1377.07		1377.07	
	Clean	Ea.	12.80	51.50		64.30	
	Paint	Ea.	99	360		459	
	Minimum Charge	Job		1025		1025	
Automatic Motor							
	Install	Ea.	525	42.50		567.50	Includes labor and material to install
	Minimum Charge	Job		211		211	electric opener motor, 1/3 HP.

Exterior Doors and Windows

Garage Door		Unit	Material	Labor	Equip.	Total	Specification
9' One Piece Wood							
	Demolish	Ea.		99		99	Includes material and labor to install
	Install	Ea.	660	106		766	one piece wood garage door 9' x 7'
	Demolish and Install	Ea.	660	205		865	primed including hardware.
	Reinstall	Ea.		84.56		84.56	
	Clean	Ea.	4.08	38.50		42.58	
	Paint	Ea.	9.20	90		99.20	
	Minimum Charge	Job		211		211	
9' Wood Sectional							
	Demolish	Ea.		99		99	Includes material and labor to install
	Install	Ea.	1000	106		1106	sectional wood garage door 9' x 7'
	Demolish and Install	Ea.	1000	205		1205	unfinished including hardware.
	Reinstall	Ea.		84.56		84.56	
	Clean	Ea.	4.08	38.50		42.58	
	Paint	Ea.	9.20	90		99.20	
	Minimum Charge	Job		211		211	
16' Wood Sectional							
	Demolish	Ea.		117		117	Includes material and labor to install
	Install	Ea.	1725	141		1866	sectional fiberglass garage door 16' x
	Demolish and Install	Ea.	1725	258		1983	7' unfinished including hardware.
	Reinstall	Ea.		112.75		112.75	
	Clean	Ea.	8.15	51.50		59.65	
	Paint	Ea.	18.35	103		121.35	
	Minimum Charge	Job		211		211	
Electric Opener							
	Demolish	Ea.		69		69	Includes material and labor to install
	Install	Ea.	340	79.50		419.50	electric opener, 1/3 HP economy.
	Demolish and Install	Ea.	340	148.50		488.50	
	Reinstall	Ea.		63.46		63.46	
	Clean	Ea.	.17	17.20		17.37	
	Minimum Charge	Job		211		211	
Springs							
	Install	Pr.	85.50	106		191.50	Includes labor and material to replace
	Minimum Charge	Job		211		211	springs for garage door.

Commercial Metal Door		Unit	Material	Labor	Equip.	Total	Specification
Pre-hung Steel Door							
2' 8" x 6' 8"							
	Demolish	Ea.		58.50		58.50	Cost includes material and labor to
	Install	Ea.	730	26.50		756.50	install 2' 8" x 6' 8" pre-hung 1-3/4"
	Demolish and Install	Ea.	730	85		815	18 gauge steel door with steel frame,
	Reinstall	Ea.		26.43		26.43	hinges, aluminum sill,
	Clean	Ea.	3.27	12.90		16.17	weather-stripped.
	Paint	Ea.	6.40	30		36.40	
	Minimum Charge	Job		211		211	
3' x 6' 8"							
	Demolish	Ea.		58.50		58.50	Cost includes material and labor to
	Install	Ea.	680	28		708	install 3' x 6' 8" pre-hung 1-3/4" 18
	Demolish and Install	Ea.	680	86.50		766.50	gauge steel door with steel frame,
	Reinstall	Ea.		28.19		28.19	hinges, aluminum sill,
	Clean	Ea.	3.27	12.90		16.17	weather-stripped.
	Paint	Ea.	6.40	30		36.40	
	Minimum Charge	Job		211		211	

Exterior Doors and Windows

Commercial Metal Door

Commercial Metal Door	Unit	Material	Labor	Equip.	Total	Specification
3′ 6″ x 6′ 8″						
Demolish	Ea.		58.50		58.50	Cost includes material and labor to
Install	Ea.	780	32.50		812.50	install 3′ 6″ x 6′ 8″ pre-hung 1-3/4″
Demolish and Install	Ea.	780	91		871	18 gauge steel door with steel frame,
Reinstall	Ea.		32.52		32.52	hinges, aluminum sill,
Clean	Ea.	3.27	12.90		16.17	weather-stripped.
Paint	Ea.	6.40	30		36.40	
Minimum Charge	Job		211		211	
4′ x 6′ 8″						
Demolish	Ea.		58.50		58.50	Cost includes material and labor to
Install	Ea.	845	42.50		887.50	install 4′ x 6′ 8″ pre-hung 1-3/4″ 18
Demolish and Install	Ea.	845	101		946	gauge steel door with steel frame,
Reinstall	Ea.		42.28		42.28	hinges, aluminum sill,
Clean	Ea.	3.27	12.90		16.17	weather-stripped.
Paint	Ea.	6.40	30		36.40	
Minimum Charge	Job		211		211	

Metal Jamb

Metal Jamb	Unit	Material	Labor	Equip.	Total	Specification
6′ 8″ High						
Demolish	Ea.		23.50		23.50	Cost includes material and labor to
Install	Ea.	168	26.50		194.50	install 18 gauge hollow metal door
Demolish and Install	Ea.	168	50		218	frame to fit a 4-1/2″ jamb 6′ 8″ high
Reinstall	Ea.		26.43		26.43	and up to 3′ 6″ opening.
Clean	Ea.	.68	6.45		7.13	
Paint	Ea.	1.77	7.20		8.97	
Minimum Charge	Job		211		211	
7′ High						
Demolish	Ea.		23.50		23.50	Cost includes material and labor to
Install	Ea.	175	28		203	install 18 gauge hollow metal door
Demolish and Install	Ea.	175	51.50		226.50	frame to fit a 4-1/2″ jamb 7′ high and
Reinstall	Ea.		28.19		28.19	up to 3′ 6″ opening.
Clean	Ea.	.68	6.45		7.13	
Paint	Ea.	1.77	7.20		8.97	
Minimum Charge	Job		211		211	
8′ High						
Demolish	Ea.		23.50		23.50	Cost includes material and labor to
Install	Ea.	216	35		251	install 18 gauge hollow metal door
Demolish and Install	Ea.	216	58.50		274.50	frame to fit a 4-1/2″ jamb 8′ high and
Reinstall	Ea.		35.23		35.23	up to 3′ 6″ opening.
Clean	Ea.	.96	7.75		8.71	
Paint	Ea.	1.98	8		9.98	
Minimum Charge	Job		211		211	
9′ High						
Demolish	Ea.		23.50		23.50	Cost includes material and labor to
Install	Ea.	278	42.50		320.50	install 18 gauge hollow metal door
Demolish and Install	Ea.	278	66		344	frame to fit a 4-1/2″ jamb 9′ high and
Reinstall	Ea.		42.28		42.28	up to 3′ 6″ opening.
Clean	Ea.	.96	7.75		8.71	
Paint	Ea.	1.98	8		9.98	
Minimum Charge	Job		211		211	

Hardware

Doorknob w / Lock	Unit	Material	Labor	Equip.	Total	Specification
Demolish	Ea.		16.80		16.80	Includes material and labor to install
Install	Ea.	163	35		198	keyed entry lock set, bored type
Demolish and Install	Ea.	163	51.80		214.80	including knob trim, strike and strike
Reinstall	Ea.		28.19		28.19	box for standard commercial
Clean	Ea.	.33	9.70		10.03	application.
Minimum Charge	Job		211		211	

Exterior Doors and Windows

Commercial Metal Door		Unit	Material	Labor	Equip.	Total	Specification
Lever Handle							
	Demolish	Ea.		16.80		16.80	Includes material and labor to install
	Install	Ea.	134	42.50		176.50	passage lockset, keyless bored type
	Demolish and Install	Ea.	134	59.30		193.30	with lever handle, non-locking
	Reinstall	Ea.		33.82		33.82	standard commercial latchset.
	Clean	Ea.	.33	9.70		10.03	
	Minimum Charge	Job		211		211	
Deadbolt							
	Demolish	Ea.		14.70		14.70	Includes material and labor to install a
	Install	Ea.	70	30		100	deadbolt.
	Demolish and Install	Ea.	70	44.70		114.70	
	Reinstall	Ea.		24.16		24.16	
	Clean	Ea.	.33	9.70		10.03	
	Minimum Charge	Job		211		211	
Panic Device							
	Demolish	Ea.		34.50		34.50	Includes material and labor to install
	Install	Ea.	400	70.50		470.50	touch bar, low profile, exit only.
	Demolish and Install	Ea.	400	105		505	
	Reinstall	Ea.		56.37		56.37	
	Clean	Ea.	.17	17.20		17.37	
	Minimum Charge	Job		211		211	
Closer							
	Demolish	Ea.		8.80		8.80	Includes material and labor to install
	Install	Ea.	285	70.50		355.50	pneumatic heavy duty model for
	Demolish and Install	Ea.	285	79.30		364.30	exterior type doors.
	Reinstall	Ea.		56.37		56.37	
	Clean	Ea.	.17	12.90		13.07	
	Minimum Charge	Job		211		211	
Peep-Hole							
	Demolish	Ea.		7.65		7.65	Includes material and labor to install
	Install	Ea.	17.40	13.20		30.60	peep-hole for door.
	Demolish and Install	Ea.	17.40	20.85		38.25	
	Reinstall	Ea.		10.57		10.57	
	Clean	Ea.	.02	3.49		3.51	
	Minimum Charge	Job		211		211	
Exit Sign							
	Demolish	Ea.		23		23	Includes material and labor to install a
	Install	Ea.	118	57.50		175.50	wall mounted interior electric exit sign.
	Demolish and Install	Ea.	118	80.50		198.50	
	Reinstall	Ea.		45.92		45.92	
	Clean	Ea.	.02	7.10		7.12	
	Minimum Charge	Job		211		211	
Kickplate							
	Demolish	Ea.		13.05		13.05	Includes material and labor to install
	Install	Ea.	54.50	28		82.50	stainless steel kickplate, 10" x 36".
	Demolish and Install	Ea.	54.50	41.05		95.55	
	Reinstall	Ea.		22.55		22.55	
	Clean	Ea.	.33	9.70		10.03	
	Minimum Charge	Job		211		211	

Aluminum Window		Unit	Material	Labor	Equip.	Total	Specification
Single Hung Two Light							
2' x 2'							
	Demolish	Ea.		21.50		21.50	Cost includes material and labor to
	Install	Ea.	189	94		283	install 2' x 2' single hung, two light,
	Demolish and Install	Ea.	189	115.50		304.50	double glazed aluminum window
	Reinstall	Ea.		75.11		75.11	including screen.
	Clean	Ea.	.03	8.10		8.13	
	Minimum Charge	Job		258		258	

Exterior Doors and Windows

Aluminum Window		Unit	Material	Labor	Equip.	Total	Specification
2' x 2' 6"							
	Demolish	Ea.		21.50		21.50	Cost includes material and labor to
	Install	Ea.	78.50	94		172.50	install 2' x 2' 6" single hung, two
	Demolish and Install	Ea.	78.50	115.50		194	light, double glazed aluminum
	Reinstall	Ea.		75.11		75.11	window including screen.
	Clean	Ea.	.03	8.10		8.13	
	Minimum Charge	Job		258		258	
3' x 1' 6"							
	Demolish	Ea.		21.50		21.50	Cost includes material and labor to
	Install	Ea.	74.50	103		177.50	install 3' x 1' 6" single hung, two
	Demolish and Install	Ea.	74.50	124.50		199	light, double glazed aluminum
	Reinstall	Ea.		82.62		82.62	window including screen.
	Clean	Ea.	.02	7.10		7.12	
	Minimum Charge	Job		258		258	
3' x 2'							
	Demolish	Ea.		21.50		21.50	Cost includes material and labor to
	Install	Ea.	89	103		192	install 3' x 2' single hung, two light,
	Demolish and Install	Ea.	89	124.50		213.50	double glazed aluminum window
	Reinstall	Ea.		82.62		82.62	including screen.
	Clean	Ea.	.02	7.10		7.12	
	Minimum Charge	Job		258		258	
3' x 2' 6"							
	Demolish	Ea.		21.50		21.50	Cost includes material and labor to
	Install	Ea.	95	103		198	install 3' x 2' 6" single hung, two
	Demolish and Install	Ea.	95	124.50		219.50	light, double glazed aluminum
	Reinstall	Ea.		82.62		82.62	window including screen.
	Clean	Ea.	.03	8.10		8.13	
	Minimum Charge	Job		258		258	
3' x 3'							
	Demolish	Ea.		21.50		21.50	Cost includes material and labor to
	Install	Ea.	105	103		208	install 3' x 3' single hung, two light,
	Demolish and Install	Ea.	105	124.50		229.50	double glazed aluminum window
	Reinstall	Ea.		82.62		82.62	including screen.
	Clean	Ea.	.03	8.10		8.13	
	Minimum Charge	Job		258		258	
3' x 3' 6"							
	Demolish	Ea.		21.50		21.50	Cost includes material and labor to
	Install	Ea.	264	103		367	install 3' x 3' 6" single hung, two
	Demolish and Install	Ea.	264	124.50		388.50	light, double glazed aluminum
	Reinstall	Ea.		82.62		82.62	window including screen.
	Clean	Ea.	.04	9.45		9.49	
	Minimum Charge	Job		258		258	
3' x 4'							
	Demolish	Ea.		21.50		21.50	Cost includes material and labor to
	Install	Ea.	276	103		379	install 3' x 4' single hung, two light,
	Demolish and Install	Ea.	276	124.50		400.50	double glazed aluminum window
	Reinstall	Ea.		82.62		82.62	including screen.
	Clean	Ea.	.04	9.45		9.49	
	Minimum Charge	Job		258		258	
3' x 5'							
	Demolish	Ea.		31.50		31.50	Cost includes material and labor to
	Install	Ea.	133	103		236	install 3' x 5' single hung, two light,
	Demolish and Install	Ea.	133	134.50		267.50	double glazed aluminum window
	Reinstall	Ea.		82.62		82.62	including screen.
	Clean	Ea.	.07	11.35		11.42	
	Minimum Charge	Job		258		258	

Aluminum Window		Unit	Material	Labor	Equip.	Total	Specification
3' x 6'							
	Demolish	Ea.		31.50		31.50	Cost includes material and labor to
	Install	Ea.	375	115		490	install 3' x 6' single hung, two light,
	Demolish and Install	Ea.	375	146.50		521.50	double glazed aluminum window
	Reinstall	Ea.		91.80		91.80	including screen.
	Clean	Ea.	.08	13.35		13.43	
	Minimum Charge	Job		258		258	
Double Hung Two Light							
2' x 2'							
	Demolish	Ea.		21.50		21.50	Cost includes material and labor to
	Install	Ea.	127	94		221	install 2' x 2' double hung, two light,
	Demolish and Install	Ea.	127	115.50		242.50	double glazed aluminum window
	Reinstall	Ea.		75.11		75.11	including screen.
	Clean	Ea.	.02	7.10		7.12	
	Minimum Charge	Job		258		258	
2' x 2' 6"							
	Demolish	Ea.		21.50		21.50	Cost includes material and labor to
	Install	Ea.	136	94		230	install 2' x 2' 6" double hung, two
	Demolish and Install	Ea.	136	115.50		251.50	light, double glazed aluminum
	Reinstall	Ea.		75.11		75.11	window including screen.
	Clean	Ea.	.02	7.10		7.12	
	Minimum Charge	Job		258		258	
3' x 1' 6"							
	Demolish	Ea.		21.50		21.50	Cost includes material and labor to
	Install	Ea.	297	103		400	install 3' x 1' 6" double hung, two
	Demolish and Install	Ea.	297	124.50		421.50	light, double glazed aluminum
	Reinstall	Ea.		82.62		82.62	window including screen.
	Clean	Ea.	.02	7.10		7.12	
	Minimum Charge	Job		258		258	
3' x 2'							
	Demolish	Ea.		21.50		21.50	Cost includes material and labor to
	Install	Ea.	345	103		448	install 3' x 2' double hung, two light,
	Demolish and Install	Ea.	345	124.50		469.50	double glazed aluminum window
	Reinstall	Ea.		82.62		82.62	including screen.
	Clean	Ea.	.02	7.10		7.12	
	Minimum Charge	Job		258		258	
3' x 2' 6"							
	Demolish	Ea.		21.50		21.50	Cost includes material and labor to
	Install	Ea.	420	103		523	install 3' x 2' 6" double hung, two
	Demolish and Install	Ea.	420	124.50		544.50	light, double glazed aluminum
	Reinstall	Ea.		82.62		82.62	window including screen.
	Clean	Ea.	.03	8.10		8.13	
	Minimum Charge	Job		258		258	
3' x 3'							
	Demolish	Ea.		21.50		21.50	Cost includes material and labor to
	Install	Ea.	470	103		573	install 3' x 3' double hung, two light,
	Demolish and Install	Ea.	470	124.50		594.50	double glazed aluminum window
	Reinstall	Ea.		82.62		82.62	including screen.
	Clean	Ea.	.03	8.10		8.13	
	Minimum Charge	Job		258		258	
3' x 3' 6"							
	Demolish	Ea.		21.50		21.50	Cost includes material and labor to
	Install	Ea.	490	103		593	install 3' x 3' 6" double hung, two
	Demolish and Install	Ea.	490	124.50		614.50	light, double glazed aluminum
	Reinstall	Ea.		82.62		82.62	window including screen.
	Clean	Ea.	.04	9.45		9.49	
	Minimum Charge	Job		258		258	

Exterior Doors and Windows

Aluminum Window		Unit	Material	Labor	Equip.	Total	Specification
3' x 4'							
	Demolish	Ea.		21.50		21.50	Cost includes material and labor to
	Install	Ea.	530	103		633	install 3' x 4' double hung, two light,
	Demolish and Install	Ea.	530	124.50		654.50	double glazed aluminum window
	Reinstall	Ea.		82.62		82.62	including screen.
	Clean	Ea.	.04	9.45		9.49	
	Minimum Charge	Job		258		258	
3' x 5'							
	Demolish	Ea.		31.50		31.50	Cost includes material and labor to
	Install	Ea.	585	103		688	install 3' x 5' double hung, two light,
	Demolish and Install	Ea.	585	134.50		719.50	double glazed aluminum window
	Reinstall	Ea.		82.62		82.62	including screen.
	Clean	Ea.	.07	11.35		11.42	
	Minimum Charge	Job		258		258	
3' x 6'							
	Demolish	Ea.		31.50		31.50	Cost includes material and labor to
	Install	Ea.	615	115		730	install 3' x 6' double hung, two light,
	Demolish and Install	Ea.	615	146.50		761.50	double glazed aluminum window
	Reinstall	Ea.		91.80		91.80	including screen.
	Clean	Ea.	.08	13.35		13.43	
	Minimum Charge	Job		258		258	
Sliding Sash							
1' 6" x 3'							
	Demolish	Ea.		21.50		21.50	Cost includes material and labor to
	Install	Ea.	239	94		333	install 1' 6" x 3' sliding sash, two
	Demolish and Install	Ea.	239	115.50		354.50	light, double glazed aluminum
	Reinstall	Ea.		75.11		75.11	window including screen.
	Clean	Ea.	.02	7.10		7.12	
	Minimum Charge	Job		258		258	
2' x 2'							
	Demolish	Ea.		21.50		21.50	Cost includes material and labor to
	Install	Ea.	64	103		167	install 2' x 2' sliding sash, two light,
	Demolish and Install	Ea.	64	124.50		188.50	double glazed aluminum window
	Reinstall	Ea.		82.62		82.62	including screen.
	Clean	Ea.	.02	7.10		7.12	
	Minimum Charge	Job		258		258	
3' x 2'							
	Demolish	Ea.		21.50		21.50	Cost includes material and labor to
	Install	Ea.	252	103		355	install 3' x 2' sliding sash, two light,
	Demolish and Install	Ea.	252	124.50		376.50	double glazed aluminum window
	Reinstall	Ea.		82.62		82.62	including screen.
	Clean	Ea.	.02	7.10		7.12	
	Minimum Charge	Job		258		258	
3' x 2' 6"							
	Demolish	Ea.		21.50		21.50	Cost includes material and labor to
	Install	Ea.	87	103		190	install 3' x 2' 6" sliding sash, two
	Demolish and Install	Ea.	87	124.50		211.50	light, double glazed aluminum
	Reinstall	Ea.		82.62		82.62	window including screen.
	Clean	Ea.	.03	8.10		8.13	
	Minimum Charge	Job		258		258	
3' x 3'							
	Demolish	Ea.		21.50		21.50	Cost includes material and labor to
	Install	Ea.	305	103		408	install 3' x 3' sliding sash, two light,
	Demolish and Install	Ea.	305	124.50		429.50	double glazed aluminum window
	Reinstall	Ea.		82.62		82.62	including screen.
	Clean	Ea.	.03	8.10		8.13	
	Minimum Charge	Job		258		258	

Exterior Doors and Windows

Aluminum Window	Unit	Material	Labor	Equip.	Total	Specification
3' x 3' 6"						
Demolish	Ea.		21.50		21.50	Cost includes material and labor to
Install	Ea.	103	103		206	install 3' x 3' 6" sliding sash, two
Demolish and Install	Ea.	103	124.50		227.50	light, double glazed aluminum
Reinstall	Ea.		82.62		82.62	window including screen.
Clean	Ea.	.04	9.45		9.49	
Minimum Charge	Job		258		258	
3' x 4'						
Demolish	Ea.		21.50		21.50	Cost includes material and labor to
Install	Ea.	112	115		227	install 3' x 4' sliding sash, two light,
Demolish and Install	Ea.	112	136.50		248.50	double glazed aluminum window
Reinstall	Ea.		91.80		91.80	including screen.
Clean	Ea.	.04	9.45		9.49	
Minimum Charge	Job		258		258	
3' x 5'						
Demolish	Ea.		31.50		31.50	Cost includes material and labor to
Install	Ea.	143	115		258	install 3' x 5' sliding sash, two light,
Demolish and Install	Ea.	143	146.50		289.50	double glazed aluminum window
Reinstall	Ea.		91.80		91.80	including screen.
Clean	Ea.	.07	11.35		11.42	
Minimum Charge	Job		258		258	
4' x 5'						
Demolish	Ea.		31.50		31.50	Cost includes material and labor to
Install	Ea.	145	148		293	install 4' x 5' sliding sash, two light,
Demolish and Install	Ea.	145	179.50		324.50	double glazed aluminum window
Reinstall	Ea.		118.03		118.03	including screen.
Clean	Ea.	.10	16.20		16.30	
Minimum Charge	Job		258		258	
5' x 5'						
Demolish	Ea.		31.50		31.50	Cost includes material and labor to
Install	Ea.	350	148		498	install 5' x 5' sliding sash, two light,
Demolish and Install	Ea.	350	179.50		529.50	double glazed aluminum window
Reinstall	Ea.		118.03		118.03	including screen.
Clean	Ea.	.10	16.20		16.30	
Minimum Charge	Job		258		258	

Awning Sash

	Unit	Material	Labor	Equip.	Total	Specification
26" x 26", 2 Light						
Demolish	Ea.		21.50		21.50	Cost includes material and labor to
Install	Ea.	320	115		435	install 26" x 26" two light awning,
Demolish and Install	Ea.	320	136.50		456.50	double glazed aluminum window
Reinstall	Ea.		91.80		91.80	including screen.
Clean	Ea.	.03	8.10		8.13	
Minimum Charge	Job		258		258	
26" x 37", 2 Light						
Demolish	Ea.		21.50		21.50	Cost includes material and labor to
Install	Ea.	360	115		475	install 26" x 37" two light awning,
Demolish and Install	Ea.	360	136.50		496.50	double glazed aluminum window
Reinstall	Ea.		91.80		91.80	including screen.
Clean	Ea.	.03	8.10		8.13	
Minimum Charge	Job		258		258	
38" x 26", 3 Light						
Demolish	Ea.		21.50		21.50	Cost includes material and labor to
Install	Ea.	395	115		510	install 38" x 26" three light awning,
Demolish and Install	Ea.	395	136.50		531.50	double glazed aluminum window
Reinstall	Ea.		91.80		91.80	including screen.
Clean	Ea.	.04	9.45		9.49	
Minimum Charge	Job		258		258	

Exterior Doors and Windows

Aluminum Window		Unit	Material	Labor	Equip.	Total	Specification
38" x 37", 3 Light							
	Demolish	Ea.		21.50		21.50	Cost includes material and labor to
	Install	Ea.	400	129		529	install 38" x 37" three light awning,
	Demolish and Install	Ea.	400	150.50		550.50	double glazed aluminum window
	Reinstall	Ea.		103.28		103.28	including screen.
	Clean	Ea.	.04	9.45		9.49	
	Minimum Charge	Job		258		258	
38" x 53", 3 Light							
	Demolish	Ea.		31.50		31.50	Cost includes material and labor to
	Install	Ea.	670	129		799	install 38" x 53" three light awning,
	Demolish and Install	Ea.	670	160.50		830.50	double glazed aluminum window
	Reinstall	Ea.		103.28		103.28	including screen.
	Clean	Ea.	.07	11.35		11.42	
	Minimum Charge	Job		258		258	
50" x 37", 4 Light							
	Demolish	Ea.		31.50		31.50	Cost includes material and labor to
	Install	Ea.	660	129		789	install 50" x 37" four light awning,
	Demolish and Install	Ea.	660	160.50		820.50	double glazed aluminum window
	Reinstall	Ea.		103.28		103.28	including screen.
	Clean	Ea.	.07	11.35		11.42	
	Minimum Charge	Job		258		258	
50" x 53", 4 Light							
	Demolish	Ea.		31.50		31.50	Cost includes material and labor to
	Install	Ea.	1100	148		1248	install 50" x 53" four light awning,
	Demolish and Install	Ea.	1100	179.50		1279.50	double glazed aluminum window
	Reinstall	Ea.		118.03		118.03	including screen.
	Clean	Ea.	.08	13.35		13.43	
	Minimum Charge	Job		258		258	
63" x 37", 5 Light							
	Demolish	Ea.		31.50		31.50	Cost includes material and labor to
	Install	Ea.	860	148		1008	install 63" x 37" five light awning,
	Demolish and Install	Ea.	860	179.50		1039.50	double glazed aluminum window
	Reinstall	Ea.		118.03		118.03	including screen.
	Clean	Ea.	.08	13.35		13.43	
	Minimum Charge	Job		258		258	
63" x 53", 5 Light							
	Demolish	Ea.		31.50		31.50	Cost includes material and labor to
	Install	Ea.	1200	148		1348	install 63" x 53" five light awning,
	Demolish and Install	Ea.	1200	179.50		1379.50	double glazed aluminum window
	Reinstall	Ea.		118.03		118.03	including screen.
	Clean	Ea.	.10	16.20		16.30	
	Minimum Charge	Job		258		258	
Mullion Bars							
	Install	Ea.	13.35	41.50		54.85	Includes labor and material to install
	Minimum Charge	Job		258		258	mullions for sliding aluminum window.
Bay							
4' 8" x 6' 4"							
	Demolish	Ea.		23		23	Cost includes material and labor to
	Install	Ea.	1600	106		1706	install 4' 8" x 6' 4" angle bay double
	Demolish and Install	Ea.	1600	129		1729	hung unit, pine frame with aluminum
	Reinstall	Ea.		84.56		84.56	cladding, picture center sash,
	Clean	Ea.	.14	18.90		19.04	hardware, screens.
	Minimum Charge	Job		258		258	

Aluminum Window	Unit	Material	Labor	Equip.	Total	Specification
4′ 8″ x 7′ 2″						
Demolish	Ea.		23		23	Cost includes material and labor to
Install	Ea.	1650	106		1756	install 4′ 8″ x 7′ 2″ angle bay double
Demolish and Install	Ea.	1650	129		1779	hung unit, pine frame with aluminum
Reinstall	Ea.		84.56		84.56	cladding, picture center sash,
Clean	Ea.	.14	18.90		19.04	hardware, screens.
Minimum Charge	Job		258		258	
4′ 8″ x 7′ 10″						
Demolish	Ea.		23		23	Cost Includes material and labor to
Install	Ea.	1675	106		1781	install 4′ 8″ x 7′ 10″ angle bay
Demolish and Install	Ea.	1675	129		1804	double hung unit, pine frame with
Reinstall	Ea.		84.56		84.56	aluminum cladding, picture center
Clean	Ea.	.14	18.90		19.04	sash, hardware, screens.
Minimum Charge	Job		258		258	
4′ 8″ x 8′ 6″						
Demolish	Ea.		26.50		26.50	Cost includes material and labor to
Install	Ea.	1650	121		1771	install 4′ 8″ x 8′ 6″ angle bay double
Demolish and Install	Ea.	1650	147.50		1797.50	hung unit, pine frame with aluminum
Reinstall	Ea.		96.64		96.64	cladding, picture center sash,
Clean	Ea.	.02	25		25.02	hardware, screens.
Minimum Charge	Job		258		258	
4′ 8″ x 8′ 10″						
Demolish	Ea.		26.50		26.50	Cost includes material and labor to
Install	Ea.	1675	121		1796	install 4′ 8″ x 8′ 10″ angle bay
Demolish and Install	Ea.	1675	147.50		1822.50	double hung unit, pine frame with
Reinstall	Ea.		96.64		96.64	aluminum cladding, picture center
Clean	Ea.	.02	25		25.02	sash, hardware, screens.
Minimum Charge	Job		258		258	
4′ 8″ x 9′ 2″						
Demolish	Ea.		26.50		26.50	Cost includes material and labor to
Install	Ea.	1725	211		1936	install 4′ 8″ x 9′ 2″ angle bay double
Demolish and Install	Ea.	1725	237.50		1962.50	hung unit, pine frame with aluminum
Reinstall	Ea.		169.12		169.12	cladding, picture center sash,
Clean	Ea.	.02	25		25.02	hardware, screens.
Minimum Charge	Job		258		258	
4′ 8″ x 9′ 10″						
Demolish	Ea.		26.50		26.50	Cost includes material and labor to
Install	Ea.	1825	211		2036	install 4′ 8″ x 9′ 10″ angle bay
Demolish and Install	Ea.	1825	237.50		2062.50	double hung unit, pine frame with
Reinstall	Ea.		169.12		169.12	aluminum cladding, picture center
Clean	Ea.	.02	25		25.02	sash, hardware, screens.
Minimum Charge	Job		258		258	
Casing						
Demolish	Ea.		9.80		9.80	Cost includes material and labor to
Install	Opng.	41.50	32.50		74	install 11/16″ x 2-1/2″ pine ranch
Demolish and Install	Opng.	41.50	42.30		83.80	style window casing.
Reinstall	Opng.		26.02		26.02	
Clean	Opng.	.46	3.10		3.56	
Paint	Ea.	5.20	7.20		12.40	
Minimum Charge	Job		258		258	
Storm Window						
2′ x 3′ 6″						
Demolish	Ea.		12.75		12.75	Includes material and labor to install
Install	Ea.	82.50	28		110.50	residential aluminum storm window
Demolish and Install	Ea.	82.50	40.75		123.25	with mill finish, 2′ x 3′ 5″ high.
Reinstall	Ea.		22.55		22.55	
Clean	Ea.	.02	7.10		7.12	
Minimum Charge	Job		258		258	

Exterior Doors and Windows

Aluminum Window

Aluminum Window		Unit	Material	Labor	Equip.	Total	Specification
2' 6" x 5'							
	Demolish	Ea.		12.75		12.75	Includes material and labor to install
	Install	Ea.	93.50	30		123.50	residential aluminum storm window
	Demolish and Install	Ea.	93.50	42.75		136.25	with mill finish, 2' 6" x 5' high.
	Reinstall	Ea.		24.16		24.16	
	Clean	Ea.	.04	9.45		9.49	
	Minimum Charge	Job		258		258	
4' x 6'							
	Demolish	Ea.		24.50		24.50	Includes material and labor to install
	Install	Ea.	108	34		142	residential aluminum storm window, 4'
	Demolish and Install	Ea.	108	58.50		166.50	x 6' high.
	Reinstall	Ea.		27.06		27.06	
	Clean	Ea.	.10	14.20		14.30	
	Minimum Charge	Job		258		258	
Hardware							
Weatherstripping							
	Install	L.F.	.55	3.52		4.07	Includes labor and material to install
	Minimum Charge	Job		141		141	vinyl V strip weatherstripping for
							double-hung window.
Trim Set							
	Demolish	Ea.		9.80		9.80	Cost includes material and labor to
	Install	Opng.	41.50	32.50		74	install 11/16" x 2-1/2" pine ranch
	Demolish and Install	Opng.	41.50	42.30		83.80	style window casing.
	Reinstall	Opng.		26.02		26.02	
	Clean	Opng.	.46	3.10		3.56	
	Paint	Ea.	5.20	7.20		12.40	
	Minimum Charge	Job		211		211	

Vinyl Window

Vinyl Window		Unit	Material	Labor	Equip.	Total	Specification
Double Hung							
2' x 2' 6"							
	Demolish	Ea.		15.65		15.65	Cost includes material and labor to
	Install	Ea.	219	56.50		275.50	install 2' x 2' 6" double hung, two
	Demolish and Install	Ea.	219	72.15		291.15	light, vinyl window with screen.
	Reinstall	Ea.		45.10		45.10	
	Clean	Ea.	.02	7.10		7.12	
	Minimum Charge	Job		141		141	
2' x 3' 6"							
	Demolish	Ea.		15.65		15.65	Cost includes material and labor to
	Install	Ea.	233	60.50		293.50	install 2' x 3' 6" double hung, two
	Demolish and Install	Ea.	233	76.15		309.15	light, vinyl window with screen.
	Reinstall	Ea.		48.32		48.32	
	Clean	Ea.	.03	8.10		8.13	
	Minimum Charge	Job		141		141	
2' 6" x 4' 6"							
	Demolish	Ea.		15.65		15.65	Cost includes material and labor to
	Install	Ea.	275	65		340	install 2' 6" x 4' 6" double hung, two
	Demolish and Install	Ea.	275	80.65		355.65	light, vinyl window with screen.
	Reinstall	Ea.		52.04		52.04	
	Clean	Ea.	.04	9.45		9.49	
	Minimum Charge	Job		141		141	
3' x 4'							
	Demolish	Ea.		15.65		15.65	Includes material and labor to install
	Install	Ea.	238	84.50		322.50	medium size double hung, 3' x 4',
	Demolish and Install	Ea.	238	100.15		338.15	two light, vinyl window with screen.
	Reinstall	Ea.		67.65		67.65	
	Clean	Ea.	.04	9.45		9.49	
	Minimum Charge	Job		141		141	

Exterior Doors and Windows

Vinyl Window

		Unit	Material	Labor	Equip.	Total	Specification
3' x 4' 6"							
	Demolish	Ea.		19.10		19.10	Cost includes material and labor to
	Install	Ea.	305	94		399	install 3' x 4' 6" double hung, two
	Demolish and Install	Ea.	305	113.10		418.10	light, vinyl window with screen.
	Reinstall	Ea.		75.16		75.16	
	Clean	Ea.	.07	11.35		11.42	
	Minimum Charge	Job		141		141	
4' x 4' 6"							
	Demolish	Ea.		19.10		19.10	Cost includes material and labor to
	Install	Ea.	320	106		426	install 4' x 4' 6" double hung, two
	Demolish and Install	Ea.	320	125.10		445.10	light, vinyl window with screen.
	Reinstall	Ea.		84.56		84.56	
	Clean	Ea.	.08	13.35		13.43	
	Minimum Charge	Job		141		141	
4' x 6'							
	Demolish	Ea.		19.10		19.10	Includes material and labor to install
	Install	Ea.	375	121		496	large double hung, 4' x 6', two light,
	Demolish and Install	Ea.	375	140.10		515.10	vinyl window with screen.
	Reinstall	Ea.		96.64		96.64	
	Clean	Ea.	.10	16.20		16.30	
	Minimum Charge	Job		141		141	
Trim Set							
	Demolish	Ea.		9.80		9.80	Cost includes material and labor to
	Install	Opng.	41.50	32.50		74	install 11/16" x 2-1/2" pine ranch
	Demolish and Install	Opng.	41.50	42.30		83.80	style window casing.
	Reinstall	Opng.		26.02		26.02	
	Clean	Opng.	.46	3.10		3.56	
	Paint	Ea.	5.20	7.20		12.40	
	Minimum Charge	Job		258		258	

Wood Window

		Unit	Material	Labor	Equip.	Total	Specification
Double Hung							
2' 2" x 3' 4"							
	Demolish	Ea.		15.65		15.65	Cost includes material and labor to
	Install	Ea.	231	56.50		287.50	install 2' 2" x 3' 4" double hung, two
	Demolish and Install	Ea.	231	72.15		303.15	light, double glazed wood window
	Reinstall	Ea.		45.10		45.10	with screen.
	Clean	Ea.	.03	8.10		8.13	
	Paint	Ea.	3.25	15.05		18.30	
	Minimum Charge	Job		211		211	
2' 2" x 4' 4"							
	Demolish	Ea.		15.65		15.65	Cost includes material and labor to
	Install	Ea.	252	60.50		312.50	install 2' 2" x 4' 4" double hung, two
	Demolish and Install	Ea.	252	76.15		328.15	light, double glazed wood window
	Reinstall	Ea.		48.32		48.32	with screen.
	Clean	Ea.	.03	8.10		8.13	
	Paint	Ea.	3.25	15.05		18.30	
	Minimum Charge	Job		211		211	
2' 6" x 3' 4"							
	Demolish	Ea.		15.65		15.65	Cost includes material and labor to
	Install	Ea.	241	65		306	install 2' 6" x 3' 4" double hung, two
	Demolish and Install	Ea.	241	80.65		321.65	light, double glazed wood window
	Reinstall	Ea.		52.04		52.04	with screen.
	Clean	Ea.	.04	9.45		9.49	
	Paint	Ea.	4.16	18.05		22.21	
	Minimum Charge	Job		211		211	

Exterior Doors and Windows

Wood Window		Unit	Material	Labor	Equip.	Total	Specification
2' 6" x 4'							
	Demolish	Ea.		15.65		15.65	Cost includes material and labor to
	Install	Ea.	252	70.50		322.50	install 2' 6" x 4' double hung, two
	Demolish and Install	Ea.	252	86.15		338.15	light, double glazed wood window
	Reinstall	Ea.		56.37		56.37	with screen.
	Clean	Ea.	.04	9.45		9.49	
	Paint	Ea.	4.16	18.05		22.21	
	Minimum Charge	Job		211		211	
2' 6" x 4' 8"							
	Demolish	Ea.		15.65		15.65	Cost includes material and labor to
	Install	Ea.	275	70.50		345.50	install 2' 6" x 4' 8" double hung, two
	Demolish and Install	Ea.	275	86.15		361.15	light, double glazed wood window
	Reinstall	Ea.		56.37		56.37	with screen.
	Clean	Ea.	.04	9.45		9.49	
	Paint	Ea.	4.16	18.05		22.21	
	Minimum Charge	Job		211		211	
2' 10" x 3' 4"							
	Demolish	Ea.		15.65		15.65	Cost includes material and labor to
	Install	Ea.	246	84.50		330.50	install 2' 10" x 3' 4" double hung, two
	Demolish and Install	Ea.	246	100.15		346.15	light, double glazed wood window
	Reinstall	Ea.		67.65		67.65	with screen.
	Clean	Ea.	.04	9.45		9.49	
	Paint	Ea.	4.16	18.05		22.21	
	Minimum Charge	Job		211		211	
2' 10" x 4'							
	Demolish	Ea.		15.65		15.65	Cost includes material and labor to
	Install	Ea.	272	84.50		356.50	install 2' 10" x 4' double hung, two
	Demolish and Install	Ea.	272	100.15		372.15	light, double glazed wood window
	Reinstall	Ea.		67.65		67.65	with screen.
	Clean	Ea.	.04	9.45		9.49	
	Paint	Ea.	4.16	18.05		22.21	
	Minimum Charge	Job		211		211	
3' 7" x 3' 4"							
	Demolish	Ea.		15.65		15.65	Cost includes material and labor to
	Install	Ea.	279	94		373	install 3' 7" x 3' 4" double hung, two
	Demolish and Install	Ea.	279	109.65		388.65	light, double glazed wood window
	Reinstall	Ea.		75.16		75.16	with screen.
	Clean	Ea.	.04	9.45		9.49	
	Paint	Ea.	4.16	18.05		22.21	
	Minimum Charge	Job		211		211	
3' 7" x 5' 4"							
	Demolish	Ea.		19.10		19.10	Cost includes material and labor to
	Install	Ea.	310	94		404	install 3' 7" x 5' 4" double hung, two
	Demolish and Install	Ea.	310	113.10		423.10	light, double glazed wood window
	Reinstall	Ea.		75.16		75.16	with screen.
	Clean	Ea.	.10	16.20		16.30	
	Paint	Ea.	6.95	30		36.95	
	Minimum Charge	Job		211		211	
4' 3" x 5' 4"							
	Demolish	Ea.		19.10		19.10	Cost includes material and labor to
	Install	Ea.	560	106		666	install 4' 3" x 5' 4" double hung, two
	Demolish and Install	Ea.	560	125.10		685.10	light, double glazed wood window
	Reinstall	Ea.		84.56		84.56	with screen.
	Clean	Ea.	.10	16.20		16.30	
	Paint	Ea.	4.61	36		40.61	
	Minimum Charge	Job		211		211	

Exterior Doors and Windows

Wood Window		Unit	Material	Labor	Equip.	Total	Specification
Casement							
2' x 3' 4"							
	Demolish	Ea.		15.65		15.65	Cost includes material and labor to
	Install	Ea.	288	42.50		330.50	install 2' x 3' 4 " casement window
	Demolish and Install	Ea.	288	58.15		346.15	with 3/4" insulated glass, screens,
	Reinstall	Ea.		33.82		33.82	weatherstripping, hardware.
	Clean	Ea.	.03	8.10		8.13	
	Paint	Ea.	3.25	15.05		18.30	
	Minimum Charge	Job		425		425	
2' x 4'							
	Demolish	Ea.		15.65		15.65	Cost includes material and labor to
	Install	Ea.	305	47		352	install 2' x 4' casement window with
	Demolish and Install	Ea.	305	62.65		367.65	3/4" insulated glass, screens,
	Reinstall	Ea.		37.58		37.58	weatherstripping, hardware.
	Clean	Ea.	.03	8.10		8.13	
	Paint	Ea.	3.25	15.05		18.30	
	Minimum Charge	Job		425		425	
2' x 5'							
	Demolish	Ea.		15.65		15.65	Cost includes material and labor to
	Install	Ea.	335	49.50		384.50	install 2' x 5' casement window with
	Demolish and Install	Ea.	335	65.15		400.15	3/4" insulated glass, screens,
	Reinstall	Ea.		39.79		39.79	weatherstripping, hardware.
	Clean	Ea.	.04	9.45		9.49	
	Paint	Ea.	4.16	18.05		22.21	
	Minimum Charge	Job		425		425	
2' x 6'							
	Demolish	Ea.		15.65		15.65	Cost includes material and labor to
	Install	Ea.	335	53		388	install 2' x 6' casement window with
	Demolish and Install	Ea.	335	68.65		403.65	3/4" insulated glass, screens,
	Reinstall	Ea.		42.28		42.28	weatherstripping, hardware.
	Clean	Ea.	.04	9.45		9.49	
	Paint	Ea.	4.16	18.05		22.21	
	Minimum Charge	Job		425		425	
4' x 3' 4"							
	Demolish	Ea.		19.10		19.10	Cost includes material and labor to
	Install	Ea.	645	56.50		701.50	install 4' x 3' 4 " casement windows
	Demolish and Install	Ea.	645	75.60		720.60	with 3/4" insulated glass, screens,
	Reinstall	Ea.		45.10		45.10	weatherstripping, hardware.
	Clean	Ea.	.04	9.45		9.49	
	Paint	Ea.	2.31	18.05		20.36	
	Minimum Charge	Job		425		425	
4' x 4'							
	Demolish	Ea.		19.10		19.10	Cost includes material and labor to
	Install	Ea.	720	56.50		776.50	install 4' x 4' casement windows with
	Demolish and Install	Ea.	720	75.60		795.60	3/4" insulated glass, screens,
	Reinstall	Ea.		45.10		45.10	weatherstripping, hardware.
	Clean	Ea.	.08	13.35		13.43	
	Paint	Ea.	2.73	22.50		25.23	
	Minimum Charge	Job		425		425	
4' x 5'							
	Demolish	Ea.		19.10		19.10	Cost includes material and labor to
	Install	Ea.	815	60.50		875.50	install 4' x 5' casement windows with
	Demolish and Install	Ea.	815	79.60		894.60	3/4" insulated glass, screens,
	Reinstall	Ea.		48.32		48.32	weatherstripping, hardware.
	Clean	Ea.	.10	16.20		16.30	
	Paint	Ea.	3.56	30		33.56	
	Minimum Charge	Job		425		425	

Wood Window		Unit	Material	Labor	Equip.	Total	Specification
4' x 6'							
	Demolish	Ea.		19.10		19.10	Cost includes material and labor to
	Install	Ea.	920	70.50		990.50	install 4' x 6' casement windows with
	Demolish and Install	Ea.	920	89.60		1009.60	3/4" insulated glass, screens,
	Reinstall	Ea.		56.37		56.37	weatherstripping, hardware.
	Clean	Ea.	.10	16.20		16.30	
	Paint	Ea.	4.19	36		40.19	
	Minimum Charge	Job		425		425	
Casement Bow							
4' 8" x 8' 1"							
	Demolish	Ea.		26.50		26.50	Includes material and labor to install
	Install	Ea.	1725	159		1884	casement bow window unit with 4
	Demolish and Install	Ea.	1725	185.50		1910.50	lights, pine frame, 1/2" insulated
	Reinstall	Ea.		126.84		126.84	glass, weatherstripping, hardware,
	Clean	Ea.	.14	18.90		19.04	screens.
	Paint	Ea.	7.75	72		79.75	
	Minimum Charge	Job		425		425	
4' 8" x 10'							
	Demolish	Ea.		26.50		26.50	Includes material and labor to install
	Install	Ea.	2125	211		2336	casement bow window unit with 5
	Demolish and Install	Ea.	2125	237.50		2362.50	lights, pine frame, 1/2" insulated
	Reinstall	Ea.		169.12		169.12	glass, weatherstripping, hardware,
	Clean	Ea.	.02	25		25.02	screens.
	Paint	Ea.	9.65	90		99.65	
	Minimum Charge	Job		425		425	
5' 4" x 8' 1"							
	Demolish	Ea.		26.50		26.50	Includes material and labor to install
	Install	Ea.	1850	159		2009	casement bow window unit with 4
	Demolish and Install	Ea.	1850	185.50		2035.50	lights, pine frame, 1/2" insulated
	Reinstall	Ea.		126.84		126.84	glass, weatherstripping, hardware,
	Clean	Ea.	.02	25		25.02	screens.
	Paint	Ea.	8.80	72		80.80	
	Minimum Charge	Job		425		425	
5' 4" x 10'							
	Demolish	Ea.		26.50		26.50	Includes material and labor to install
	Install	Ea.	2100	211		2311	casement bow window unit with 5
	Demolish and Install	Ea.	2100	237.50		2337.50	lights, pine frame, 1/2" insulated
	Reinstall	Ea.		169.12		169.12	glass, weatherstripping, hardware,
	Clean	Ea.	.02	25		25.02	screens.
	Paint	Ea.	11.10	90		101.10	
	Minimum Charge	Job		425		425	
6' x 8' 1"							
	Demolish	Ea.		26.50		26.50	Includes material and labor to install
	Install	Ea.	2125	211		2336	casement bow window unit with 5
	Demolish and Install	Ea.	2125	237.50		2362.50	lights, pine frame, 1/2" insulated
	Reinstall	Ea.		169.12		169.12	glass, weatherstripping, hardware,
	Clean	Ea.	.02	25		25.02	screens.
	Paint	Ea.	10.05	90		100.05	
	Minimum Charge	Job		425		425	
6' x 10'							
	Demolish	Ea.		26.50		26.50	Includes material and labor to install
	Install	Ea.	2200	211		2411	casement bow window unit with 5
	Demolish and Install	Ea.	2200	237.50		2437.50	lights, pine frame, 1/2" insulated
	Reinstall	Ea.		169.12		169.12	glass, weatherstripping, hardware,
	Clean	Ea.	.02	25		25.02	screens.
	Paint	Ea.	12.60	120		132.60	
	Minimum Charge	Job		425		425	

Exterior Doors and Windows

Wood Window		Unit	Material	Labor	Equip.	Total	Specification
Casing							
	Demolish	Ea.		9.80		9.80	Cost includes material and labor to
	Install	Opng.	41.50	32.50		74	install 11/16" x 2-1/2" pine ranch
	Demolish and Install	Opng.	41.50	42.30		83.80	style window casing.
	Reinstall	Opng.		26.02		26.02	
	Clean	Opng.	.46	3.10		3.56	
	Paint	Ea.	5.20	7.20		12.40	
	Minimum Charge	Job		258		258	
Trim Set							
	Demolish	Ea.		7.45		7.45	Cost includes material and labor to
	Install	Opng.	41.50	32.50		74	install 11/16" x 2-1/2" pine ranch
	Demolish and Install	Opng.	41.50	39.95		81.45	style window casing.
	Reinstall	Opng.		26.02		26.02	
	Clean	Opng.	.46	3.10		3.56	
	Paint	Ea.	5.20	7.20		12.40	
	Minimum Charge	Job		211		211	
Wood Picture Window							
3' 6" x 4'							
	Demolish	Ea.		19.10		19.10	Includes material and labor to install a
	Install	Ea.	480	70.50		550.50	wood framed picture window
	Demolish and Install	Ea.	480	89.60		569.60	including exterior trim.
	Clean	Ea.	.10	16.20		16.30	
	Paint	Ea.	3.47	28		31.47	
	Minimum Charge	Job		425		425	
4' x 4' 6"							
	Demolish	Ea.		19.10		19.10	Includes material and labor to install a
	Install	Ea.	585	77		662	wood framed picture window
	Demolish and Install	Ea.	585	96.10		681.10	including exterior trim.
	Clean	Ea.	.10	16.20		16.30	
	Paint	Ea.	3.47	28		31.47	
	Minimum Charge	Job		425		425	
5' x 4'							
	Demolish	Ea.		19.10		19.10	Includes material and labor to install a
	Install	Ea.	665	77		742	wood framed picture window
	Demolish and Install	Ea.	665	96.10		761.10	including exterior trim.
	Clean	Ea.	.10	16.20		16.30	
	Paint	Ea.	7.45	45		52.45	
	Minimum Charge	Job		425		425	
6' x 4' 6"							
	Demolish	Ea.		19.10		19.10	Includes material and labor to install a
	Install	Ea.	700	84.50		784.50	wood framed picture window
	Demolish and Install	Ea.	700	103.60		803.60	including exterior trim.
	Clean	Ea.	.10	16.20		16.30	
	Paint	Ea.	7.45	45		52.45	
	Minimum Charge	Job		425		425	

Window Screens		Unit	Material	Labor	Equip.	Total	Specification
Small							
	Demolish	Ea.		3.58		3.58	Includes material and labor to install
	Install	Ea.	19.95	12.80		32.75	small window screen, 4 S.F.
	Demolish and Install	Ea.	19.95	16.38		36.33	
	Reinstall	Ea.		10.25		10.25	
	Clean	Ea.	.33	6.45		6.78	
	Minimum Charge	Job		106		106	

Exterior Doors and Windows

Window Screens		Unit	Material	Labor	Equip.	Total	Specification
Standard Size							
	Demolish	Ea.		3.58		3.58	Includes material and labor to install
	Install	Ea.	48.50	12.80		61.30	medium window screen, 12 S.F.
	Demolish and Install	Ea.	48.50	16.38		64.88	
	Reinstall	Ea.		10.25		10.25	
	Clean	Ea.	.33	6.45		6.78	
	Minimum Charge	Job		106		106	
Large							
	Demolish	Ea.		3.58		3.58	Includes material and labor to install
	Install	Ea.	65	12.80		77.80	large window screen, 16 S.F.
	Demolish and Install	Ea.	65	16.38		81.38	
	Reinstall	Ea.		10.25		10.25	
	Clean	Ea.	.33	6.45		6.78	
	Minimum Charge	Job		106		106	

Window Glass		Unit	Material	Labor	Equip.	Total	Specification
Plate, Clear, 1/4"							
	Demolish	S.F.		1.72		1.72	Includes material and labor to install
	Install	S.F.	9.20	6.80		16	clear glass.
	Demolish and Install	S.F.	9.20	8.52		17.72	
	Reinstall	S.F.		5.42		5.42	
	Clean	S.F.	.02	.18		.20	
	Minimum Charge	Job		203		203	
Insulated							
	Demolish	S.F.		1.72		1.72	Includes material and labor to install
	Install	S.F.	15.05	11.60		26.65	insulated glass.
	Demolish and Install	S.F.	15.05	13.32		28.37	
	Reinstall	S.F.		9.30		9.30	
	Clean	S.F.	.02	.18		.20	
	Minimum Charge	Job		203		203	
Tinted							
	Demolish	S.F.		1.72		1.72	Includes material and labor to install
	Install	S.F.	8.40	6.25		14.65	tinted glass.
	Demolish and Install	S.F.	8.40	7.97		16.37	
	Reinstall	S.F.		5.01		5.01	
	Clean	S.F.	.02	.18		.20	
	Minimum Charge	Job		203		203	
Obscure / Florentine							
	Demolish	S.F.		1.72		1.72	Includes material and labor to install
	Install	S.F.	12.60	5.80		18.40	obscure glass.
	Demolish and Install	S.F.	12.60	7.52		20.12	
	Reinstall	S.F.		4.65		4.65	
	Clean	S.F.	.02	.18		.20	
	Minimum Charge	Job		203		203	
Laminated Safety							
	Demolish	S.F.		1.72		1.72	Includes material and labor to install
	Install	S.F.	13.15	9.05		22.20	clear laminated glass.
	Demolish and Install	S.F.	13.15	10.77		23.92	
	Reinstall	S.F.		7.23		7.23	
	Clean	S.F.	.02	.18		.20	
	Minimum Charge	Job		203		203	

Exterior Doors and Windows

Window Glass		Unit	Material	Labor	Equip.	Total	Specification
Wired							
	Demolish	S.F.		1.72		1.72	Includes material and labor to install
	Install	S.F.	25	6.05		31.05	rough obscure wire glass.
	Demolish and Install	S.F.	25	7.77		32.77	
	Reinstall	S.F.		4.82		4.82	
	Clean	S.F.	.02	.18		.20	
	Minimum Charge	Job		203		203	
Tempered							
	Demolish	S.F.		1.72		1.72	Includes material and labor to install
	Install	S.F.	9.20	6.80		16	clear glass.
	Demolish and Install	S.F.	9.20	8.52		17.72	
	Reinstall	S.F.		5.42		5.42	
	Clean	S.F.	.02	.18		.20	
	Minimum Charge	Job		203		203	
Stained							
	Demolish	S.F.		1.72		1.72	Includes material and labor to install
	Install	S.F.	67	6.80		73.80	stained glass.
	Demolish and Install	S.F.	67	8.52		75.52	
	Reinstall	S.F.		5.42		5.42	
	Clean	S.F.	.02	.18		.20	
	Minimum Charge	Job		203		203	
Etched							
	Demolish	S.F.		1.72		1.72	Includes material and labor to install
	Install	S.F.	55	6.80		61.80	etched glass.
	Demolish and Install	S.F.	55	8.52		63.52	
	Reinstall	S.F.		5.42		5.42	
	Clean	S.F.	.02	.18		.20	
	Minimum Charge	Job		203		203	
Wall Mirror							
	Demolish	S.F.		1.72		1.72	Includes material and labor to install
	Install	S.F.	10.45	6.50		16.95	distortion-free float glass mirror, 1/4"
	Demolish and Install	S.F.	10.45	8.22		18.67	thick, cut and polished edges.
	Reinstall	S.F.		5.21		5.21	
	Clean	S.F.	.02	.18		.20	
	Minimum Charge	Job		203		203	
Bullet Resistant							
	Demolish	S.F.		1.72		1.72	Includes material and labor to install
	Install	S.F.	125	81.50		206.50	bullet resistant clear laminated glass.
	Demolish and Install	S.F.	125	83.22		208.22	
	Reinstall	S.F.		65.09		65.09	
	Clean	S.F.	.02	.18		.20	
	Minimum Charge	Job		203		203	

Storefront System		Unit	Material	Labor	Equip.	Total	Specification
Clear Glass							
Stub Wall to 8'							
	Demolish	S.F.		2.69		2.69	Includes material and labor to install
	Install	S.F.	24	11.25		35.25	commercial grade 4-1/2" section,
	Demolish and Install	S.F.	24	13.94		37.94	anodized aluminum with clear glass in
	Reinstall	S.F.		8.99		8.99	upper section and safety glass in
	Clean	S.F.		.38		.38	lower section.
	Minimum Charge	Job		258		258	

Exterior Doors and Windows

Storefront System		Unit	Material	Labor	Equip.	Total	Specification
Floor to 10'							Includes material and labor to install
	Demolish	S.F.		2.69		2.69	commercial grade 4-1/2" section,
	Install	S.F.	25	11.25		36.25	anodized aluminum with clear glass in
	Demolish and Install	S.F.	25	13.94		38.94	upper section and safety glass in
	Reinstall	S.F.		8.99		8.99	lower section.
	Clean	S.F.		.38		.38	
	Minimum Charge	Job		258		258	
Polished Plate Glass							
Stub Wall to 9'							
	Demolish	S.F.		2.69		2.69	Cost includes material and labor to
	Install	S.F.	15.95	11.25		27.20	install 4' x 9' polished plate glass
	Demolish and Install	S.F.	15.95	13.94		29.89	window built on 8" stub wall.
	Reinstall	S.F.		8.99		8.99	
	Clean	S.F.		.38		.38	
	Minimum Charge	Job		258		258	
Stub Wall to 13'							
	Demolish	S.F.		2.69		2.69	Cost includes material and labor to
	Install	S.F.	17.25	11.25		28.50	install 4' x 13' polished plate glass
	Demolish and Install	S.F.	17.25	13.94		31.19	window built on 8" stub wall.
	Reinstall	S.F.		8.99		8.99	
	Clean	S.F.		.38		.38	
	Minimum Charge	Job		258		258	
Floor to 9'							
	Demolish	S.F.		2.69		2.69	Cost includes material and labor to
	Install	S.F.	17.80	11.25		29.05	install 4' x 9' polished plate glass
	Demolish and Install	S.F.	17.80	13.94		31.74	window built on floor.
	Reinstall	S.F.		8.99		8.99	
	Clean	S.F.		.38		.38	
	Minimum Charge	Job		258		258	
Floor to 13'							
	Demolish	S.F.		2.69		2.69	Cost includes material and labor to
	Install	S.F.	19.15	11.25		30.40	install 4' x 13' polished plate glass
	Demolish and Install	S.F.	19.15	13.94		33.09	window built on floor.
	Reinstall	S.F.		8.99		8.99	
	Clean	S.F.		.38		.38	
	Minimum Charge	Job		258		258	
Clear Plate Glass							
Floor to 9'							
	Demolish	S.F.		2.69		2.69	Cost includes material and labor to
	Install	S.F.	25.50	11.25		36.75	install 4' x 9' clear plate glass window
	Demolish and Install	S.F.	25.50	13.94		39.44	built on floor.
	Reinstall	S.F.		8.99		8.99	
	Clean	S.F.		.38		.38	
	Minimum Charge	Job		258		258	
Floor to 11'							
	Demolish	S.F.		2.69		2.69	Cost includes material and labor to
	Install	S.F.	25	11.25		36.25	install 4' x 11' clear plate glass
	Demolish and Install	S.F.	25	13.94		38.94	window built on floor.
	Reinstall	S.F.		8.99		8.99	
	Clean	S.F.		.38		.38	
	Minimum Charge	Job		258		258	
Floor to 13'							
	Demolish	S.F.		2.69		2.69	Cost includes material and labor to
	Install	S.F.	25	11.25		36.25	install 4' x 13' clear plate glass
	Demolish and Install	S.F.	25	13.94		38.94	window built on floor.
	Reinstall	S.F.		8.99		8.99	
	Clean	S.F.		.38		.38	
	Minimum Charge	Job		258		258	

Exterior Doors and Windows

Storefront System		Unit	Material	Labor	Equip.	Total	Specification
Entrance							
Narrow Stile 3′ x 7″							
	Demolish	Ea.		23.50		23.50	Cost includes material and labor to
	Install	Ea.	980	282		1262	install 3′ x 7′ narrow stile door, center
	Demolish and Install	Ea.	980	305.50		1285.50	pivot concealed closer.
	Reinstall	Ea.		225.49		225.49	
	Clean	Ea.	4.08	38.50		42.58	
	Minimum Charge	Job		258		258	
Tempered Glass							
	Demolish	Ea.		23.50		23.50	Cost includes material and labor to
	Install	Ea.	3375	560		3935	install 3′ x 7′ tempered glass door,
	Demolish and Install	Ea.	3375	583.50		3958.50	center pivot concealed closer,.
	Reinstall	Ea.		446.62		446.62	
	Clean	Ea.	4.08	38.50		42.58	
	Minimum Charge	Job		258		258	
Heavy Section							
	Demolish	Ea.		23.50		23.50	Includes material and labor to install
	Install	Ea.	1325	515		1840	wide stile door with 3′ 6″ x 7′
	Demolish and Install	Ea.	1325	538.50		1863.50	opening.
	Reinstall	Ea.		413.12		413.12	
	Clean	Ea.	.81	12.90		13.71	
	Minimum Charge	Job		815		815	

Curtain Wall System		Unit	Material	Labor	Equip.	Total	Specification
Regular Weight							
Bronze Anodized							
	Demolish	S.F.		2.69		2.69	Includes material and labor to install
	Install	S.F.	39.50	11.25		50.75	structural curtain wall system with
	Demolish and Install	S.F.	39.50	13.94		53.44	bronze anodized frame of heavy
	Clean	S.F.		.38		.38	weight, float glass.
	Minimum Charge	Job		211		211	
Black Anodized							
	Demolish	S.F.		2.69		2.69	Includes material and labor to install
	Install	S.F.	43	11.25		54.25	structural curtain wall system with
	Demolish and Install	S.F.	43	13.94		56.94	black anodized frame of regular
	Clean	S.F.		.38		.38	weight, float glass.
	Minimum Charge	Job		211		211	
Heavy Weight							
Bronze Anodized							
	Demolish	S.F.		2.69		2.69	Includes material and labor to install
	Install	S.F.	39.50	11.25		50.75	structural curtain wall system with
	Demolish and Install	S.F.	39.50	13.94		53.44	bronze anodized frame of heavy
	Clean	S.F.		.38		.38	weight, float glass.
	Minimum Charge	Job		211		211	
Black Anodized							
	Demolish	S.F.		2.69		2.69	Includes material and labor to install
	Install	S.F.	40.50	11.25		51.75	structural curtain wall system with
	Demolish and Install	S.F.	40.50	13.94		54.44	black anodized frame of heavy
	Clean	S.F.		.38		.38	weight, float glass.
	Minimum Charge	Job		211		211	
Glass							
1″ Insulated							
	Install	S.F.	17.80	10.85		28.65	Cost includes labor and material to
	Minimum Charge	Job		203		203	install 1″ insulated glass.

Exterior Doors and Windows

Skylight		Unit	Material	Labor	Equip.	Total	Specification
Dome							
22″ x 22″							
	Demolish	Ea.		80.50		80.50	Cost includes material and labor to
	Install	Ea.	231	130		361	install 22″ x 22″ fixed dome skylight.
	Demolish and Install	Ea.	231	210.50		441.50	
	Reinstall	Ea.		103.79		103.79	
	Clean	Ea.	.36	11.05		11.41	
	Minimum Charge	Job		780		780	
22″ x 46″							
	Demolish	Ea.		80.50		80.50	Cost includes material and labor to
	Install	Ea.	255	156		411	install 22″ x 46″ fixed dome skylight.
	Demolish and Install	Ea.	255	236.50		491.50	
	Reinstall	Ea.		124.54		124.54	
	Clean	Ea.		17.20		17.20	
	Minimum Charge	Job		780		780	
30″ x 30″							
	Demolish	Ea.		80.50		80.50	Cost includes material and labor to
	Install	Ea.	259	130		389	install 30″ x 30″ fixed dome skylight.
	Demolish and Install	Ea.	259	210.50		469.50	
	Reinstall	Ea.		103.79		103.79	
	Clean	Ea.		17.20		17.20	
	Minimum Charge	Job		780		780	
30″ x 46″							
	Demolish	Ea.		80.50		80.50	Cost includes material and labor to
	Install	Ea.	335	156		491	install 30″ x 46″ fixed dome skylight.
	Demolish and Install	Ea.	335	236.50		571.50	
	Reinstall	Ea.		124.54		124.54	
	Clean	Ea.		17.20		17.20	
	Minimum Charge	Job		780		780	
Fixed							
22″ x 27″							
	Demolish	Ea.		80.50		80.50	Includes material and labor to install
	Install	Ea.	242	130		372	double glazed skylight, 22″ x 27″,
	Demolish and Install	Ea.	242	210.50		452.50	fixed.
	Reinstall	Ea.		103.79		103.79	
	Clean	Ea.	.36	11.05		11.41	
	Minimum Charge	Job		780		780	
22″ x 46″							
	Demolish	Ea.		80.50		80.50	Includes material and labor to install
	Install	Ea.	299	156		455	double glazed skylight, 22″ x 46″,
	Demolish and Install	Ea.	299	236.50		535.50	fixed.
	Reinstall	Ea.		124.54		124.54	
	Clean	Ea.		17.20		17.20	
	Minimum Charge	Job		780		780	
44″ x 46″							
	Demolish	Ea.		80.50		80.50	Includes material and labor to install
	Install	Ea.	450	156		606	double glazed skylight, 44″ x 46″,
	Demolish and Install	Ea.	450	236.50		686.50	fixed.
	Reinstall	Ea.		124.54		124.54	
	Clean	Ea.		19.35		19.35	
	Minimum Charge	Job		780		780	
Operable							
22″ x 27″							
	Demolish	Ea.		80.50		80.50	Includes material and labor to install
	Install	Ea.	345	130		475	double glazed skylight, 22″ x 27″,
	Demolish and Install	Ea.	345	210.50		555.50	operable.
	Reinstall	Ea.		103.79		103.79	
	Clean	Ea.	.36	11.05		11.41	
	Minimum Charge	Job		780		780	

Exterior Doors and Windows

Skylight

Skylight		Unit	Material	Labor	Equip.	Total	Specification
22" x 46"							
	Demolish	Ea.		80.50		80.50	Includes material and labor to install
	Install	Ea.	405	156		561	double glazed skylight, 22" x 46",
	Demolish and Install	Ea.	405	236.50		641.50	operable.
	Reinstall	Ea.		124.54		124.54	
	Clean	Ea.		17.20		17.20	
	Minimum Charge	Job		780		780	
44" x 46"							
	Demolish	Ea.		80.50		80.50	Includes material and labor to install
	Install	Ea.	895	156		1051	double glazed skylight, 44" x 46",
	Demolish and Install	Ea.	895	236.50		1131.50	operable.
	Reinstall	Ea.		124.54		124.54	
	Clean	Ea.		19.35		19.35	
	Minimum Charge	Job		780		780	

Industrial Steel Window

Industrial Steel Window		Unit	Material	Labor	Equip.	Total	Specification
100% Fixed							
	Demolish	S.F.		2.21		2.21	Includes material and labor to install
	Install	S.F.	31.50	5.15		36.65	commercial grade steel windows,
	Demolish and Install	S.F.	31.50	7.36		38.86	including glazing.
	Reinstall	S.F.		4.13		4.13	
	Clean	Ea.	.14	1.18		1.32	
	Paint	S.F.	.18	.80		.98	
	Minimum Charge	Job		258		258	
50% Vented							
	Demolish	S.F.		2.21		2.21	Includes material and labor to install
	Install	S.F.	58.50	5.15		63.65	commercial grade steel windows,
	Demolish and Install	S.F.	58.50	7.36		65.86	including glazing.
	Reinstall	S.F.		4.13		4.13	
	Clean	Ea.	.14	1.18		1.32	
	Paint	S.F.	.18	.80		.98	
	Minimum Charge	Job		258		258	

Brick Veneer **Concrete Masonry Units**

Masonry Wall	Unit	Material	Labor	Equip.	Total	Specification
Unreinforced CMU						
4" x 8" x 16"						
Demolish	S.F.		1.91		1.91	Includes material and labor to install
Install	S.F.	1.75	4.15		5.90	normal weight 4" x 8" x 16" block.
Demolish and Install	S.F.	1.75	6.06		7.81	
Clean	S.F.	.11	.48		.59	
Paint	S.F.	.14	.82		.96	
Minimum Charge	Job		207		207	
6" x 8" x 16"						
Demolish	S.F.		2.02		2.02	Includes material and labor to install
Install	S.F.	2.43	4.34		6.77	normal weight 6" x 8" x 16" block.
Demolish and Install	S.F.	2.43	6.36		8.79	
Clean	S.F.	.11	.48		.59	
Paint	S.F.	.14	.82		.96	
Minimum Charge	Job		207		207	
8" x 8" x 16"						
Demolish	S.F.		2.29		2.29	Includes material and labor to install
Install	S.F.	2.60	4.78		7.38	normal weight 8" x 8" x 16" block.
Demolish and Install	S.F.	2.60	7.07		9.67	
Clean	S.F.	.11	.48		.59	
Paint	S.F.	.14	.82		.96	
Minimum Charge	Job		207		207	
12" x 8" x 16"						
Demolish	S.F.		2.29		2.29	Includes material and labor to install
Install	S.F.	3.95	7.25		11.20	normal weight 12" x 8" x 16" block.
Demolish and Install	S.F.	3.95	9.54		13.49	
Clean	S.F.	.11	.48		.59	
Paint	S.F.	.14	.82		.96	
Minimum Charge	Job		207		207	
Reinforced CMU						
4" x 8" x 16"						
Demolish	S.F.		1.91	1.19	3.10	Includes material and labor to install
Install	S.F.	1.91	4.24		6.15	normal weight 4" x 8" x 16" block,
Demolish and Install	S.F.	1.91	6.15	1.19	9.25	mortar and 1/2" re-bar installed
Clean	S.F.	.11	.48		.59	vertically every 24" O.C.
Paint	S.F.	.14	.82		.96	
Minimum Charge	Job		375		375	
6" x 8" x 16"						
Demolish	S.F.		2.01	.60	2.61	Includes material and labor to install
Install	S.F.	2.60	4.44		7.04	normal weight 6" x 8" x 16" block,
Demolish and Install	S.F.	2.60	6.45	.60	9.65	mortar and 1/2" re-bar installed
Clean	S.F.	.11	.48		.59	vertically every 24" O.C.
Paint	S.F.	.14	.82		.96	
Minimum Charge	Job		375		375	

Masonry

Masonry Wall		Unit	Material	Labor	Equip.	Total	Specification
8" x 8" x 16"							
	Demolish	S.F.		2.12	.63	2.75	Includes material and labor to install
	Install	S.F.	2.77	4.84		7.61	normal weight 8" x 8" x 16" block,
	Demolish and Install	S.F.	2.77	6.96	.63	10.36	mortar and 1/2" re-bar installed
	Clean	S.F.	.11	.48		.59	vertically every 24" O.C.
	Paint	S.F.	.14	.82		.96	
	Minimum Charge	Job		375		375	
12" x 8" x 16"							
	Demolish	S.F.		2.19	.65	2.84	Includes material and labor to install
	Install	S.F.	4.13	7.50		11.63	normal weight 12" x 8" x 16" block,
	Demolish and Install	S.F.	4.13	9.69	.65	14.47	mortar and 1/2" re-bar installed
	Clean	S.F.	.11	.48		.59	vertically every 24" O.C.
	Paint	S.F.	.14	.82		.96	
	Minimum Charge	Job		375		375	
Structural Brick							
4" Single Row							
	Demolish	S.F.		2.75		2.75	Cost includes material and labor to
	Install	S.F.	4.37	8.90		13.27	install 4" red brick wall, running bond
	Demolish and Install	S.F.	4.37	11.65		16.02	with 3/8" concave joints, mortar and
	Clean	S.F.	.11	.41		.52	ties.
	Paint	S.F.	.14	.82		.96	
	Minimum Charge	Job		375		375	
8" Double Row							
	Demolish	S.F.		2.75		2.75	Cost includes material and labor to
	Install	S.F.	6.60	14.15		20.75	install 8" red brick wall, running bond,
	Demolish and Install	S.F.	6.60	16.90		23.50	with 3/8" concave joints, mortar and
	Clean	S.F.	.11	.41		.52	wall ties.
	Paint	S.F.	.14	.82		.96	
	Minimum Charge	Job		375		375	
12" Triple Row							
	Demolish	S.F.		3.82		3.82	Cost includes material and labor to
	Install	S.F.	9.95	20		29.95	install 12" red brick wall, running
	Demolish and Install	S.F.	9.95	23.82		33.77	bond, with 3/8" concave joints,
	Clean	S.F.	.11	.41		.52	mortar and wall ties.
	Paint	S.F.	.14	.82		.96	
	Minimum Charge	Job		375		375	
Brick Veneer							
4" Red Common							
	Demolish	S.F.		2.75		2.75	Cost includes material and labor to
	Install	S.F.	3.09	8.30		11.39	install 4" common red face brick,
	Demolish and Install	S.F.	3.09	11.05		14.14	running bond with 3/8" concave
	Clean	S.F.	.11	.41		.52	joints, mortar and wall ties.
	Paint	S.F.	.14	.82		.96	
	Minimum Charge	Job		207		207	
4" Used							
	Demolish	S.F.		2.75		2.75	Cost includes material and labor to
	Install	S.F.	4.37	8.90		13.27	install 4" red brick wall, running bond
	Demolish and Install	S.F.	4.37	11.65		16.02	with 3/8" concave joints, mortar and
	Clean	S.F.	.11	.41		.52	ties.
	Paint	S.F.	.14	.82		.96	
	Minimum Charge	Job		207		207	

Masonry

Masonry Wall

	Unit	Material	Labor	Equip.	Total	Specification
6" Jumbo						
Demolish	S.F.		2.75		2.75	Cost includes material and labor to
Install	S.F.	4.98	4.39		9.37	install 6" red jumbo brick veneer,
Demolish and Install	S.F.	4.98	7.14		12.12	running bond with 3/8" concave
Clean	S.F.	.11	.41		.52	joints, mortar and ties.
Paint	S.F.	.14	.82		.96	
Minimum Charge	Job		207		207	
Norman (11-1/2")						
Demolish	S.F.		2.75		2.75	Cost includes material and labor to
Install	S.F.	6.25	5.95		12.20	install 4" norman red brick wall,
Demolish and Install	S.F.	6.25	8.70		14.95	running bond with 3/8" concave
Clean	S.F.	.11	.41		.52	joints, mortar and ties.
Paint	S.F.	.14	.82		.96	
Minimum Charge	Job		207		207	
Glazed						
Demolish	S.F.		2.75		2.75	Includes material, labor and
Install	S.F.	18.35	9.10		27.45	equipment to install glazed brick.
Demolish and Install	S.F.	18.35	11.85		30.20	
Clean	S.F.	.11	.41		.52	
Paint	S.F.	.14	.82		.96	
Minimum Charge	Job		207		207	
Slumpstone						
6" x 4"						
Demolish	S.F.		1.91		1.91	Cost includes material and labor to
Install	S.F.	5.35	4.67		10.02	install 6" x 4" x 16" concrete slump
Demolish and Install	S.F.	5.35	6.58		11.93	block.
Clean	S.F.	.11	.48		.59	
Paint	S.F.	.14	.82		.96	
Minimum Charge	Job		207		207	
6" x 6"						
Demolish	S.F.		2.02		2.02	Cost includes material and labor to
Install	S.F.	4.97	4.82		9.79	install 6" x 6" x 16" concrete slump
Demolish and Install	S.F.	4.97	6.84		11.81	block.
Clean	S.F.	.11	.48		.59	
Paint	S.F.	.14	.82		.96	
Minimum Charge	Job		207		207	
8" x 6"						
Demolish	S.F.		2.02		2.02	Cost includes material and labor to
Install	S.F.	7.30	4.98		12.28	install 8" x 6" x 16" concrete slump
Demolish and Install	S.F.	7.30	7		14.30	block.
Clean	S.F.	.11	.48		.59	
Paint	S.F.	.14	.82		.96	
Minimum Charge	Job		207		207	
12" x 4"						
Demolish	S.F.		1.91		1.91	Cost includes material and labor to
Install	S.F.	11.20	5.75		16.95	install 12" x 4" x 16" concrete slump
Demolish and Install	S.F.	11.20	7.66		18.86	block.
Clean	S.F.	.11	.48		.59	
Paint	S.F.	.14	.82		.96	
Minimum Charge	Job		207		207	
12" x 6"						
Demolish	S.F.		2.02		2.02	Cost includes material and labor to
Install	S.F.	11.75	6.25		18	install 12" x 6" x 16" concrete slump
Demolish and Install	S.F.	11.75	8.27		20.02	block.
Clean	S.F.	.11	.48		.59	
Paint	S.F.	.14	.82		.96	
Minimum Charge	Job		207		207	

Masonry

Masonry Wall		Unit	Material	Labor	Equip.	Total	Specification
Stucco							
Three Coat							
	Demolish	S.Y.		2.37		2.37	Includes material and labor to install
	Install	S.Y.	3.89	9	.78	13.67	stucco on exterior masonry walls.
	Demolish and Install	S.Y.	3.89	11.37	.78	16.04	
	Clean	S.Y.		4.30		4.30	
	Paint	S.F.	.28	.59	.	.87	
	Minimum Charge	Job		190		190	
Three Coat w / Wire Mesh							
	Demolish	S.Y.		4.30		4.30	Includes material and labor to install
	Install	S.Y.	6.60	34.50	2.48	43.58	stucco on wire mesh over wood
	Demolish and Install	S.Y.	6.60	38.80	2.48	47.88	framing.
	Clean	S.Y.		4.30		4.30	
	Paint	S.F.	.28	.59		.87	
	Minimum Charge	Job		190		190	
Fill Cracks							
	Install	S.F.	.43	3.04		3.47	Includes labor and material to fill-in
	Minimum Charge	Job		190		190	hairline cracks up to 1/8" with filler,
							including sand and fill per S.F.
Furring							
1" x 2"							
	Demolish	L.F.		.14		.14	Cost includes material and labor to
	Install	L.F.	.28	.85		1.13	install 1" x 2" furring on masonry wall.
	Demolish and Install	L.F.	.28	.99		1.27	
	Minimum Charge	Job		211		211	
2" x 2"							
	Demolish	L.F.		.15		.15	Cost includes material and labor to
	Install	L.F.	.24	.94		1.18	install 2" x 2" furring on masonry wall.
	Demolish and Install	L.F.	.24	1.09		1.33	
	Minimum Charge	Job		211		211	
2" x 4"							
	Demolish	L.F.		.16		.16	Cost includes material and labor to
	Install	L.F.	.40	1.07		1.47	install 2" x 4" furring on masonry wall.
	Demolish and Install	L.F.	.40	1.23		1.63	
	Reinstall	L.F.		.86		.86	
	Minimum Charge	Job		211		211	
Maintenance / Cleaning							
Repoint							
	Install	S.F.	.58	5.20		5.78	Includes labor and material to cut and
	Minimum Charge	Hr.		52		52	repoint brick, hard mortar, running
							bond.
Acid Etch							
	Install	S.F.	.07	.74		.81	Includes labor and material to acid
	Minimum Charge	Job		155	46	201	wash smooth brick.
Pressure Wash							
	Clean	S.F.		.83	.20	1.03	Includes labor and material to high
	Minimum Charge	Job		155	46	201	pressure water wash masonry,
							average.
Waterproof							
	Install	S.F.	.34	.77		1.11	Includes labor and material to install
	Minimum Charge	Job		192		192	masory waterproofing.

Masonry

Chimney

Chimney		Unit	Material	Labor	Equip.	Total	Specification
Demo & Haul Exterior							
	Demolish	C.F.		5.65		5.65	Includes minimum labor and
	Minimum Charge	Job		172		172	equipment to remove masonry and haul debris to truck or dumpster.
Demo & Haul Interior							
	Demolish	C.F.		7.75		7.75	Includes minimum labor and
	Minimum Charge	Job		172		172	equipment to remove masonry and haul debris to truck or dumpster.
Install New (w / flue)							
Brick							
	Install	V.L.F.	42.50	46.50		89	Includes material and labor to install
	Minimum Charge	Job		207		207	brick chimney with flue from ground level up. Foundation not included.
Natural Stone							
	Install	L.F.	97.50	75		172.50	Includes labor and material to install
	Minimum Charge	Job		207		207	natural stone chimney.
Install From Roof Up							
Brick							
	Install	V.L.F.	42.50	46.50		89	Includes material and labor to install
	Minimum Charge	Job		207		207	brick chimney with flue from ground level up. Foundation not included.
Natural Stone							
	Install	L.F.	97.50	62.50		160	Includes labor and material to install
	Minimum Charge	Job		207		207	natural stone chimney.
Install Base Pad							
	Install	Ea.	270	230	1.49	501.49	Includes labor and material to replace
	Minimum Charge	Job		980	36	1016	spread footing, 5' square x 16" deep.
Metal Cap							
	Install	Ea.	58.50	26		84.50	Includes labor and materials to install
	Minimum Charge	Job		211		211	bird and squirrel screens in a chimney flue.

Fireplace

Fireplace		Unit	Material	Labor	Equip.	Total	Specification
Common Brick							
30"							
	Demolish	Ea.		1225		1225	Includes material and labor to install a
	Install	Ea.	2025	3750		5775	complete unit which includes common
	Demolish and Install	Ea.	2025	4975		7000	brick, foundation, damper, flue lining
	Minimum Charge	Job		207		207	stack to 15' high and a 30" fire box.
36"							
	Demolish	Ea.		1225		1225	Includes material and labor to install a
	Install	Ea.	2150	4150		6300	complete unit which includes common
	Demolish and Install	Ea.	2150	5375		7525	brick, foundation, damper, flue lining
	Minimum Charge	Job		207		207	stack to 15' high and a 36" fire box.
42"							
	Demolish	Ea.		1225		1225	Includes material and labor to install a
	Install	Ea.	2300	4675		6975	complete unit which includes common
	Demolish and Install	Ea.	2300	5900		8200	brick, foundation, damper, flue lining
	Minimum Charge	Job		207		207	stack to 15' high and a 42" fire box.

125

Masonry

Fireplace		Unit	Material	Labor	Equip.	Total	Specification
48″							Includes material and labor to install a complete unit which includes common brick, foundation, damper, flue lining stack to 15′ high and a 48″ fire box.
	Demolish	Ea.		1225		1225	
	Install	Ea.	2450	5325		7775	
	Demolish and Install	Ea.	2450	6550		9000	
	Minimum Charge	Job		207		207	
Raised Hearth							
Brick Hearth							Includes material and labor to install a brick fireplace hearth based on 30″ x 29″ opening.
	Demolish	S.F.		5.40		5.40	
	Install	S.F.	18.45	31		49.45	
	Demolish and Install	S.F.	18.45	36.40		54.85	
	Clean	S.F.	.11	.41		.52	
	Minimum Charge	Job		207		207	
Marble Hearth / Facing							Includes material and labor to install marble fireplace hearth.
	Demolish	Ea.		34.50		34.50	
	Install	Ea.	660	126		786	
	Demolish and Install	Ea.	660	160.50		820.50	
	Reinstall	Ea.		100.78		100.78	
	Clean	Ea.	39.50	6.20		45.70	
	Minimum Charge	Job		207		207	
Breast (face) Brick							Includes material and labor to install standard red brick, running bond.
	Demolish	S.F.		5.40		5.40	
	Install	S.F.	4.44	8.70		13.14	
	Demolish and Install	S.F.	4.44	14.10		18.54	
	Clean	S.F.		.83	.20	1.03	
	Minimum Charge	Hr.		52		52	
Fire Box							Includes material and labor to install fireplace box only (110 brick).
	Demolish	Ea.		275		275	
	Install	Ea.	168	375		543	
	Demolish and Install	Ea.	168	650		818	
	Clean	Ea.	.12	14.20		14.32	
	Minimum Charge	Hr.		52		52	
Natural Stone							Includes material and labor to install high priced stone, split or rock face.
	Demolish	S.F.		5.40		5.40	
	Install	S.F.	21	15.90		36.90	
	Demolish and Install	S.F.	21	21.30		42.30	
	Clean	S.F.	.09	.93		1.02	
	Minimum Charge	Job		375		375	
Slumpstone Veneer							Cost includes material and labor to install 8″ x 6″ x 16″ concrete slump block.
	Demolish	S.F.		2.02		2.02	
	Install	S.F.	7.30	4.98		12.28	
	Demolish and Install	S.F.	7.30	7		14.30	
	Clean	S.F.	.11	.48		.59	
	Paint	S.F.	.14	.82		.96	
	Minimum Charge	Job		207		207	
Prefabricated							
Firebox							Includes material and labor to install up to 43″ radiant heat, zero clearance, prefabricated fireplace. Flues, doors or blowers not included.
	Demolish	Ea.		49		49	
	Install	Ea.	2275	385		2660	
	Demolish and Install	Ea.	2275	434		2709	
	Reinstall	Ea.		307.49		307.49	
	Clean	Ea.	1.63	19.35		20.98	
	Minimum Charge	Job		211		211	

Masonry

Fireplace		Unit	Material	Labor	Equip.	Total	Specification
Flue							
	Demolish	V.L.F.		5.75		5.75	Includes material and labor to install
	Install	V.L.F.	18.10	16.20		34.30	prefabricated chimney vent, 8"
	Demolish and Install	V.L.F.	18.10	21.95		40.05	exposed metal flue.
	Reinstall	V.L.F.		12.98		12.98	
	Clean	L.F.	.65	3.87		4.52	
	Minimum Charge	Job		420		420	
Blower							
	Demolish	Ea.		4.59		4.59	Includes material and labor to install
	Install	Ea.	160	29.50		189.50	blower for metal fireplace flue.
	Demolish and Install	Ea.	160	34.09		194.09	
	Reinstall	Ea.		23.44		23.44	
	Clean	Ea.	3.27	38.50		41.77	
	Minimum Charge	Job		420		420	
Accessories							
Wood Mantle							
	Demolish	L.F.		5.75		5.75	Includes material and labor to install
	Install	L.F.	12	12.10		24.10	wood fireplace mantel.
	Demolish and Install	L.F.	12	17.85		29.85	
	Reinstall	L.F.		12.08		12.08	
	Clean	Ea.	.02	.87		.89	
	Paint	L.F.	.11	9.80		9.91	
	Minimum Charge	Job		211		211	
Colonial Style							
	Demolish	Ea.		52		52	Includes material and labor to install
	Install	Opng.	1375	211		1586	wood fireplace mantel.
	Demolish and Install	Opng.	1375	263		1638	
	Reinstall	Opng.		211.40		211.40	
	Clean	Ea.		.63		.63	
	Paint	Ea.	19.35	135		154.35	
	Minimum Charge	Job		211		211	
Modern Design							
	Demolish	Ea.		52		52	Includes material and labor to install
	Install	Opng.	450	84.50		534.50	wood fireplace mantel.
	Demolish and Install	Opng.	450	136.50		586.50	
	Reinstall	Opng.		84.56		84.56	
	Clean	Ea.		.63		.63	
	Paint	Ea.	19.35	135		154.35	
	Minimum Charge	Job		211		211	
Glass Door Unit							
	Demolish	Ea.		7.15		7.15	Includes material and labor to install a
	Install	Ea.	335	8.65		343.65	fireplace glass door unit.
	Demolish and Install	Ea.	335	15.80		350.80	
	Reinstall	Ea.		8.64		8.64	
	Clean	Ea.	.40	11.90		12.30	
	Minimum Charge	Job		211		211	
Fire Screen							
	Demolish	Ea.		3.58		3.58	Includes material and labor to install a
	Install	Ea.	135	4.32		139.32	fireplace fire screen.
	Demolish and Install	Ea.	135	7.90		142.90	
	Reinstall	Ea.		4.32		4.32	
	Clean	Ea.	.20	10.30		10.50	
	Minimum Charge	Job		211		211	

Siding

Shingles

Board & Batten

Texture 1-11

Exterior Siding		Unit	Material	Labor	Equip.	Total	Specification
Trim							
1″ x 2″							
	Demolish	L.F.		.29		.29	Cost includes material and labor to
	Install	L.F.	.29	1.28		1.57	install 1″ x 2″ pine trim.
	Demolish and Install	L.F.	.29	1.57		1.86	
	Clean	L.F.	.01	.21		.22	
	Paint	L.F.	.02	.56		.58	
	Minimum Charge	Job		211		211	
1″ x 3″							
	Demolish	L.F.		.29		.29	Cost includes material and labor to
	Install	L.F.	.26	1.46		1.72	install 1″ x 3″ pine trim.
	Demolish and Install	L.F.	.26	1.75		2.01	
	Clean	L.F.	.01	.21		.22	
	Paint	L.F.	.02	.56		.58	
	Minimum Charge	Job		211		211	
1″ x 4″							
	Demolish	L.F.		.29		.29	Cost includes material and labor to
	Install	L.F.	.48	1.69		2.17	install 1″ x 4″ pine trim.
	Demolish and Install	L.F.	.48	1.98		2.46	
	Clean	L.F.	.01	.21		.22	
	Paint	L.F.	.02	.56		.58	
	Minimum Charge	Job		211		211	
1″ x 6″							
	Demolish	L.F.		.29		.29	Cost includes material and labor to
	Install	L.F.	.94	1.69		2.63	install 1″ x 6″ pine trim.
	Demolish and Install	L.F.	.94	1.98		2.92	
	Clean	L.F.	.01	.21		.22	
	Paint	L.F.	.08	.56		.64	
	Minimum Charge	Job		211		211	
1″ x 8″							
	Demolish	L.F.		.31		.31	Cost includes material and labor to
	Install	L.F.	1.28	2.11		3.39	install 1″ x 8″ pine trim.
	Demolish and Install	L.F.	1.28	2.42		3.70	
	Clean	L.F.	.02	.22		.24	
	Paint	L.F.	.08	.56		.64	
	Minimum Charge	Job		211		211	
Aluminum							
8″ Non-Insulated							
	Demolish	S.F.		.77		.77	Includes material and labor to install
	Install	S.F.	2.65	1.68		4.33	horizontal, colored clapboard, 8″ to
	Demolish and Install	S.F.	2.65	2.45		5.10	10″ wide, plain.
	Reinstall	S.F.		1.68		1.68	
	Clean	S.F.		.31		.31	
	Minimum Charge	Job		211		211	

Siding

Exterior Siding	Unit	Material	Labor	Equip.	Total	Specification
8" Insulated						
Demolish	S.F.		.77		.77	Includes material and labor to install
Install	S.F.	2.83	1.68		4.51	horizontal, colored insulated
Demolish and Install	S.F.	2.83	2.45		5.28	clapboard, 8" wide, plain.
Reinstall	S.F.		1.68		1.68	
Clean	S.F.		.31		.31	
Minimum Charge	Job		211		211	
12" Non-Insulated						
Demolish	S.F.		.77		.77	Includes material and labor to install
Install	S.F.	2.65	1.44		4.09	horizontal, colored clapboard, plain.
Demolish and Install	S.F.	2.65	2.21		4.86	
Reinstall	S.F.		1.44		1.44	
Clean	S.F.		.31		.31	
Minimum Charge	Job		211		211	
12" Insulated						
Demolish	S.F.		.77		.77	Includes material and labor to install
Install	S.F.	2.83	1.44		4.27	horizontal, colored insulated
Demolish and Install	S.F.	2.83	2.21		5.04	clapboard, plain.
Reinstall	S.F.		1.44		1.44	
Clean	S.F.		.31		.31	
Minimum Charge	Job		211		211	
Clapboard 8" to 10"						
Demolish	S.F.		.77		.77	Includes material and labor to install
Install	S.F.	2.65	1.68		4.33	horizontal, colored clapboard, 8" to
Demolish and Install	S.F.	2.65	2.45		5.10	10" wide, plain.
Reinstall	S.F.		1.68		1.68	
Clean	S.F.		.31		.31	
Minimum Charge	Job		211		211	
Corner Strips						
Demolish	L.F.		.40		.40	Includes material and labor to install
Install	V.L.F.	3.89	1.68		5.57	corners for horizontal lapped
Demolish and Install	V.L.F.	3.89	2.08		5.97	aluminum siding.
Reinstall	V.L.F.		1.68		1.68	
Clean	L.F.		.21		.21	
Minimum Charge	Job		211		211	
Vinyl						
Non-Insulated						
Demolish	S.F.		.61		.61	Cost includes material and labor to
Install	S.F.	1.43	1.54		2.97	install 10" non-insulated solid vinyl
Demolish and Install	S.F.	1.43	2.15		3.58	siding.
Reinstall	S.F.		1.23		1.23	
Clean	S.F.		.31		.31	
Minimum Charge	Job		211		211	
Insulated						
Demolish	S.F.		.61		.61	Cost includes material and labor to
Install	S.F.	3.15	2.11		5.26	install insulated solid vinyl siding
Demolish and Install	S.F.	3.15	2.72		5.87	panels.
Reinstall	S.F.		1.69		1.69	
Clean	S.F.		.31		.31	
Minimum Charge	Job		211		211	
Corner Strips						
Demolish	L.F.		.38		.38	Cost includes material and labor to
Install	L.F.	2	1.21		3.21	install corner posts for vinyl siding
Demolish and Install	L.F.	2	1.59		3.59	panels.
Reinstall	L.F.		.97		.97	
Clean	L.F.		.21		.21	
Minimum Charge	Job		211		211	

Siding

Exterior Siding	Unit	Material	Labor	Equip.	Total	Specification
Fiber Cement Siding						
4' x 8' Panel						
Demolish	S.F.		.64		.64	Cost includes material and labor to
Install	S.F.	1.27	.93		2.20	install 4' x 8' fiber cement 7/16"
Demolish and Install	S.F.	1.27	1.57		2.84	thick.
Clean	S.F.		.31		.31	
Paint	S.F.	.32	1.21		1.53	
Minimum Charge	Job		211		211	
Lap						
Demolish	S.F.		.63		.63	Cost includes material and labor to
Install	S.F.	1.08	1.67		2.75	install 8" or 12" fiber cement lap
Demolish and Install	S.F.	1.08	2.30		3.38	siding 7/16" thick with wood grain
Clean	S.F.		.31		.31	finish.
Paint	S.F.	.32	1.21		1.53	
Minimum Charge	Job		211		211	
Plywood Panel						
Douglas Fir (T-1-11)						
Demolish	S.F.		.47		.47	Includes material and labor to install
Install	S.F.	1.38	1.28		2.66	Douglas fir 4' x 8' plywood exterior
Demolish and Install	S.F.	1.38	1.75		3.13	siding plain or patterns.
Reinstall	S.F.		1.28		1.28	
Clean	S.F.		.31		.31	
Paint	S.F.	.32	1.21		1.53	
Minimum Charge	Job		211		211	
Redwood						
Demolish	S.F.		.47		.47	Includes material and labor to install
Install	S.F.	2.11	1.28		3.39	redwood 4' x 8' plywood exterior
Demolish and Install	S.F.	2.11	1.75		3.86	siding plain or patterns.
Reinstall	S.F.		1.28		1.28	
Clean	S.F.		.31		.31	
Paint	S.F.	.32	1.21		1.53	
Minimum Charge	Job		211		211	
Cedar						
Demolish	S.F.		.47		.47	Cost includes material and labor to
Install	S.F.	2.67	1.28		3.95	install 4' x 8' cedar plywood exterior
Demolish and Install	S.F.	2.67	1.75		4.42	siding plain or patterns.
Reinstall	S.F.		1.28		1.28	
Clean	S.F.		.31		.31	
Paint	S.F.	.32	1.21		1.53	
Minimum Charge	Job		211		211	
Tongue and Groove						
Fir, Tongue and Groove						
Demolish	S.F.		.86		.86	Includes material and labor to install
Install	S.F.	5.15	1.28		6.43	fir tongue and groove exterior siding
Demolish and Install	S.F.	5.15	2.14		7.29	1" x 8".
Reinstall	S.F.		1.02		1.02	
Clean	S.F.		.31		.31	
Paint	S.F.	.32	1.21		1.53	
Minimum Charge	Job		211		211	
Pine, Tongue and Groove						
Demolish	S.F.		.86		.86	Includes labor and material to install
Install	S.F.	2.29	1.28		3.57	Pine tongue and groove exterior
Demolish and Install	S.F.	2.29	2.14		4.43	siding 1" x 8".
Reinstall	S.F.		1.28		1.28	
Clean	S.F.		.31		.31	
Paint	S.F.	.32	1.21		1.53	
Minimum Charge	Job		211		211	

Siding

Exterior Siding	Unit	Material	Labor	Equip.	Total	Specification
Cedar, Tongue and Groove						
Demolish	S.F.		.86		.86	Includes labor and material to install
Install	S.F.	6.40	1.92		8.32	cedar tongue and groove exterior
Demolish and Install	S.F.	6.40	2.78		9.18	siding 1" x 8".
Reinstall	S.F.		1.54		1.54	
Clean	S.F.		.31		.31	
Paint	S.F.	.32	1.21		1.53	
Minimum Charge	Job		211		211	
Redwood, Tongue and Groove						
Demolish	S.F.		.86		.86	Includes labor and material to install
Install	S.F.	2.79	1.92		4.71	redwood tongue and groove exterior
Demolish and Install	S.F.	2.79	2.78		5.57	siding 1" x 4".
Reinstall	S.F.		1.92		1.92	
Clean	S.F.		.31		.31	
Paint	S.F.	.32	1.21		1.53	
Minimum Charge	Job		211		211	
Board and Batten						
Fir						
Demolish	S.F.		.86		.86	Cost includes material and labor to
Install	S.F.	6.15	1.28		7.43	install 1" x 8" fir boards with 1" x 3"
Demolish and Install	S.F.	6.15	2.14		8.29	battens.
Reinstall	S.F.		1.02		1.02	
Clean	S.F.		.31		.31	
Paint	S.F.	.32	1.21		1.53	
Minimum Charge	Job		211		211	
Pine						
Demolish	S.F.		.86		.86	Cost includes material and labor to
Install	S.F.	3.50	1.13		4.63	install 1" x 10" pine boards with 1" x
Demolish and Install	S.F.	3.50	1.99		5.49	3" battens.
Reinstall	S.F.		.90		.90	
Clean	S.F.		.31		.31	
Paint	S.F.	.32	1.21		1.53	
Minimum Charge	Job		211		211	
Cedar						
Demolish	S.F.		.86		.86	Cost includes material and labor to
Install	S.F.	4.88	1.13		6.01	install 1" x 10" cedar boards with 1" x
Demolish and Install	S.F.	4.88	1.99		6.87	3" battens.
Reinstall	S.F.		.90		.90	
Clean	S.F.		.31		.31	
Paint	S.F.	.32	1.21		1.53	
Minimum Charge	Job		211		211	
Redwood						
Demolish	S.F.		.86		.86	Cost includes material and labor to
Install	S.F.	5.80	1.13		6.93	install 1" x 10" redwood boards with
Demolish and Install	S.F.	5.80	1.99		7.79	1" x 3" battens.
Reinstall	S.F.		.90		.90	
Clean	S.F.		.31		.31	
Paint	S.F.	.32	1.21		1.53	
Minimum Charge	Job		211		211	
Wood Shingle						
Demolish	Sq.		79.50		79.50	Includes material and labor to install
Install	Sq.	212	188		400	wood shingles/shakes.
Demolish and Install	Sq.	212	267.50		479.50	
Clean	Sq.		32.50		32.50	
Paint	Sq.	20.50	108		128.50	
Minimum Charge	Job		211		211	

Siding

Exterior Siding	Unit	Material	Labor	Equip.	Total	Specification
Synthetic Stucco System						
Demolish	S.F.		2.87		2.87	Cost includes material and labor to
Install	S.F.	2.88	6.70	.59	10.17	install 2 coats of adhesive mixture,
Demolish and Install	S.F.	2.88	9.57	.59	13.04	glass fiber mesh, 1" insulation board,
Clean	S.F.		.28		.28	and textured finish coat.
Paint	S.F.	.28	.59		.87	
Minimum Charge	Job		900	78.50	978.50	

Front-end Loader

Bulldozer

Grading

Grading		Unit	Material	Labor	Equip.	Total	Specification
Hand (fine)							
	Install	S.F.		.62		.62	Includes labor and material to install
	Minimum Charge	Job		155		155	fine grade for slab on grade, hand grade.
Machine (fine)							
	Install	S.Y.		.55	.58	1.13	Includes labor and material to install
	Minimum Charge	Job		370	390	760	fine grade for roadway, base or leveling course.

Asphalt Paving

Asphalt Paving		Unit	Material	Labor	Equip.	Total	Specification
Seal Coat							
	Install	S.F.	.15	.11		.26	Includes labor and material to install
	Minimum Charge	Job		155		155	sealcoat for a small area.
Wearing Course							
	Install	S.F.	.18	.34	.28	.80	Cost includes labor and material to
	Minimum Charge	Job		155		155	install 1" thick wearing course.

Masonry Pavers

Masonry Pavers		Unit	Material	Labor	Equip.	Total	Specification
Natural Concrete							
Sand Base							
	Demolish	S.F.		3.44		3.44	Includes material and labor to install
	Install	S.F.	2.82	6.90		9.72	concrete pavers on sand base.
	Demolish and Install	S.F.	2.82	10.34		13.16	
	Reinstall	S.F.		5.53		5.53	
	Clean	S.F.		1.24		1.24	
	Minimum Charge	Job		207		207	
Mortar Base							
	Demolish	S.F.		5.75	2.25	8	Includes material and labor to install
	Install	S.F.	3.70	5.75		9.45	concrete pavers on mortar base.
	Demolish and Install	S.F.	3.70	11.50	2.25	17.45	
	Reinstall	S.F.		4.61		4.61	
	Clean	S.F.		1.13		1.13	
	Minimum Charge	Job		207		207	

Masonry Pavers	Unit	Material	Labor	Equip.	Total	Specification
Adobe						
Sand Base						
Demolish	S.F.		3.44		3.44	Includes material and labor to install
Install	S.F.	2.31	4.61		6.92	adobe pavers on sand base.
Demolish and Install	S.F.	2.31	8.05		10.36	
Reinstall	S.F.		3.69		3.69	
Clean	S.F.		1.24		1.24	
Minimum Charge	Job		207		207	
Mortar Base						
Demolish	S.F.		5.75	2.25	8	Includes material and labor to install
Install	S.F.	2.97	5.75		8.72	adobe pavers on mortar base.
Demolish and Install	S.F.	2.97	11.50	2.25	16.72	
Reinstall	S.F.		4.61		4.61	
Clean	S.F.		1.13		1.13	
Minimum Charge	Job		207		207	
Standard Brick						
Sand Base						
Demolish	S.F.		3.44		3.44	Includes material and labor to install
Install	S.F.	3.27	7.50		10.77	standard brick pavers on sand base.
Demolish and Install	S.F.	3.27	10.94		14.21	
Reinstall	S.F.		5.98		5.98	
Clean	S.F.		1.24		1.24	
Minimum Charge	Job		207		207	
Mortar Base						
Demolish	S.F.		5.75	2.25	8	Includes material and labor to install
Install	S.F.	3.30	10.35		13.65	standard brick pavers on mortar base.
Demolish and Install	S.F.	3.30	16.10	2.25	21.65	
Reinstall	S.F.		8.30		8.30	
Clean	S.F.		1.13		1.13	
Minimum Charge	Job		207		207	
Deluxe Brick						
Sand Base						
Demolish	S.F.		3.44		3.44	Includes material and labor to install
Install	S.F.	3.85	5.75		9.60	deluxe brick pavers on sand base.
Demolish and Install	S.F.	3.85	9.19		13.04	
Reinstall	S.F.		4.61		4.61	
Clean	S.F.		1.24		1.24	
Minimum Charge	Job		207		207	
Mortar Base						
Demolish	S.F.		5.75	2.25	8	Includes material and labor to install
Install	S.F.	3.96	10.35		14.31	deluxe brick pavers on mortar base.
Demolish and Install	S.F.	3.96	16.10	2.25	22.31	
Reinstall	S.F.		8.30		8.30	
Clean	S.F.		1.13		1.13	
Minimum Charge	Job		207		207	

Fencing	Unit	Material	Labor	Equip.	Total	Specification
Block w / Footing						
4″						
Demolish	S.F.		1.91		1.91	Cost includes material and labor to
Install	S.F.	1.65	4.44		6.09	install 4″ non-reinforced concrete
Demolish and Install	S.F.	1.65	6.35		8	block.
Clean	S.F.	.24	1.03	.25	1.52	
Paint	S.F.	.31	.69		1	
Minimum Charge	Job		207		207	

Fencing	Unit	Material	Labor	Equip.	Total	Specification
4" Reinforced						
Demolish	S.F.		1.91	1.19	3.10	Cost includes material and labor to
Install	S.F.	1.82	4.49		6.31	install 4" reinforced concrete block.
Demolish and Install	S.F.	1.82	6.40	1.19	9.41	
Clean	S.F.	.24	1.03	.25	1.52	
Paint	S.F.	.31	.69		1	
Minimum Charge	Job		207		207	
6"						
Demolish	S.F.		2.02		2.02	Cost includes material and labor to
Install	S.F.	2.33	4.78		7.11	install 6" non-reinforced concrete
Demolish and Install	S.F.	2.33	6.80		9.13	block.
Clean	S.F.	.24	1.03	.25	1.52	
Paint	S.F.	.31	.69		1	
Minimum Charge	Job		207		207	
6" Reinforced						
Demolish	S.F.		2.01	.60	2.61	Cost includes material and labor to
Install	S.F.	2.50	4.84		7.34	install 6" reinforced concrete block.
Demolish and Install	S.F.	2.50	6.85	.60	9.95	
Clean	S.F.	.24	1.03	.25	1.52	
Paint	S.F.	.31	.69		1	
Minimum Charge	Job		207		207	
Cap for Block Wall						
Demolish	L.F.		2.15		2.15	Includes labor and equipment to install
Install	L.F.	19.95	9.95		29.90	precast concrete coping.
Demolish and Install	L.F.	19.95	12.10		32.05	
Clean	L.F.	.02	3.22		3.24	
Paint	S.F.	.31	.69		1	
Minimum Charge	Job		207		207	
Block Pilaster						
Demolish	Ea.		148		148	Includes material and labor to install
Install	Ea.	75.50	150		225.50	block pilaster for fence.
Demolish and Install	Ea.	75.50	298		373.50	
Reinstall	Ea.		149.54		149.54	
Paint	Ea.	4.53	16.05		20.58	
Minimum Charge	Job		207		207	
Block w / o Footing						
4"						
Demolish	S.F.		1.91		1.91	Cost includes material and labor to
Install	S.F.	1.79	3.40		5.19	install 4" non-reinforced hollow
Demolish and Install	S.F.	1.79	5.31		7.10	lightweight block.
Clean	S.F.	.24	1.03	.25	1.52	
Paint	S.F.	.31	.69		1	
Minimum Charge	Job		207		207	
4" Reinforced						
Demolish	S.F.		1.91	1.19	3.10	Cost includes material and labor to
Install	S.F.	1.93	4.39		6.32	install 4" reinforced hollow lightweight
Demolish and Install	S.F.	1.93	6.30	1.19	9.42	block.
Clean	S.F.	.24	1.03	.25	1.52	
Paint	S.F.	.31	.69		1	
Minimum Charge	Job		207		207	
6"						
Demolish	S.F.		2.02		2.02	Cost includes material and labor to
Install	S.F.	2.60	3.74		6.34	install 6" non-reinforced hollow
Demolish and Install	S.F.	2.60	5.76		8.36	lightweight block.
Clean	S.F.	.24	1.03	.25	1.52	
Paint	S.F.	.31	.69		1	
Minimum Charge	Job		207		207	

Site Work

Fencing	Unit	Material	Labor	Equip.	Total	Specification
6" Reinforced						
Demolish	S.F.		2.01	.60	2.61	Cost includes material and labor to
Install	S.F.	2.74	4.72		7.46	install 6" reinforced hollow lightweight
Demolish and Install	S.F.	2.74	6.73	.60	10.07	block.
Clean	S.F.	.24	1.03	.25	1.52	
Paint	S.F.	.31	.69		1	
Minimum Charge	Job		207		207	
Cap for Block Wall						
Demolish	L.F.		2.15		2.15	Includes labor and equipment to install
Install	L.F.	19.95	9.95		29.90	precast concrete coping.
Demolish and Install	L.F.	19.95	12.10		32.05	
Clean	L.F.	.02	3.22		3.24	
Paint	S.F.	.31	.69		1	
Minimum Charge	Job		207		207	
Block Pilaster						
Demolish	Ea.		148		148	Includes material and labor to install
Install	Ea.	75.50	150		225.50	block pilaster for fence.
Demolish and Install	Ea.	75.50	298		373.50	
Reinstall	Ea.		149.54		149.54	
Paint	Ea.	4.53	16.05		20.58	
Minimum Charge	Job		207		207	
Chain Link						
3' High w / o Top Rail						
Demolish	L.F.		2.71	1.06	3.77	Includes material and labor to install
Install	L.F.	6.05	1.82	.57	8.44	residential galvanized fence 3' high
Demolish and Install	L.F.	6.05	4.53	1.63	12.21	with 1-3/8" steel post 10' O.C. set
Reinstall	L.F.		1.46	.46	1.92	without concrete.
Clean	L.F.	.10	1.13		1.23	
Minimum Charge	Job		480		480	
4' High w / o Top Rail						
Demolish	L.F.		2.71	1.06	3.77	Includes material and labor to install
Install	L.F.	6.80	2.25	.71	9.76	residential galvanized fence 4' high
Demolish and Install	L.F.	6.80	4.96	1.77	13.53	with 1-3/8" steel post 10' O.C. set
Reinstall	L.F.		1.80	.57	2.37	without concrete.
Clean	L.F.	.13	1.24		1.37	
Minimum Charge	Job		480		480	
5' High w / o Top Rail						
Demolish	L.F.		2.71	1.06	3.77	Includes material and labor to install
Install	L.F.	9	2.95	.93	12.88	residential galvanized fence 5' high
Demolish and Install	L.F.	9	5.66	1.99	16.65	with 1-3/8" steel post 10' O.C. set
Reinstall	L.F.		2.36	.74	3.10	without concrete.
Clean	L.F.	.17	1.38		1.55	
Minimum Charge	Job		480		480	
3' High w / Top Rail						
Demolish	L.F.		2.88	1.12	4	Includes material and labor to install
Install	L.F.	2.31	1.92	.60	4.83	residential galvanized fence 3' high
Demolish and Install	L.F.	2.31	4.80	1.72	8.83	with 1-3/8" steel post 10' O.C. set
Reinstall	L.F.		1.53	.48	2.01	without concrete.
Clean	L.F.	.10	1.15		1.25	
Minimum Charge	Job		480		480	
4' High w / Top Rail						
Demolish	L.F.		2.88	1.12	4	Includes material and labor to install
Install	L.F.	7.90	2.39	.75	11.04	residential galvanized fence 4' high
Demolish and Install	L.F.	7.90	5.27	1.87	15.04	with 1-5/8" steel post 10' O.C. set
Reinstall	L.F.		1.92	.60	2.52	without concrete and 1-5/8" top rail
Clean	L.F.	.13	1.26		1.39	with sleeves.
Minimum Charge	Job		480		480	

Site Work

Fencing	Unit	Material	Labor	Equip.	Total	Specification
5' High w / Top Rail						
Demolish	L.F.		2.88	1.12	4	Includes material and labor to install
Install	L.F.	9	3.19	1	13.19	residential galvanized fence 5' high
Demolish and Install	L.F.	9	6.07	2.12	17.19	with 1-5/8" steel post 10' O.C. set
Reinstall	L.F.		2.55	.80	3.35	without concrete and 1-5/8" top rail
Clean	L.F.	.17	1.41		1.58	with sleeves.
Minimum Charge	Job		480		480	
6' High w / Top Rail						
Demolish	L.F.		2.88	1.12	4	Includes material and labor to install
Install	L.F.	11.20	4.79	1.51	17.50	residential galvanized fence 6' high
Demolish and Install	L.F.	11.20	7.67	2.63	21.50	with 2" steel post 10' O.C. set without
Reinstall	L.F.		3.83	1.20	5.03	concrete and 1-3/8" top rail with
Clean	L.F.	.19	1.59		1.78	sleeves.
Minimum Charge	Job		480		480	
Barbed Wire						
Demolish	L.F.		1.60		1.60	Cost includes material and labor to
Install	L.F.	.40	.62	.24	1.26	install 3 strand galvanized barbed
Demolish and Install	L.F.	.40	2.22	.24	2.86	wire fence.
Reinstall	L.F.		.50	.19	.69	
Minimum Charge	Job		480	415	895	
6' Redwood						
Demolish	L.F.		10.60		10.60	Cost includes material and labor to
Install	L.F.	26.50	7.65	2.41	36.56	install 6' high redwood fencing.
Demolish and Install	L.F.	26.50	18.25	2.41	47.16	
Reinstall	L.F.		6.13	1.93	8.06	
Clean	L.F.	.19	1.44		1.63	
Paint	L.F.	.84	3.80		4.64	
Minimum Charge	Job		480		480	
6' Red Cedar						
Demolish	L.F.		10.60		10.60	Cost includes material and labor to
Install	L.F.	18.10	7.65	2.41	28.16	install 6' cedar fence.
Demolish and Install	L.F.	18.10	18.25	2.41	38.76	
Reinstall	L.F.		6.13	1.93	8.06	
Clean	L.F.	.19	1.44		1.63	
Paint	L.F.	.84	3.80		4.64	
Minimum Charge	Job		480		480	
6' Shadowbox						
Demolish	L.F.		10.60		10.60	Cost includes material and labor to
Install	L.F.	15.30	6.40	2.01	23.71	install 6' pressure treated pine fence.
Demolish and Install	L.F.	15.30	17	2.01	34.31	
Reinstall	L.F.		5.11	1.61	6.72	
Clean	L.F.	.19	1.44		1.63	
Paint	L.F.	.84	3.80		4.64	
Minimum Charge	Job		480		480	
3' Picket						
Demolish	L.F.		10.60		10.60	Includes material, labor and
Install	L.F.	5.90	6.85	2.15	14.90	equipment to install pressure treated
Demolish and Install	L.F.	5.90	17.45	2.15	25.50	pine picket fence.
Reinstall	L.F.		5.47	1.72	7.19	
Clean	L.F.	.10	1.24		1.34	
Paint	L.F.	.42	2.06		2.48	
Minimum Charge	Job		480		480	

Fencing	Unit	Material	Labor	Equip.	Total	Specification
Wrought Iron						
Demolish	L.F.		23		23	Cost includes material and labor to
Install	L.F.	39.50	12.75		52.25	install 5' high wrought iron fence with
Demolish and Install	L.F.	39.50	35.75		75.25	pickets at 4-1/2" O.C.
Reinstall	L.F.		10.20		10.20	
Clean	L.F.		1.24		1.24	
Paint	L.F.	3.99	4.24		8.23	
Minimum Charge	Job		480		480	
Post						
Dirt Based Wood						
Demolish	Ea.		26.50		26.50	Includes material and labor to install
Install	Ea.	25	119		144	wood post set in earth.
Demolish and Install	Ea.	25	145.50		170.50	
Reinstall	Ea.		95.59		95.59	
Clean	Ea.		3.87		3.87	
Paint	Ea.	.92	1.44		2.36	
Minimum Charge	Job		480		480	
Concrete Based Wood						
Demolish	Ea.		38.50	14.95	53.45	Includes material and labor to install
Install	Ea.	15.40	159		174.40	wood post set in concrete.
Demolish and Install	Ea.	15.40	197.50	14.95	227.85	
Clean	Ea.		3.87		3.87	
Paint	Ea.	.92	1.44		2.36	
Minimum Charge	Job		480		480	
Block Pilaster						
Demolish	Ea.		148		148	Includes material and labor to install
Install	Ea.	75.50	150		225.50	block pilaster for fence.
Demolish and Install	Ea.	75.50	298		373.50	
Reinstall	Ea.		149.54		149.54	
Paint	Ea.	4.53	16.05		20.58	
Minimum Charge	Job		207		207	
Gate						
Chain Link						
Demolish	Ea.		35.50		35.50	Cost includes material and labor to
Install	Ea.	107	53		160	install 3' wide chain link pass gate
Demolish and Install	Ea.	107	88.50		195.50	and hardware.
Reinstall	Ea.		42.49		42.49	
Clean	Ea.		4.30		4.30	
Minimum Charge	Job		480		480	
Redwood						
Demolish	Ea.		30.50		30.50	Includes material, labor and
Install	Ea.	217	53		270	equipment to install redwood pass
Demolish and Install	Ea.	217	83.50		300.50	gate and hardware.
Reinstall	Ea.		42.49		42.49	
Clean	Ea.		4.30		4.30	
Paint	Ea.	1.39	7.20		8.59	
Minimum Charge	Job		480		480	
Red Cedar						
Demolish	Ea.		30.50		30.50	Cost includes material and labor to
Install	Ea.	97.50	53		150.50	install 3' wide red cedar pass gate
Demolish and Install	Ea.	97.50	83.50		181	and hardware.
Reinstall	Ea.		42.49		42.49	
Clean	Ea.		4.30		4.30	
Paint	Ea.	1.39	7.20		8.59	
Minimum Charge	Job		480		480	

Site Work

Fencing	Unit	Material	Labor	Equip.	Total	Specification
Wrought Iron						
Demolish	Ea.		23		23	Includes material, labor and
Install	Ea.	276	119		395	equipment to install wrought iron gate
Demolish and Install	Ea.	276	142		418	with hardware.
Reinstall	Ea.		95.59		95.59	
Clean	Ea.		4.30		4.30	
Paint	Ea.	8	7.20		15.20	
Minimum Charge	Job		480		480	

Rough Mechanical

Fire-extinguishing System	Space Heater	Hot-air Furnace

Plumbing	Unit	Material	Labor	Equip.	Total	Specification
Components						
3/4″ Supply						
Demolish	L.F.		2.70		2.70	Includes material and labor to install
Install	L.F.	7.15	6.40		13.55	type L copper tubing including
Demolish and Install	L.F.	7.15	9.10		16.25	couplings and hangers at 10′ O.C.
Reinstall	L.F.		6.38		6.38	
Minimum Charge	Job		243		243	
1/2″ Fixture Supply						
Install	Ea.	78	162		240	Includes labor and materials to install supply pipe only to a plumbing fixture.
2″ Single Vent Stack						
Demolish	Ea.		45		45	Includes material and labor to install
Install	Ea.	31.50	243		274.50	PVC vent stack pipe with couplings at
Demolish and Install	Ea.	31.50	288		319.50	10′ O.C.
Reinstall	Ea.		242.60		242.60	
Minimum Charge	Job		243		243	
4″ Drain						
Demolish	L.F.		3.59		3.59	Includes material and labor to install
Install	L.F.	10.50	7.95		18.45	DVW type PVC pipe with couplings at
Demolish and Install	L.F.	10.50	11.54		22.04	10′ O.C. and 3 strap hangers per 10′
Reinstall	L.F.		7.94		7.94	run.
Minimum Charge	Job		243		243	
1-1/2″ Drain						
Demolish	L.F.		2.70		2.70	Includes material and labor to install
Install	L.F.	3.42	5.15		8.57	DVW type PVC pipe with couplings at
Demolish and Install	L.F.	3.42	7.85		11.27	10′ O.C. and 3 strap hangers per 10′
Reinstall	L.F.		5.16		5.16	run.
Minimum Charge	Job		243		243	
Maintenance						
Flush Out / Unclog						
Clean	Job		243		243	Minimum labor and equipment charge
Minimum Charge	Job		243		243	to install water cooler.
Vent Cap						
Large						
Demolish	Ea.		12.25		12.25	Includes labor and material to install a
Install	Ea.	41.50	22		63.50	PVC plumbing pipe vent cap.
Demolish and Install	Ea.	41.50	34.25		75.75	
Clean	Ea.	.45	7.05		7.50	
Paint	Ea.	.19	8.20		8.39	
Minimum Charge	Job		121		121	

143

Rough Mechanical

Plumbing

Plumbing	Unit	Material	Labor	Equip.	Total	Specification
Medium						
Demolish	Ea.		12.25		12.25	Includes labor and material to install a
Install	Ea.	36	21		57	PVC plumbing pipe vent cap.
Demolish and Install	Ea.	36	33.25		69.25	
Clean	Ea.	.45	7.05		7.50	
Paint	Ea.	.19	8.20		8.39	
Minimum Charge	Job		121		121	
Small						
Demolish	Ea.		12.25		12.25	Includes material and labor to install a
Install	Ea.	9	20		29	PVC plumbing pipe vent cap.
Demolish and Install	Ea.	9	32.25		41.25	
Clean	Ea.	.45	7.05		7.50	
Paint	Ea.	.19	8.20		8.39	
Minimum Charge	Job		121		121	

Plumbing

Plumbing	Unit	Material	Labor	Equip.	Total	Specification
Sump Pump						
1/4 HP						
Demolish	Ea.		48.50		48.50	Includes material and labor to install a
Install	Ea.	143	81		224	submersible sump pump.
Demolish and Install	Ea.	143	129.50		272.50	
Reinstall	Ea.		64.69		64.69	
Clean	Ea.	1.14	19.35		20.49	
Minimum Charge	Job		243		243	
1/3 HP						
Demolish	Ea.		48.50		48.50	Includes material and labor to install a
Install	Ea.	215	97		312	submersible sump pump.
Demolish and Install	Ea.	215	145.50		360.50	
Reinstall	Ea.		77.63		77.63	
Clean	Ea.	1.14	19.35		20.49	
Minimum Charge	Job		243		243	
1/2 HP						
Demolish	Ea.		48.50		48.50	Includes material and labor to install a
Install	Ea.	254	97		351	submersible sump pump.
Demolish and Install	Ea.	254	145.50		399.50	
Reinstall	Ea.		77.63		77.63	
Clean	Ea.	1.14	19.35		20.49	
Minimum Charge	Job		243		243	
Clean and Service						
Clean	Ea.		30.50		30.50	Includes labor to clean and service a
Minimum Charge	Job		243		243	sump pump.

Water Heater

Water Heater	Unit	Material	Labor	Equip.	Total	Specification
30 Gallon						
Demolish	Ea.		152		152	Includes material and labor to install a
Install	Ea.	765	221		986	30 gallon electric water heater.
Demolish and Install	Ea.	765	373		1138	
Reinstall	Ea.		176.44		176.44	
Clean	Ea.	1.14	19.35		20.49	
Minimum Charge	Job		243		243	

Rough Mechanical

Water Heater		Unit	Material	Labor	Equip.	Total	Specification
40 Gallon							
	Demolish	Ea.		152		152	Includes material and labor to install a
	Install	Ea.	815	243		1058	40 gallon electric water heater.
	Demolish and Install	Ea.	815	395		1210	
	Reinstall	Ea.		194.08		194.08	
	Clean	Ea.	1.14	19.35		20.49	
	Minimum Charge	Job		243		243	
52 Gallon							
	Demolish	Ea.		152		152	Includes material and labor to install a
	Install	Ea.	915	243		1158	52 gallon electric water heater.
	Demolish and Install	Ea.	915	395		1310	
	Reinstall	Ea.		194.08		194.08	
	Clean	Ea.	1.14	19.35		20.49	
	Minimum Charge	Job		243		243	
82 Gallon							
	Demolish	Ea.		152		152	Includes material and labor to install
	Install	Ea.	1375	305		1680	an 82 gallon electric water heater.
	Demolish and Install	Ea.	1375	457		1832	
	Reinstall	Ea.		242.60		242.60	
	Clean	Ea.	1.14	19.35		20.49	
	Minimum Charge	Job		243		243	
Gas-fired							
30 Gallon Gas-fired							
	Demolish	Ea.		187		187	Includes material and labor to install a
	Install	Ea.	985	243		1228	30 gallon gas water heater.
	Demolish and Install	Ea.	985	430		1415	
	Reinstall	Ea.		194.08		194.08	
	Clean	Ea.	1.14	19.35		20.49	
	Minimum Charge	Job		243		243	
30 Gallon Gas-fired, Flue							
	Install	Ea.	61	69.50		130.50	Includes labor and materials to install
	Minimum Charge	Job		243		243	a flue for a water heater (up to 50 gallon size).
40 Gallon Gas-fired							
	Demolish	Ea.		187		187	Includes material and labor to install a
	Install	Ea.	985	255		1240	40 gallon gas water heater.
	Demolish and Install	Ea.	985	442		1427	
	Reinstall	Ea.		204.29		204.29	
	Clean	Ea.	1.14	19.35		20.49	
	Minimum Charge	Job		243		243	
40 Gallon Gas-fired, Flue							
	Install	Ea.	61	69.50		130.50	Includes labor and materials to install
	Minimum Charge	Job		243		243	a flue for a water heater (up to 50 gallon size).
50 Gallon Gas-fired							
	Demolish	Ea.		187		187	Includes material and labor to install a
	Install	Ea.	1025	270		1295	50 gallon gas water heater.
	Demolish and Install	Ea.	1025	457		1482	
	Reinstall	Ea.		215.64		215.64	
	Clean	Ea.	1.14	19.35		20.49	
	Minimum Charge	Job		243		243	
50 Gallon Gas-fired, Flue							
	Install	Ea.	61	69.50		130.50	Includes labor and materials to install
	Minimum Charge	Job		243		243	a flue for a water heater (up to 50 gallon size).

Rough Mechanical

Water Heater	Unit	Material	Labor	Equip.	Total	Specification
Solar Hot Water System						
Water Heater						
Demolish	Ea.		121		121	Includes material and labor to install a
Install	Ea.	1625	325		1950	solar hot water heat exchanger
Demolish and Install	Ea.	1625	446		2071	storage tank, 100 gallon.
Reinstall	Ea.		258.77		258.77	
Minimum Charge	Job		246		246	
Solar Panel						
Demolish	S.F.		1.77		1.77	Includes material and labor to install
Install	S.F.	35	3.19		38.19	solar panel.
Demolish and Install	S.F.	35	4.96		39.96	
Reinstall	S.F.		2.55		2.55	
Minimum Charge	Job		246		246	
Repair Plumbing						
Install	Job		246		246	Includes 4 hour minimum labor charge
Minimum Charge	Job		246		246	for a steamfitter or pipefitter.
Insulating Wrap						
Demolish	Ea.		5.60		5.60	Includes material and labor to install
Install	Ea.	19.65	15.15		34.80	insulating wrap around a 100 gallon
Demolish and Install	Ea.	19.65	20.75		40.40	solar hot water heat exchanger
Reinstall	Ea.		12.13		12.13	storage tank.
Minimum Charge	Job		243		243	
Water Softener						
Demolish	Ea.		90		90	Includes material and labor to install a
Install	Ea.	1650	218		1868	14 GPM water softener.
Demolish and Install	Ea.	1650	308		1958	
Reinstall	Ea.		174.72		174.72	
Clean	Ea.	1.14	19.35		20.49	
Minimum Charge	Ea.		243		243	

Ductwork	Unit	Material	Labor	Equip.	Total	Specification
Galvanized, Square or Rectangular						
6"						
Demolish	L.F.		2.17		2.17	Includes material and labor to install
Install	L.F.	1.78	10.40		12.18	rectangular ductwork with a greatest
Demolish and Install	L.F.	1.78	12.57		14.35	dimension of 6".
Reinstall	L.F.		8.33		8.33	
Clean	L.F.		2.50		2.50	
Minimum Charge	Job		234		234	
8"						
Demolish	L.F.		2.17		2.17	Includes material and labor to install
Install	L.F.	2.37	13.40		15.77	rectangular ductwork with a greatest
Demolish and Install	L.F.	2.37	15.57		17.94	dimension of 8".
Reinstall	L.F.		10.72		10.72	
Clean	L.F.		3.33		3.33	
Minimum Charge	Job		234		234	
10"						
Demolish	L.F.		2.17		2.17	Includes material and labor to install
Install	L.F.	2.96	17.35		20.31	rectangular ductwork with a greatest
Demolish and Install	L.F.	2.96	19.52		22.48	dimension of 10".
Reinstall	L.F.		13.89		13.89	
Clean	L.F.		4.17		4.17	
Minimum Charge	Job		234		234	

Rough Mechanical

Ductwork	Unit	Material	Labor	Equip.	Total	Specification
Galvanized, Spiral						
4″						
Demolish	L.F.		2.60		2.60	Cost includes material and labor to
Install	L.F.	2.16	2.34		4.50	install 4″ diameter spiral pre-formed
Demolish and Install	L.F.	2.16	4.94		7.10	steel galvanized ductwork.
Reinstall	L.F.		1.87		1.87	
Clean	L.F.		1.31		1.31	
Minimum Charge	Job		234		234	
6″						
Demolish	L.F.		2.60		2.60	Cost includes material and labor to
Install	L.F.	2.16	3.01		5.17	install 6″ diameter spiral pre-formed
Demolish and Install	L.F.	2.16	5.61		7.77	steel galvanized ductwork.
Reinstall	L.F.		2.41		2.41	
Clean	L.F.		1.96		1.96	
Minimum Charge	Job		234		234	
8″						
Demolish	L.F.		2.60		2.60	Cost includes material and labor to
Install	L.F.	2.87	4.22		7.09	install 8″ diameter spiral pre-formed
Demolish and Install	L.F.	2.87	6.82		9.69	steel galvanized ductwork.
Reinstall	L.F.		3.37		3.37	
Clean	L.F.		2.62		2.62	
Minimum Charge	Job		234		234	
10″						
Demolish	L.F.		3.26		3.26	Cost includes material and labor to
Install	L.F.	3.59	5.25		8.84	install 10″ diameter spiral pre-formed
Demolish and Install	L.F.	3.59	8.51		12.10	steel galvanized ductwork.
Reinstall	L.F.		4.22		4.22	
Clean	L.F.		3.27		3.27	
Minimum Charge	Job		234		234	
12″						
Demolish	L.F.		4.34		4.34	Cost includes material and labor to
Install	L.F.	4.29	7.05		11.34	install 12″ diameter spiral pre-formed
Demolish and Install	L.F.	4.29	11.39		15.68	steel galvanized ductwork.
Reinstall	L.F.		5.62		5.62	
Clean	L.F.		3.93		3.93	
Minimum Charge	Job		234		234	
16″						
Demolish	L.F.		8.70		8.70	Cost includes material and labor to
Install	L.F.	6.65	14.05		20.70	install 16″ diameter spiral pre-formed
Demolish and Install	L.F.	6.65	22.75		29.40	steel galvanized ductwork.
Reinstall	L.F.		11.25		11.25	
Clean	L.F.		5.25		5.25	
Minimum Charge	Job		234		234	
Flexible Fiberglass						
6″ Non-insulated						
Demolish	L.F.		2		2	Cost includes material and labor to
Install	L.F.	1.54	3.01		4.55	install 6″ diameter non-insulated
Demolish and Install	L.F.	1.54	5.01		6.55	flexible fiberglass fabric on corrosion
Reinstall	L.F.		2.41		2.41	resistant metal ductwork.
Clean	L.F.		1.96		1.96	
Minimum Charge	Job		234		234	
8″ Non-insulated						
Demolish	L.F.		2.89		2.89	Cost includes material and labor to
Install	L.F.	1.94	4.22		6.16	install 8″ diameter non-insulated
Demolish and Install	L.F.	1.94	7.11		9.05	flexible fiberglass fabric on corrosion
Reinstall	L.F.		3.37		3.37	resistant metal ductwork.
Clean	L.F.		2.62		2.62	
Minimum Charge	Job		234		234	

Rough Mechanical

Ductwork

	Unit	Material	Labor	Equip.	Total	Specification
10" Non-insulated						
Demolish	L.F.		3.72		3.72	Cost includes material and labor to
Install	L.F.	2.51	5.25		7.76	install 10" diameter non-insulated
Demolish and Install	L.F.	2.51	8.97		11.48	flexible fiberglass fabric on corrosion
Reinstall	L.F.		4.22		4.22	resistant metal ductwork.
Clean	L.F.		3.27		3.27	
Minimum Charge	Job		234		234	
12" Non-insulated						
Demolish	L.F.		5.20		5.20	Cost includes material and labor to
Install	L.F.	3	7.05		10.05	install 12" diameter non-insulated
Demolish and Install	L.F.	3	12.25		15.25	flexible fiberglass fabric on corrosion
Reinstall	L.F.		5.62		5.62	resistant metal ductwork.
Clean	L.F.		3.93		3.93	
Minimum Charge	Job		234		234	
4" Insulated						
Demolish	L.F.		1.53		1.53	Cost includes material and labor to
Install	L.F.	2.44	2.48		4.92	install 4" diameter insulated flexible
Demolish and Install	L.F.	2.44	4.01		6.45	fiberglass fabric on corrosion resistant
Reinstall	L.F.		1.99		1.99	metal ductwork.
Clean	L.F.		1.31		1.31	
Minimum Charge	Job		234		234	
6" Insulated						
Demolish	L.F.		2		2	Cost includes material and labor to
Install	L.F.	2.75	3.25		6	install 6" diameter insulated flexible
Demolish and Install	L.F.	2.75	5.25		8	fiberglass fabric on corrosion resistant
Reinstall	L.F.		2.60		2.60	metal ductwork.
Clean	L.F.		1.96		1.96	
Minimum Charge	Job		234		234	
8" Insulated						
Demolish	L.F.		2.89		2.89	Cost includes material and labor to
Install	L.F.	3.31	4.69		8	install 8" diameter insulated flexible
Demolish and Install	L.F.	3.31	7.58		10.89	fiberglass fabric on corrosion resistant
Reinstall	L.F.		3.75		3.75	metal ductwork.
Clean	L.F.		2.62		2.62	
Minimum Charge	Job		234		234	
10" Insulated						
Demolish	L.F.		3.72		3.72	Cost includes material and labor to
Install	L.F.	4.03	6.05		10.08	install 10" diameter insulated flexible
Demolish and Install	L.F.	4.03	9.77		13.80	fiberglass fabric on corrosion resistant
Reinstall	L.F.		4.82		4.82	metal ductwork.
Clean	L.F.		3.27		3.27	
Minimum Charge	Job		234		234	
12" Insulated						
Demolish	L.F.		5.20		5.20	Cost includes material and labor to
Install	L.F.	4.62	8.45		13.07	install 12" diameter insulated flexible
Demolish and Install	L.F.	4.62	13.65		18.27	fiberglass fabric on corrosion resistant
Reinstall	L.F.		6.75		6.75	metal ductwork.
Clean	L.F.		3.93		3.93	
Minimum Charge	Job		234		234	

Rough Mechanical

Heating	Unit	Material	Labor	Equip.	Total	Specification
Hot Water Baseboard						
Demolish	L.F.		8.40		8.40	Includes material and labor to install
Install	L.F.	8.40	14.75		23.15	baseboard radiation.
Demolish and Install	L.F.	8.40	23.15		31.55	
Reinstall	L.F.		11.81		11.81	
Clean	L.F.	1.14	4.84		5.98	
Minimum Charge	Job		243		243	
3'						
Demolish	Ea.		27		27	Includes material and labor to install
Install	Ea.	28	49		77	fin tube baseboard radiation in 3'
Demolish and Install	Ea.	28	76		104	sections.
Reinstall	Ea.		39.38		39.38	
Clean	Ea.	.33	14.75		15.08	
Minimum Charge	Job		243		243	
4'						
Demolish	Ea.		36		36	Includes material and labor to install
Install	Ea.	37	55.50		92.50	fin tube baseboard radiation in 4'
Demolish and Install	Ea.	37	91.50		128.50	sections.
Reinstall	Ea.		44.30		44.30	
Clean	Ea.	.65	19.35		20	
Minimum Charge	Job		243		243	
5'						
Demolish	Ea.		45		45	Includes material and labor to install
Install	Ea.	46	63.50		109.50	fin tube baseboard radiation in 5'
Demolish and Install	Ea.	46	108.50		154.50	sections.
Reinstall	Ea.		50.63		50.63	
Clean	Ea.	.65	24		24.65	
Minimum Charge	Job		243		243	
6'						
Demolish	Ea.		54		54	Includes material and labor to install
Install	Ea.	55.50	74		129.50	fin tube baseboard radiation in 6'
Demolish and Install	Ea.	55.50	128		183.50	sections.
Reinstall	Ea.		59.07		59.07	
Clean	Ea.	.65	28		28.65	
Minimum Charge	Job		243		243	
7'						
Demolish	Ea.		60		60	Includes material and labor to install
Install	Ea.	65	97		162	fin tube baseboard radiation in 7'
Demolish and Install	Ea.	65	157		222	sections.
Reinstall	Ea.		77.63		77.63	
Clean	Ea.	.81	34.50		35.31	
Minimum Charge	Job		243		243	
9'						
Demolish	Ea.		90		90	Includes material and labor to install
Install	Ea.	83.50	139		222.50	fin tube baseboard radiation in 9'
Demolish and Install	Ea.	83.50	229		312.50	sections.
Reinstall	Ea.		110.90		110.90	
Clean	Ea.	1.31	44		45.31	
Minimum Charge	Job		243		243	
Cabinet Unit Heater						
50 MBH						
Demolish	Ea.		44.50		44.50	Includes material and labor to install a
Install	Ea.	1300	127		1427	cabinet space heater with an output of
Demolish and Install	Ea.	1300	171.50		1471.50	50 MBH.
Reinstall	Ea.		101.27		101.27	
Clean	Ea.	.81	38.50		39.31	
Minimum Charge	Job		246		246	

Rough Mechanical

Heating		Unit	Material	Labor	Equip.	Total	Specification
75 MBH							
	Demolish	Ea.		44.50		44.50	Includes material and labor to install a
	Install	Ea.	1450	148		1598	cabinet space heater with an output of
	Demolish and Install	Ea.	1450	192.50		1642.50	75 MBH.
	Reinstall	Ea.		118.14		118.14	
	Clean	Ea.	.81	38.50		39.31	
	Minimum Charge	Job		246		246	
125 MBH							
	Demolish	Ea.		53		53	Includes material and labor to install a
	Install	Ea.	1825	177		2002	cabinet space heater with an output of
	Demolish and Install	Ea.	1825	230		2055	125 MBH.
	Reinstall	Ea.		141.77		141.77	
	Clean	Ea.	.81	38.50		39.31	
	Minimum Charge	Job		246		246	
175 MBH							
	Demolish	Ea.		66.50		66.50	Includes material and labor to install a
	Install	Ea.	2175	253		2428	cabinet space heater with an output of
	Demolish and Install	Ea.	2175	319.50		2494.50	175 MBH.
	Reinstall	Ea.		202.53		202.53	
	Clean	Ea.	.81	38.50		39.31	
	Minimum Charge	Job		246		246	

Furnace Unit		Unit	Material	Labor	Equip.	Total	Specification
Wall (gas-fired)							
25 MBH w / Thermostat							
	Demolish	Ea.		105		105	Includes material and labor to install a
	Install	Ea.	670	169		839	gas fired hot air furnace rated at 25
	Demolish and Install	Ea.	670	274		944	MBH.
	Reinstall	Ea.		134.99		134.99	
	Clean	Ea.	1.63	86		87.63	
	Minimum Charge	Job		246		246	
35 MBH w / Thermostat							
	Demolish	Ea.		105		105	Includes material and labor to install a
	Install	Ea.	860	187		1047	gas fired hot air furnace rated at 35
	Demolish and Install	Ea.	860	292		1152	MBH.
	Reinstall	Ea.		149.99		149.99	
	Clean	Ea.	1.63	86		87.63	
	Minimum Charge	Job		246		246	
50 MBH w / Thermostat							
	Demolish	Ea.		105		105	Includes material and labor to install a
	Install	Ea.	1000	211		1211	gas fired hot air furnace rated at 50
	Demolish and Install	Ea.	1000	316		1316	MBH.
	Reinstall	Ea.		168.74		168.74	
	Clean	Ea.	1.63	86		87.63	
	Minimum Charge	Job		246		246	
60 MBH w / Thermostat							
	Demolish	Ea.		105		105	Includes material and labor to install a
	Install	Ea.	1125	234		1359	gas fired hot air furnace rated at 60
	Demolish and Install	Ea.	1125	339		1464	MBH.
	Reinstall	Ea.		187.48		187.48	
	Clean	Ea.	1.63	86		87.63	
	Minimum Charge	Job		246		246	

Furnace Unit	Unit	Material	Labor	Equip.	Total	Specification
Floor (gas-fired)						
30 MBH w / Thermostat						Includes material and labor to install a
Demolish	Ea.		105		105	gas fired hot air floor furnace rated at
Install	Ea.	890	169		1059	30 MBH.
Demolish and Install	Ea.	890	274		1164	
Reinstall	Ea.		134.99		134.99	
Clean	Ea.	1.63	86		87.63	
Minimum Charge	Job		246		246	
50 MBH w / Thermostat						Includes material and labor to install a
Demolish	Ea.		105		105	gas fired hot air floor furnace rated at
Install	Ea.	965	211		1176	50 MBH.
Demolish and Install	Ea.	965	316		1281	
Reinstall	Ea.		168.74		168.74	
Clean	Ea.	1.63	86		87.63	
Minimum Charge	Job		246		246	
Gas Forced Air						
50 MBH						Includes material and labor to install
Demolish	Ea.		105		105	an up flow gas fired hot air furnace
Install	Ea.	990	211		1201	rated at 50 MBH.
Demolish and Install	Ea.	990	316		1306	
Reinstall	Ea.		168.74		168.74	
Clean	Ea.	1.63	86		87.63	
Minimum Charge	Job		246		246	
80 MBH						Includes material and labor to install
Demolish	Ea.		105		105	an up flow gas fired hot air furnace
Install	Ea.	950	234		1184	rated at 80 MBH.
Demolish and Install	Ea.	950	339		1289	
Reinstall	Ea.		187.48		187.48	
Clean	Ea.	1.63	86		87.63	
Minimum Charge	Job		246		246	
100 MBH						Includes material and labor to install
Demolish	Ea.		105		105	an up flow gas fired hot air furnace
Install	Ea.	1100	264		1364	rated at 100 MBH.
Demolish and Install	Ea.	1100	369		1469	
Reinstall	Ea.		210.92		210.92	
Clean	Ea.	1.63	86		87.63	
Minimum Charge	Job		246		246	
120 MBH						Includes material and labor to install
Demolish	Ea.		105		105	an up flow gas fired hot air furnace
Install	Ea.	1200	281		1481	rated at 120 MBH.
Demolish and Install	Ea.	1200	386		1586	
Reinstall	Ea.		224.98		224.98	
Clean	Ea.	1.63	86		87.63	
Minimum Charge	Job		246		246	
Electric Forced Air						
31.4 MBH						Includes material and labor to install
Demolish	Ea.		105		105	an electric hot air furnace rated at
Install	Ea.	1400	238		1638	31.4 MBH.
Demolish and Install	Ea.	1400	343		1743	
Reinstall	Ea.		190.79		190.79	
Clean	Ea.	1.63	86		87.63	
Minimum Charge	Job		246		246	

Furnace Unit		Unit	Material	Labor	Equip.	Total	Specification
47.1 MBH							
	Demolish	Ea.		105		105	Includes material and labor to install
	Install	Ea.	1725	250		1975	an electric hot air furnace rated at
	Demolish and Install	Ea.	1725	355		2080	47.1 MBH.
	Reinstall	Ea.		199.67		199.67	
	Clean	Ea.	1.63	86		87.63	
	Minimum Charge	Job		246		246	
62.9 MBH							
	Demolish	Ea.		105		105	Includes material and labor to install
	Install	Ea.	1850	268		2118	an electric hot air furnace rated at
	Demolish and Install	Ea.	1850	373		2223	62.9 MBH.
	Reinstall	Ea.		214.64		214.64	
	Clean	Ea.	1.63	86		87.63	
	Minimum Charge	Job		246		246	
78.5 MBH							
	Demolish	Ea.		105		105	Includes material and labor to install
	Install	Ea.	2250	275		2525	an electric hot air furnace rated at
	Demolish and Install	Ea.	2250	380		2630	78.5 MBH.
	Reinstall	Ea.		220.14		220.14	
	Clean	Ea.	1.63	86		87.63	
	Minimum Charge	Job		246		246	
110 MBH							
	Demolish	Ea.		105		105	Includes material and labor to install
	Install	Ea.	3100	290		3390	an electric hot air furnace rated at
	Demolish and Install	Ea.	3100	395		3495	110 MBH.
	Reinstall	Ea.		232.04		232.04	
	Clean	Ea.	1.63	86		87.63	
	Minimum Charge	Job		246		246	
A/C Junction Box							
	Install	Ea.	61.50	115		176.50	Includes labor and materials to install
	Minimum Charge	Job		246		246	a heating system emergency shut-off switch. Cost includes up to 40 feet of non-metallic cable.

Boiler		Unit	Material	Labor	Equip.	Total	Specification
Gas-Fired							
60 MBH							
	Demolish	Ea.		895		895	Includes material and labor to install a
	Install	Ea.	2275	1600		3875	gas fired cast iron boiler rated at 60
	Demolish and Install	Ea.	2275	2495		4770	MBH.
	Reinstall	Ea.		1288.84		1288.84	
	Clean	Ea.	3.27	51.50		54.77	
	Minimum Charge	Job		246		246	
Gas-Fired							
125 MBH							
	Demolish	Ea.		2450		2450	Includes material and labor to install a
	Install	Ea.	4025	1975		6000	gas fired cast iron boiler rated at 125
	Demolish and Install	Ea.	4025	4425		8450	MBH.
	Reinstall	Ea.		1575.25		1575.25	
	Clean	Ea.	3.27	51.50		54.77	
	Minimum Charge	Job		246		246	

Rough Mechanical

Boiler

Boiler	Unit	Material	Labor	Equip.	Total	Specification
Gas-Fired						
400 MBH						
Demolish	Ea.		2450		2450	Includes material and labor to install a
Install	Ea.	5125	2775		7900	gas fired cast iron boiler rated at 400
Demolish and Install	Ea.	5125	5225		10350	MBH.
Reinstall	Ea.		2769		2769	
Clean	Ea.	3.27	51.50		54.77	
Minimum Charge	Job		246		246	

Heat Pump

Heat Pump	Unit	Material	Labor	Equip.	Total	Specification
2 Ton						
Demolish	Ea.		246		246	Includes material and labor to install a
Install	Ea.	2450	445		2895	2 ton heat pump.
Demolish and Install	Ea.	2450	691		3141	
Reinstall	Ea.		354.43		354.43	
Clean	Ea.	3.27	38.50		41.77	
Minimum Charge	Job		246		246	
3 Ton						
Demolish	Ea.		410		410	Includes material and labor to install a
Install	Ea.	2975	740		3715	3 ton heat pump.
Demolish and Install	Ea.	2975	1150		4125	
Reinstall	Ea.		590.72		590.72	
Clean	Ea.	3.27	38.50		41.77	
Minimum Charge	Job		246		246	
4 Ton						
Demolish	Ea.		615		615	Includes material and labor to install a
Install	Ea.	3525	1100		4625	4 ton heat pump.
Demolish and Install	Ea.	3525	1715		5240	
Reinstall	Ea.		886.08		886.08	
Clean	Ea.	3.27	38.50		41.77	
Minimum Charge	Job		246		246	
5 Ton						
Demolish	Ea.		985		985	Includes material and labor to install a
Install	Ea.	3850	1775		5625	5 ton heat pump.
Demolish and Install	Ea.	3850	2760		6610	
Reinstall	Ea.		1417.73		1417.73	
Clean	Ea.	3.27	38.50		41.77	
Minimum Charge	Job		246		246	
Thermostat						
Demolish	Ea.		13.70		13.70	Includes material and labor to install
Install	Ea.	177	61.50		238.50	an electric thermostat, 2 set back.
Demolish and Install	Ea.	177	75.20		252.20	
Reinstall	Ea.		49.24		49.24	
Clean	Ea.	.07	9.70		9.77	
Minimum Charge	Job		246		246	
A/C Junction Box						
Install	Ea.	595	355		950	Includes labor and materials to install
Minimum Charge	Job		230		230	a heat pump circuit including up to 40
						feet of non-metallic cable.

Rough Mechanical

Heating		Unit	Material	Labor	Equip.	Total	Specification
Electric Baseboard							
	Demolish	L.F.		6.40		6.40	Includes material and labor to install
	Install	L.F.	11.20	15.30		26.50	electric baseboard heating elements.
	Demolish and Install	L.F.	11.20	21.70		32.90	
	Reinstall	L.F.		12.25		12.25	
	Clean	L.F.	1.14	4.84		5.98	
	Minimum Charge	Job		230		230	
2' 6"							
	Demolish	Ea.		21.50		21.50	Includes material and labor to install a
	Install	Ea.	33	57.50		90.50	2' 6" long section of electric
	Demolish and Install	Ea.	33	79		112	baseboard heating elements.
	Reinstall	Ea.		45.92		45.92	
	Clean	Ea.	.33	11.90		12.23	
	Minimum Charge	Job		230		230	
3'							
	Demolish	Ea.		23		23	Includes material and labor to install a
	Install	Ea.	50.50	57.50		108	3' long section of electric baseboard
	Demolish and Install	Ea.	50.50	80.50		131	heating elements.
	Reinstall	Ea.		45.92		45.92	
	Clean	Ea.	.33	14.75		15.08	
	Minimum Charge	Job		230		230	
4'							
	Demolish	Ea.		25.50		25.50	Includes material and labor to install a
	Install	Ea.	45	68.50		113.50	4' long section of electric baseboard
	Demolish and Install	Ea.	45	94		139	heating elements.
	Reinstall	Ea.		54.83		54.83	
	Clean	Ea.	.13	3.22		3.35	
	Minimum Charge	Job		230		230	
5'							
	Demolish	Ea.		28.50		28.50	Includes material and labor to install a
	Install	Ea.	56	80.50		136.50	5' long section of electric baseboard
	Demolish and Install	Ea.	56	109		165	heating elements.
	Reinstall	Ea.		64.45		64.45	
	Clean	Ea.	.65	24		24.65	
	Minimum Charge	Job		230		230	
6'							
	Demolish	Ea.		32		32	Includes material and labor to install a
	Install	Ea.	61.50	92		153.50	6' long section of electric baseboard
	Demolish and Install	Ea.	61.50	124		185.50	heating elements.
	Reinstall	Ea.		73.47		73.47	
	Clean	Ea.	.13	3.87		4	
	Minimum Charge	Job		230		230	
8'							
	Demolish	Ea.		42.50		42.50	Includes material and labor to install
	Install	Ea.	76.50	115		191.50	an 8' long section of electric
	Demolish and Install	Ea.	76.50	157.50		234	baseboard heating elements.
	Reinstall	Ea.		91.84		91.84	
	Clean	Ea.	.13	4.84		4.97	
	Minimum Charge	Job		230		230	
10'							
	Demolish	Ea.		51		51	Includes material and labor to install a
	Install	Ea.	212	139		351	10' long section of electric baseboard
	Demolish and Install	Ea.	212	190		402	heating elements.
	Reinstall	Ea.		111.32		111.32	
	Clean	Ea.	1.31	48.50		49.81	
	Minimum Charge	Job		230		230	

Rough Mechanical

Heating

	Unit	Material	Labor	Equip.	Total	Specification
Electric Radiant Wall Panel						Includes material and labor to install
Demolish	Ea.		34		34	an 18" x 23" electric radiant heat wall
Install	Ea.	445	92		537	panel.
Demolish and Install	Ea.	445	126		571	
Reinstall	Ea.		73.47		73.47	
Clean	Ea.	.17	9.70		9.87	
Minimum Charge	Job		230		230	

Heating Control

	Unit	Material	Labor	Equip.	Total	Specification
Thermostat						Includes material and labor to install
Demolish	Ea.		13.70		13.70	an electric thermostat, 2 set back.
Install	Ea.	177	61.50		238.50	
Demolish and Install	Ea.	177	75.20		252.20	
Reinstall	Ea.		49.24		49.24	
Clean	Ea.	.07	9.70		9.77	
Minimum Charge	Job		246		246	

Air Conditioning

	Unit	Material	Labor	Equip.	Total	Specification
Self-Contained						
2 Ton						
Demolish	Ea.		450		450	Includes material and labor to install a
Install	Ea.	2250	765		3015	rooftop air conditioning unit, 2 ton.
Demolish and Install	Ea.	2250	1215		3465	No hoisting is included.
Reinstall	Ea.		764.52		764.52	
Clean	Ea.	3.27	38.50		41.77	
Minimum Charge	Job		246		246	
3 Ton						
Demolish	Ea.		450		450	Includes material and labor to install a
Install	Ea.	2375	1000		3375	rooftop air conditioning unit, 3 ton.
Demolish and Install	Ea.	2375	1450		3825	No hoisting is included.
Reinstall	Ea.		1008.05		1008.05	
Clean	Ea.	3.27	38.50		41.77	
Minimum Charge	Job		246		246	
4 Ton						
Demolish	Ea.		450		450	Includes material and labor to install a
Install	Ea.	2800	1175		3975	rooftop air conditioning unit, 4 ton.
Demolish and Install	Ea.	2800	1625		4425	No hoisting is included.
Reinstall	Ea.		1168.97		1168.97	
Clean	Ea.	3.27	38.50		41.77	
Minimum Charge	Job		246		246	
5 Ton						
Demolish	Ea.		450		450	Includes material and labor to install a
Install	Ea.	2975	1250		4225	rooftop air conditioning unit, 5 ton.
Demolish and Install	Ea.	2975	1700		4675	No hoisting is included.
Reinstall	Ea.		1262.22		1262.22	
Clean	Ea.	3.27	38.50		41.77	
Minimum Charge	Job		246		246	
Compressor						
2 Ton						
Demolish	Ea.		410		410	Includes material and labor to install a
Install	Ea.	284	246		530	refrigeration compressor.
Demolish and Install	Ea.	284	656		940	
Reinstall	Ea.		196.96		196.96	
Clean	Ea.	3.27	38.50		41.77	
Minimum Charge	Job		246		246	

Rough Mechanical

Air Conditioning

	Unit	Material	Labor	Equip.	Total	Specification
2.5 Ton						
Demolish	Ea.		410		410	Includes material and labor to install a
Install	Ea.	335	267		602	refrigeration compressor.
Demolish and Install	Ea.	335	677		1012	
Reinstall	Ea.		213.39		213.39	
Clean	Ea.	3.27	38.50		41.77	
Minimum Charge	Job		246		246	
3 Ton						
Demolish	Ea.		410		410	Includes material and labor to install a
Install	Ea.	370	278		648	refrigeration compressor.
Demolish and Install	Ea.	370	688		1058	
Reinstall	Ea.		222.68		222.68	
Clean	Ea.	3.27	38.50		41.77	
Minimum Charge	Job		246		246	
4 Ton						
Demolish	Ea.		410		410	Includes material and labor to install a
Install	Ea.	470	224		694	refrigeration compressor.
Demolish and Install	Ea.	470	634		1104	
Reinstall	Ea.		179.05		179.05	
Clean	Ea.	3.27	38.50		41.77	
Minimum Charge	Job		246		246	
5 Ton						
Demolish	Ea.		410		410	Includes material and labor to install a
Install	Ea.	515	305		820	refrigeration compressor.
Demolish and Install	Ea.	515	715		1230	
Reinstall	Ea.		243.91		243.91	
Clean	Ea.	3.27	38.50		41.77	
Minimum Charge	Job		246		246	

Condensing Unit

	Unit	Material	Labor	Equip.	Total	Specification
2 Ton Air-cooled						
Demolish	Ea.		117		117	Includes material and labor to install a
Install	Ea.	1550	420		1970	2 ton condensing unit.
Demolish and Install	Ea.	1550	537		2087	
Reinstall	Ea.		337.55		337.55	
Clean	Ea.	3.27	38.50		41.77	
Minimum Charge	Job		246		246	
2.5 Ton Air-cooled						
Demolish	Ea.		137		137	Includes material and labor to install a
Install	Ea.	1625	520		2145	2.5 ton condensing unit.
Demolish and Install	Ea.	1625	657		2282	
Reinstall	Ea.		416.98		416.98	
Clean	Ea.	3.27	38.50		41.77	
Minimum Charge	Job		246		246	
3 Ton Air-cooled						
Demolish	Ea.		164		164	Includes material and labor to install a
Install	Ea.	1625	680		2305	3 ton condensing unit.
Demolish and Install	Ea.	1625	844		2469	
Reinstall	Ea.		545.28		545.28	
Clean	Ea.	3.27	38.50		41.77	
Minimum Charge	Job		246		246	
4 Ton Air-cooled						
Demolish	Ea.		246		246	Includes material and labor to install a
Install	Ea.	2125	985		3110	4 ton condensing unit.
Demolish and Install	Ea.	2125	1231		3356	
Reinstall	Ea.		787.63		787.63	
Clean	Ea.	3.27	38.50		41.77	
Minimum Charge	Job		246		246	

Rough Mechanical

Air Conditioning	Unit	Material	Labor	Equip.	Total	Specification
5 Ton Air-cooled						
Demolish	Ea.		370		370	Includes material and labor to install a
Install	Ea.	2625	1475		4100	5 ton condensing unit.
Demolish and Install	Ea.	2625	1845		4470	
Reinstall	Ea.		1181.44		1181.44	
Clean	Ea.	3.27	38.50		41.77	
Minimum Charge	Job		246		246	
Air Handler						
1,200 CFM						
Demolish	Ea.		296		296	Includes material and labor to install a
Install	Ea.	2900	680		3580	1200 CFM capacity air handling unit.
Demolish and Install	Ea.	2900	976		3876	
Reinstall	Ea.		545.28		545.28	
Clean	Ea.	3.27	38.50		41.77	
Minimum Charge	Job		246		246	
2,000 CFM						
Demolish	Ea.		296		296	Includes material and labor to install a
Install	Ea.	4825	805		5630	2000 CFM capacity air handling unit.
Demolish and Install	Ea.	4825	1101		5926	
Reinstall	Ea.		644.42		644.42	
Clean	Ea.	3.27	38.50		41.77	
Minimum Charge	Job		246		246	
3,000 CFM						
Demolish	Ea.		330		330	Includes material and labor to install a
Install	Ea.	6025	885		6910	3000 CFM capacity air handling unit.
Demolish and Install	Ea.	6025	1215		7240	
Reinstall	Ea.		708.86		708.86	
Clean	Ea.	3.27	38.50		41.77	
Minimum Charge	Job		246		246	
Evaporative Cooler						
4,300 CFM, 1/3 HP						
Demolish	Ea.		120		120	Includes material and labor to install
Install	Ea.	790	264		1054	an evaporative cooler.
Demolish and Install	Ea.	790	384		1174	
Reinstall	Ea.		210.92		210.92	
Clean	Ea.	1.31	38.50		39.81	
Minimum Charge	Job		246		246	
4,700 CFM, 1/2 HP						
Demolish	Ea.		120		120	Includes material and labor to install
Install	Ea.	790	272		1062	an evaporative cooler.
Demolish and Install	Ea.	790	392		1182	
Reinstall	Ea.		217.72		217.72	
Clean	Ea.	1.31	38.50		39.81	
Minimum Charge	Job		246		246	
5,600 CFM, 3/4 HP						
Demolish	Ea.		120		120	Includes material and labor to install
Install	Ea.	1400	281		1681	an evaporative cooler.
Demolish and Install	Ea.	1400	401		1801	
Reinstall	Ea.		224.98		224.98	
Clean	Ea.	1.31	38.50		39.81	
Minimum Charge	Job		246		246	
Master Cool Brand						
Demolish	Ea.		120		120	Includes material and labor to install
Install	Ea.	790	291		1081	an evaporative cooler.
Demolish and Install	Ea.	790	411		1201	
Reinstall	Ea.		232.74		232.74	
Clean	Ea.	1.31	38.50		39.81	
Minimum Charge	Job		246		246	

157

Rough Mechanical

Air Conditioning	Unit	Material	Labor	Equip.	Total	Specification
Motor						
Demolish	Ea.		67.50		67.50	Includes material and labor to install
Install	Ea.	197	205		402	an evaporative cooler motor.
Demolish and Install	Ea.	197	272.50		469.50	
Reinstall	Ea.		164.06		164.06	
Clean	Ea.	1.31	38.50		39.81	
Minimum Charge	Job		246		246	
Water Pump						
Demolish	Ea.		67.50		67.50	Includes material and labor to install
Install	Ea.	55.50	117		172.50	an evaporative cooler water pump.
Demolish and Install	Ea.	55.50	184.50		240	
Reinstall	Ea.		93.76		93.76	
Clean	Ea.	1.31	38.50		39.81	
Minimum Charge	Job		246		246	
Fan Belt						
Demolish	Ea.		13.50		13.50	Includes material and labor to install
Install	Ea.	39.50	23.50		63	an evaporative cooler fan belt.
Demolish and Install	Ea.	39.50	37		76.50	
Reinstall	Ea.		18.75		18.75	
Minimum Charge	Job		246		246	
Pads (4 ea.)						
Demolish	Ea.		13.50		13.50	Includes material and labor to install
Install	Ea.	22.50	58.50		81	pre-fabricated roof pads for an
Demolish and Install	Ea.	22.50	72		94.50	evaporative cooler.
Reinstall	Ea.		46.88		46.88	
Clean	Ea.	1.31	38.50		39.81	
Minimum Charge	Job		246		246	
Service call						
Install	Job		234		234	Includes 4 hour minimum labor charge
						for an sheetmetal worker.
Window Mounted						
Low Output BTU						
Install	Ea.	360	53		413	Includes material and labor to install a
Reinstall	Ea.		42.28		42.28	window air conditioning unit.
Clean	Ea.	.50	9.70		10.20	
Minimum Charge	Job		246		246	
Mid-range Output BTU						
Demolish	Ea.		37		37	Includes material and labor to install a
Install	Ea.	490	70.50		560.50	window air conditioning unit.
Demolish and Install	Ea.	490	107.50		597.50	
Reinstall	Ea.		56.37		56.37	
Clean	Ea.	.50	9.70		10.20	
Minimum Charge	Job		246		246	
High Output BTU						
Demolish	Ea.		47		47	Includes material and labor to install a
Install	Ea.	625	70.50		695.50	window air conditioning unit.
Demolish and Install	Ea.	625	117.50		742.50	
Reinstall	Ea.		56.37		56.37	
Clean	Ea.	.50	9.70		10.20	
Minimum Charge	Job		246		246	
Hotel-type Hot / Cold						
Demolish	Ea.		78.50		78.50	Includes material and labor to install
Install	Ea.	1150	295		1445	packaged air conditioning unit,
Demolish and Install	Ea.	1150	373.50		1523.50	15000 BTU cooling, 13900 BTU
Reinstall	Ea.		236.29		236.29	heating.
Clean	Ea.	.65	19.35		20	
Minimum Charge	Job		246		246	

Rough Mechanical

Air Conditioning		Unit	Material	Labor	Equip.	Total	Specification
Thermostat							
	Demolish	Ea.		13.70		13.70	Includes material and labor to install
	Install	Ea.	177	61.50		238.50	an electric thermostat, 2 set back.
	Demolish and Install	Ea.	177	75.20		252.20	
	Reinstall	Ea.		49.24		49.24	
	Clean	Ea.	.07	9.70		9.77	
	Minimum Charge	Job		246		246	
Concrete Pad							
	Demolish	S.F.		2.15		2.15	Includes material and labor to install
	Install	S.F.	2.48	4.87		7.35	air conditioner pad.
	Demolish and Install	S.F.	2.48	7.02		9.50	
	Reinstall	S.F.		3.89		3.89	
	Clean	S.F.	.07	.62		.69	
	Paint	S.F.	.18	.37		.55	
	Minimum Charge	Job		246		246	
A/C Junction Box							
	Install	Ea.	222	131		353	Includes labor and materials to install
	Minimum Charge	Job		230		230	an air conditioning circuit including up to 40 feet of non-metallic cable.

Sprinkler System		Unit	Material	Labor	Equip.	Total	Specification
Exposed							
Wet Type							
	Demolish	SF Flr.		.22		.22	Includes material and labor to install
	Install	SF Flr.	2.31	2.08		4.39	an exposed wet pipe fire sprinkler
	Demolish and Install	SF Flr.	2.31	2.30		4.61	system.
	Reinstall	SF Flr.		2.09		2.09	
	Minimum Charge	Job		243		243	
Dry Type							
	Demolish	SF Flr.		.22		.22	Includes material and labor to install
	Install	SF Flr.	2.66	2.08		4.74	an exposed dry pipe fire sprinkler
	Demolish and Install	SF Flr.	2.66	2.30		4.96	system.
	Reinstall	SF Flr.		2.09		2.09	
	Minimum Charge	Job		243		243	
Concealed							
Wet Type							
	Demolish	SF Flr.		.43		.43	Includes material and labor to install a
	Install	SF Flr.	2.31	2.85		5.16	concealed wet pipe fire sprinkler
	Demolish and Install	SF Flr.	2.31	3.28		5.59	system.
	Reinstall	SF Flr.		2.85		2.85	
	Minimum Charge	Job		243		243	
Dry Type							
	Demolish	SF Flr.		.43		.43	Includes material and labor to install a
	Install	SF Flr.	.02	2.85		2.87	concealed dry pipe fire sprinkler
	Demolish and Install	SF Flr.	.02	3.28		3.30	system.
	Reinstall	SF Flr.		2.85		2.85	
	Minimum Charge	Job		243		243	

| Junction Boxes | | | Cable | | | Switch Box | |

Electrical Per S.F.	Unit	Material	Labor	Equip.	Total	Specification
Residential						
Wiring, Lighting and Fixtures						
Install	S.F.	1.74	1.91		3.65	Includes labor and material to replace house wiring, lighting fixtures, and built-in appliances. Apply S.F. cost to floor area including garages but not basements.
Minimum Charge	Job		230		230	
Wiring Only						
Install	S.F.	.21	.75		.96	Includes labor and material to replace house wiring only. Apply S.F. cost to floor area including garages but not basements.

Main Service	Unit	Material	Labor	Equip.	Total	Specification
Complete Panel						
100 Amp.						
Demolish	Ea.		189		189	Cost includes material and labor to install 100 Amp. residential service panel including 12 breaker interior panel, 10 single pole breakers and hook-up. Wiring is not included.
Install	Ea.	610	385		995	
Demolish and Install	Ea.	610	574		1184	
Reinstall	Ea.		308.71		308.71	
Clean	Ea.	.33	77.50		77.83	
Minimum Charge	Job		230		230	
150 Amp.						
Demolish	Ea.		189		189	Cost includes material and labor to install 150 Amp. residential service, incl. 20 breaker exterior panel, main switch, 1 GFI and 14 single pole breakers, meter socket including hookup. Wiring is not included.
Install	Ea.	945	445		1390	
Demolish and Install	Ea.	945	634		1579	
Reinstall	Ea.		356.66		356.66	
Clean	Ea.	.33	77.50		77.83	
Minimum Charge	Job		230		230	
200 Amp.						
Demolish	Ea.		189		189	Cost includes material and labor to install 200 Amp. residential service, incl. 20 breaker exterior panel, main switch, 1 GFI and 18 single pole breakers, meter socket including hookup. Wiring is not included.
Install	Ea.	1250	510		1760	
Demolish and Install	Ea.	1250	699		1949	
Reinstall	Ea.		408.18		408.18	
Clean	Ea.	.33	77.50		77.83	
Minimum Charge	Job		230		230	

Rough Electrical

Main Service	Unit	Material	Labor	Equip.	Total	Specification
Masthead						
Demolish	Ea.		32		32	Includes material and labor to install
Install	Ea.	149	191		340	galvanized steel conduit including
Demolish and Install	Ea.	149	223		372	masthead.
Reinstall	Ea.		153.07		153.07	
Minimum Charge	Job		230		230	
Allowance to Repair						
Install	Job		230		230	Minimum labor charge for residential
Minimum Charge	Job		230		230	wiring.

Insulation

Wall Insulation

Floor Insulation		Unit	Material	Labor	Equip.	Total	Specification
Kraft Faced Batt							
3-1/2″ (R-13)							
	Demolish	S.F.		.14		.14	Cost includes material and labor to
	Install	S.F.	.42	.60		1.02	install 3-1/2″ (R-13) kraft paper or foil
	Demolish and Install	S.F.	.42	.74		1.16	faced roll and batt fiberglass
	Minimum Charge	Job		211		211	insulation, joists 16″ O.C.
6″ (R-19)							
	Demolish	S.F.		.14		.14	Cost includes material and labor to
	Install	S.F.	.50	.70		1.20	install 6″ (R-19) kraft paper or foil
	Demolish and Install	S.F.	.50	.84		1.34	faced roll and batt fiberglass
	Minimum Charge	Job		211		211	insulation, joists 16″ O.C.
10″ (R-30)							
	Demolish	S.F.		.14		.14	Cost includes material and labor to
	Install	S.F.	.65	.85		1.50	install 10″ (R-30) kraft paper-faced roll
	Demolish and Install	S.F.	.65	.99		1.64	and batt fiberglass insulation, joists
	Minimum Charge	Job		211		211	16″ O.C.
12″ (R-38)							
	Demolish	S.F.		.14		.14	Cost includes material and labor to
	Install	S.F.	.79	.89		1.68	install 12″ (R-38) kraft paper-faced roll
	Demolish and Install	S.F.	.79	1.03		1.82	and batt fiberglass insulation, joists
	Minimum Charge	Job		211		211	16″ O.C.
Foil Faced Batt							
3-1/2″ (R-13)							
	Demolish	S.F.		.14		.14	Cost includes material and labor to
	Install	S.F.	.42	.60		1.02	install 3-1/2″ (R-13) kraft paper or foil
	Demolish and Install	S.F.	.42	.74		1.16	faced roll and batt fiberglass
	Minimum Charge	Job		211		211	insulation, joists 16″ O.C.
6″ (R-19)							
	Demolish	S.F.		.14		.14	Cost includes material and labor to
	Install	S.F.	.50	.70		1.20	install 6″ (R-19) kraft paper or foil
	Demolish and Install	S.F.	.50	.84		1.34	faced roll and batt fiberglass
	Minimum Charge	Job		211		211	insulation, joists 16″ O.C.
Unfaced Batt							
3-1/2″ (R-13)							
	Demolish	S.F.		.14		.14	Cost includes material and labor to
	Install	S.F.	.33	.70		1.03	install 3-1/2″ (R-13) unfaced roll and
	Demolish and Install	S.F.	.33	.84		1.17	batt fiberglass insulation, joists 16″
	Minimum Charge	Job		211		211	O.C.

Insulation

Floor Insulation

Floor Insulation		Unit	Material	Labor	Equip.	Total	Specification
6" (R-19)							
	Demolish	S.F.		.14		.14	Cost includes material and labor to
	Install	S.F.	.44	.85		1.29	install 6" (R-19) unfaced roll and batt
	Demolish and Install	S.F.	.44	.99		1.43	fiberglass insulation, joists 16" O.C.
	Minimum Charge	Job		211		211	
10" (R-30)							
	Demolish	S.F.		.14		.14	Cost includes material and labor to
	Install	S.F.	.66	.94		1.60	install 10" (R-30) unfaced roll and batt
	Demolish and Install	S.F.	.66	1.08		1.74	fiberglass insulation, joists 16" O.C.
	Minimum Charge	Job		211		211	
12" (R-38)							
	Demolish	S.F.		.14		.14	Cost includes material and labor to
	Install	S.F.	.83	.99		1.82	install 12" (R-38) unfaced roll and batt
	Demolish and Install	S.F.	.83	1.13		1.96	fiberglass insulation, joists 16" O.C.
	Minimum Charge	Job		211		211	
Rigid Board							
1/2"							
	Demolish	S.F.		.14		.14	Includes material and labor to install
	Install	S.F.	.32	.53		.85	rigid foam insulation board foil faced
	Demolish and Install	S.F.	.32	.67		.99	two sides.
	Minimum Charge	Job		211		211	
3/4"							
	Demolish	S.F.		.14		.14	Includes material and labor to install
	Install	S.F.	.36	.53		.89	rigid foam insulation board foil faced
	Demolish and Install	S.F.	.36	.67		1.03	two sides.
	Minimum Charge	Job		211		211	
1-1/2"							
	Demolish	S.F.		.14		.14	Includes material and labor to install
	Install	S.F.	.69	.56		1.25	rigid foam insulation board foil faced
	Demolish and Install	S.F.	.69	.70		1.39	two sides.
	Minimum Charge	Job		211		211	
Support Wire							
	Demolish	S.F.		.06		.06	Includes material and labor to install
	Install	S.F.	.02	.12		.14	friction fit support wire.
	Demolish and Install	S.F.	.02	.18		.20	
	Minimum Charge	Job		211		211	
Vapor Barrier							
	Demolish	S.F.		.08		.08	Cost includes material and labor to
	Install	S.F.	.03	.11		.14	install 4 mil polyethylene vapor barrier
	Demolish and Install	S.F.	.03	.19		.22	10' wide sheets with 6" overlaps.
	Minimum Charge	Job		211		211	

Wall Insulation

Wall Insulation		Unit	Material	Labor	Equip.	Total	Specification
Kraft Faced Batt							
3-1/2" (R-13)							
	Demolish	S.F.		.14		.14	Cost includes material and labor to
	Install	S.F.	.35	.31		.66	install 3-1/2" (R-13) kraft paper-faced
	Demolish and Install	S.F.	.35	.45		.80	roll and batt fiberglass insulation, joist
	Minimum Charge	Job		211		211	or studs 16" O.C.
6" (R-19)							
	Demolish	S.F.		.14		.14	Cost includes material and labor to
	Install	S.F.	.43	.31		.74	install 6" (R-19) kraft paper-faced roll
	Demolish and Install	S.F.	.43	.45		.88	and batt fiberglass insulation, joist or
	Minimum Charge	Job		211		211	studs 16" O.C.

Insulation

Wall Insulation		Unit	Material	Labor	Equip.	Total	Specification
10" (R-30)							
	Demolish	S.F.		.14		.14	Cost includes material and labor to
	Install	S.F.	.65	.37		1.02	install 10" (R-30) kraft paper-faced roll
	Demolish and Install	S.F.	.65	.51		1.16	and batt fiberglass insulation, joist or
	Minimum Charge	Job		211		211	studs 16" O.C.
12" (R-38)							
	Demolish	S.F.		.14		.14	Cost includes material and labor to
	Install	S.F.	.79	.37		1.16	install 12" (R-38) kraft paper-faced roll
	Demolish and Install	S.F.	.79	.51		1.30	and batt fiberglass insulation, joist or
	Minimum Charge	Job		211		211	studs 16" O.C.
Foil Faced Batt							
3-1/2" (R-13)							
	Demolish	S.F.		.14		.14	Cost includes material and labor to
	Install	S.F.	.41	.31		.72	install 3-1/2" (R-13) foil faced roll and
	Demolish and Install	S.F.	.41	.45		.86	batt fiberglass insulation, joist or studs
	Minimum Charge	Job		211		211	16" O.C.
6" (R-19)							
	Demolish	S.F.		.14		.14	Cost includes material and labor to
	Install	S.F.	.48	.31		.79	install 6" (R-19) foil faced roll and batt
	Demolish and Install	S.F.	.48	.45		.93	fiberglass insulation, joist or studs 16"
	Minimum Charge	Job		211		211	O.C.
Unfaced Batt							
3-1/2" (R-13)							
	Demolish	S.F.		.14		.14	Cost includes material and labor to
	Install	S.F.	.33	.31		.64	install 3-1/2" (R-13) unfaced roll and
	Demolish and Install	S.F.	.33	.45		.78	batt fiberglass insulation, joist or studs
	Minimum Charge	Job		211		211	16" O.C.
6" (R-19)							
	Demolish	S.F.		.14		.14	Cost includes material and labor to
	Install	S.F.	.44	.37		.81	install 6" (R-19) unfaced roll and batt
	Demolish and Install	S.F.	.44	.51		.95	fiberglass insulation, joist or studs 16"
	Minimum Charge	Job		211		211	O.C.
10" (R-30)							
	Demolish	S.F.		.14		.14	Cost includes material and labor to
	Install	S.F.	.66	.37		1.03	install 10" (R-30) unfaced roll and batt
	Demolish and Install	S.F.	.66	.51		1.17	fiberglass insulation, joist or studs 16"
	Minimum Charge	Job		211		211	O.C.
12" (R-38)							
	Demolish	S.F.		.14		.14	Cost includes material and labor to
	Install	S.F.	.83	.42		1.25	install 12" (R-38) unfaced roll and batt
	Demolish and Install	S.F.	.83	.56		1.39	fiberglass insulation, joist or studs 16"
	Minimum Charge	Job		211		211	O.C.
Blown							
R-13							
	Demolish	S.F.		.14		.14	Includes material and labor to install
	Install	S.F.	.23	1.39	.68	2.30	fiberglass blown-in insulation in wall
	Demolish and Install	S.F.	.23	1.53	.68	2.44	with wood siding.
	Minimum Charge	Job		480	233	713	
R-19							
	Demolish	S.F.		.14		.14	Includes material and labor to install
	Install	S.F.	.35	1.59	.78	2.72	fiberglass blown-in insulation in wall
	Demolish and Install	S.F.	.35	1.73	.78	2.86	with wood siding.
	Minimum Charge	Job		480	233	713	

Insulation

Wall Insulation		Unit	Material	Labor	Equip.	Total	Specification
Rigid Board							
1/2"							
	Demolish	S.F.		.17		.17	Includes material and labor to install
	Install	S.F.	.32	.53		.85	rigid foam insulation board, foil faced,
	Demolish and Install	S.F.	.32	.70		1.02	two sides.
	Minimum Charge	Job		211		211	
3/4"							
	Demolish	S.F.		.17		.17	Includes material and labor to install
	Install	S.F.	.36	.53		.89	rigid foam insulation board, foil faced,
	Demolish and Install	S.F.	.36	.70		1.06	two sides.
	Minimum Charge	Job		211		211	
1-1/2"							
	Demolish	S.F.		.17		.17	Includes material and labor to install
	Install	S.F.	.69	.58		1.27	rigid foam insulation board, foil faced,
	Demolish and Install	S.F.	.69	.75		1.44	two sides.
	Minimum Charge	Job		211		211	
Vapor Barrier							
	Demolish	S.F.		.05		.05	Cost includes material and labor to
	Install	S.F.	.03	.11		.14	install 4 mil polyethylene vapor barrier
	Demolish and Install	S.F.	.03	.16		.19	10' wide sheets with 6" overlaps.
	Minimum Charge	Job		211		211	

Ceiling Insulation		Unit	Material	Labor	Equip.	Total	Specification
Kraft Faced Batt							
3-1/2" (R-13)							
	Demolish	S.F.		.14		.14	Cost includes material and labor to
	Install	S.F.	.35	.31		.66	install 3-1/2" (R-13) kraft paper-faced
	Demolish and Install	S.F.	.35	.45		.80	roll and batt fiberglass insulation, joist
	Minimum Charge	Job		211		211	or studs 16" O.C.
6" (R-19)							
	Demolish	S.F.		.14		.14	Cost includes material and labor to
	Install	S.F.	.43	.31		.74	install 6" (R-19) kraft paper-faced roll
	Demolish and Install	S.F.	.43	.45		.88	and batt fiberglass insulation, joist or
	Minimum Charge	Job		211		211	studs 16" O.C.
10" (R-30)							
	Demolish	S.F.		.14		.14	Cost includes material and labor to
	Install	S.F.	.65	.37		1.02	install 10" (R-30) kraft paper-faced roll
	Demolish and Install	S.F.	.65	.51		1.16	and batt fiberglass insulation, joist or
	Minimum Charge	Job		211		211	studs 16" O.C.
12" (R-38)							
	Demolish	S.F.		.14		.14	Cost includes material and labor to
	Install	S.F.	.79	.37		1.16	install 12" (R-38) kraft paper-faced roll
	Demolish and Install	S.F.	.79	.51		1.30	and batt fiberglass insulation, joist or
	Minimum Charge	Job		211		211	studs 16" O.C.
Foil Faced Batt							
3-1/2" (R-13)							
	Demolish	S.F.		.14		.14	Cost includes material and labor to
	Install	S.F.	.41	.31		.72	install 3-1/2" (R-13) foil faced roll and
	Demolish and Install	S.F.	.41	.45		.86	batt fiberglass insulation, joist or studs
	Minimum Charge	Job		211		211	16" O.C.

Insulation

Ceiling Insulation		Unit	Material	Labor	Equip.	Total	Specification
6" (R-19)							
	Demolish	S.F.		.14		.14	Cost includes material and labor to
	Install	S.F.	.48	.31		.79	install 6" (R-19) foil faced roll and batt
	Demolish and Install	S.F.	.48	.45		.93	fiberglass insulation, joist or studs 16"
	Minimum Charge	Job		211		211	O.C.
Unfaced Batt							
3-1/2" (R-13)							
	Demolish	S.F.		.14		.14	Cost includes material and labor to
	Install	S.F.	.33	.31		.64	install 3-1/2" (R-13) unfaced roll and
	Demolish and Install	S.F.	.33	.45		.78	batt fiberglass insulation, joist or studs
	Minimum Charge	Job		211		211	16" O.C.
6" (R-19)							
	Demolish	S.F.		.14		.14	Cost includes material and labor to
	Install	S.F.	.44	.37		.81	install 6" (R-19) unfaced roll and batt
	Demolish and Install	S.F.	.44	.51		.95	fiberglass insulation, joist or studs 16"
	Minimum Charge	Job		211		211	O.C.
10" (R-30)							
	Demolish	S.F.		.14		.14	Cost includes material and labor to
	Install	S.F.	.66	.37		1.03	install 10" (R-30) unfaced roll and batt
	Demolish and Install	S.F.	.66	.51		1.17	fiberglass insulation, joist or studs 16"
	Minimum Charge	Job		211		211	O.C.
12" (R-38)							
	Demolish	S.F.		.14		.14	Cost includes material and labor to
	Install	S.F.	.83	.42		1.25	install 12" (R-38) unfaced roll and batt
	Demolish and Install	S.F.	.83	.56		1.39	fiberglass insulation, joist or studs 16"
	Minimum Charge	Job		211		211	O.C.
Blown							
5"							
	Demolish	S.F.		.43		.43	Includes material and labor to install
	Install	S.F.	.17	.25	.12	.54	fiberglass blown-in insulation.
	Demolish and Install	S.F.	.17	.68	.12	.97	
	Minimum Charge	Job		480	233	713	
6"							
	Demolish	S.F.		.43		.43	Includes material and labor to install
	Install	S.F.	.23	.32	.16	.71	fiberglass blown-in insulation.
	Demolish and Install	S.F.	.23	.75	.16	1.14	
	Minimum Charge	Job		480	233	713	
9"							
	Demolish	S.F.		.43		.43	Includes material and labor to install
	Install	S.F.	.29	.43	.21	.93	fiberglass blown-in insulation.
	Demolish and Install	S.F.	.29	.86	.21	1.36	
	Minimum Charge	Job		480	233	713	
12"							
	Demolish	S.F.		.43		.43	Includes material and labor to install
	Install	S.F.	.40	.64	.31	1.35	fiberglass blown-in insulation.
	Demolish and Install	S.F.	.40	1.07	.31	1.78	
	Minimum Charge	Job		480	233	713	
Rigid Board							
1/2"							
	Demolish	S.F.		.24		.24	Includes material and labor to install
	Install	S.F.	.32	.53		.85	rigid foam insulation board, foil faced,
	Demolish and Install	S.F.	.32	.77		1.09	two sides.
	Minimum Charge	Job		211		211	

Insulation

Ceiling Insulation

	Unit	Material	Labor	Equip.	Total	Specification
3/4″						
Demolish	S.F.		.24		.24	Includes material and labor to install
Install	S.F.	.36	.53		.89	rigid foam insulation board, foil faced,
Demolish and Install	S.F.	.36	.77		1.13	two sides.
Minimum Charge	Job		211		211	
1-1/2″						
Demolish	S.F.		.24		.24	Includes material and labor to install
Install	S.F.	.69	.58		1.27	rigid foam insulation board, foil faced,
Demolish and Install	S.F.	.69	.82		1.51	two sides.
Minimum Charge	Job		211		211	

Exterior Insulation

	Unit	Material	Labor	Equip.	Total	Specification
Rigid Board						
1/2″						
Demolish	S.F.		.13		.13	Includes material and labor to install
Install	S.F.	.32	.53		.85	rigid foam insulation board foil faced
Demolish and Install	S.F.	.32	.66		.98	two sides.
Minimum Charge	Job		211		211	
3/4″						
Demolish	S.F.		.13		.13	Includes material and labor to install
Install	S.F.	.36	.53		.89	rigid foam insulation board foil faced
Demolish and Install	S.F.	.36	.66		1.02	two sides.
Minimum Charge	Job		211		211	
1-1/2″						
Demolish	S.F.		.13		.13	Includes material and labor to install
Install	S.F.	.69	.56		1.25	rigid foam insulation board foil faced
Demolish and Install	S.F.	.69	.69		1.38	two sides.
Minimum Charge	Job		211		211	
Vapor Barrier						
Demolish	S.F.		.08		.08	Cost includes material and labor to
Install	S.F.	.03	.11		.14	install 4 mil polyethylene vapor barrier
Demolish and Install	S.F.	.03	.19		.22	10′ wide sheets with 6″ overlaps.
Minimum Charge	Job		211		211	

Roof Insulation Board

	Unit	Material	Labor	Equip.	Total	Specification
Fiberglass						
3/4″ Fiberglass						
Demolish	Sq.		45		45	Includes material and labor to install
Install	Sq.	61.50	24		85.50	fiberglass board insulation 3/4″ thick,
Demolish and Install	Sq.	61.50	69		130.50	R-2.80 and C-0.36 values.
Minimum Charge	Job		211		211	
1″ Fiberglass						
Demolish	Sq.		45		45	Includes material and labor to install
Install	Sq.	87	27		114	fiberglass board insulation 1″ thick,
Demolish and Install	Sq.	87	72		159	R-4.20 and C-0.24 values.
Minimum Charge	Job		211		211	
1-3/8″ Fiberglass						
Demolish	Sq.		45		45	Includes material and labor to install
Install	Sq.	139	29.50		168.50	fiberglass board insulation 1-3/8″
Demolish and Install	Sq.	139	74.50		213.50	thick, R-5.30 and C-0.19 values.
Minimum Charge	Job		211		211	

Insulation

Roof Insulation Board		Unit	Material	Labor	Equip.	Total	Specification
1-5/8″ Fiberglass							
	Demolish	Sq.		45		45	Includes material and labor to install
	Install	Sq.	143	31		174	fiberglass board insulation 1-5/8″
	Demolish and Install	Sq.	143	76		219	thick, R-6.70 and C-0.15 values.
	Minimum Charge	Job		211		211	
2-1/4″ Fiberglass							
	Demolish	Sq.		45		45	Includes material and labor to install
	Install	Sq.	169	34		203	fiberglass board insulation 2-1/4″
	Demolish and Install	Sq.	169	79		248	thick, R-8.30 and C-0.12 values.
	Minimum Charge	Job		211		211	
Perlite							
1″ Perlite							
	Demolish	Sq.		45		45	Includes material and labor to install
	Install	Sq.	53	27		80	perlite board insulation 1″ thick,
	Demolish and Install	Sq.	53	72		125	R-2.80 and C-0.36 values.
	Minimum Charge	Job		211		211	
4″ Perlite							
	Demolish	Sq.		45		45	Includes material and labor to install
	Install	Sq.	211	52		263	perlite board insulation 4″ thick,
	Demolish and Install	Sq.	211	97		308	R-10.0 and C-0.10 values.
	Minimum Charge	Job		211		211	
1-1/2″ Perlite							
	Demolish	Sq.		45		45	Includes material and labor to install
	Install	Sq.	79	29.50		108.50	perlite board insulation 1-1/2″ thick,
	Demolish and Install	Sq.	79	74.50		153.50	R-4.20 and C-0.24 values.
	Minimum Charge	Job		211		211	
2″ Perlite							
	Demolish	Sq.		45		45	Includes material and labor to install
	Install	Sq.	106	36		142	perlite board insulation 2″ thick,
	Demolish and Install	Sq.	106	81		187	R-5.30 and C-0.19 values.
	Minimum Charge	Job		211		211	
2-1/2″ Perlite							
	Demolish	Sq.		45		45	Includes material and labor to install
	Install	Sq.	134	40.50		174.50	perlite board insulation 2-1/2″ thick,
	Demolish and Install	Sq.	134	85.50		219.50	R-6.70 and C-0.15 values.
	Minimum Charge	Job		211		211	
3″ Perlite							
	Demolish	Sq.		45		45	Includes material and labor to install
	Install	Sq.	158	43.50		201.50	perlite board insulation 3″ thick,
	Demolish and Install	Sq.	158	88.50		246.50	R-8.30 and C-0.12 values.
	Minimum Charge	Job		211		211	
Polystyrene							
2″ Polystyrene							
	Demolish	Sq.		45		45	Includes material and labor to install
	Install	Sq.	53	34		87	extruded polystyrene board, R5.0/ 2″
	Demolish and Install	Sq.	53	79		132	in thickness @25PSI compressive
	Minimum Charge	Job		211		211	strength.
1″ Polystyrene							
	Demolish	Sq.		45		45	Includes material and labor to install
	Install	Sq.	26.50	23		49.50	extruded polystyrene board, R5.0/ 1″
	Demolish and Install	Sq.	26.50	68		94.50	in thickness @25PSI compressive
	Minimum Charge	Job		211		211	strength.
1-1/2″ Polystyrene							
	Demolish	Sq.		45		45	Includes material and labor to install
	Install	Sq.	39.50	27		66.50	extruded polystyrene board, R5.0/
	Demolish and Install	Sq.	39.50	72		111.50	1-1/2″ in thickness @25PSI
	Minimum Charge	Job		211		211	compressive strength.

Insulation

Roof Insulation Board		Unit	Material	Labor	Equip.	Total	Specification
Urethane							
3/4″ Urethane							
	Demolish	Sq.		45		45	Includes material and labor to install
	Install	Sq.	48.50	20.50		69	urethane board insulation 3/4″ thick,
	Demolish and Install	Sq.	48.50	65.50		114	R-5.30 and C-0.19 values.
	Minimum Charge	Job		211		211	
1″ Urethane							
	Demolish	Sq.		45		45	Includes material and labor to install
	Install	Sq.	55	21.50		76.50	urethane board insulation 1″ thick,
	Demolish and Install	Sq.	55	66.50		121.50	R-6.70 and C-0.15 values.
	Minimum Charge	Job		211		211	
1-1/2″ Urethane							
	Demolish	Sq.		45		45	Includes material and labor to install
	Install	Sq.	71.50	25		96.50	urethane board insulation 1-1/2″
	Demolish and Install	Sq.	71.50	70		141.50	thick, R-10.1 and C-0.09 values.
	Minimum Charge	Job		211		211	
2″ Urethane							
	Demolish	Sq.		45		45	Includes material and labor to install
	Install	Sq.	87	29.50		116.50	urethane board insulation 2″ thick,
	Demolish and Install	Sq.	87	74.50		161.50	R-13.4 and C-0.06 values.
	Minimum Charge	Job		211		211	

Gypsum Wallboard   **Tile Ceiling**   **Gypsum Wallboard Ceiling**

Wall Plaster	Unit	Material	Labor	Equip.	Total	Specification
Interior						
Cement						
Demolish	S.Y.		5.75		5.75	Includes material and labor to install
Install	S.Y.	14.15	24.50	2.12	40.77	cement plaster on interior wood
Demolish and Install	S.Y.	14.15	30.25	2.12	46.52	framing with metal lath.
Clean	S.Y.		2.81		2.81	
Paint	S.Y.	2.05	4.81		6.86	
Minimum Charge	Job		190		190	
Acoustic						
Demolish	S.Y.		5.75		5.75	Includes material and labor to install
Install	S.Y.	11.15	29	2.11	42.26	acoustical cement plaster on interior
Demolish and Install	S.Y.	11.15	34.75	2.11	48.01	wood framing with metal lath.
Clean	S.Y.		2.81		2.81	
Paint	S.Y.	2.05	4.81		6.86	
Minimum Charge	Job		190		190	
Gypsum						
Demolish	S.Y.		5.75		5.75	Includes material and labor to install
Install	S.Y.	9.65	25	1.80	36.45	gypsum wall or ceiling plaster on
Demolish and Install	S.Y.	9.65	30.75	1.80	42.20	interior wood framing with metal lath.
Clean	S.Y.		2.81		2.81	
Paint	S.Y.	2.05	4.81		6.86	
Minimum Charge	Job		190		190	
Stucco						
Demolish	S.Y.		13.75		13.75	Includes material and labor to install
Install	S.Y.	6.60	34.50	2.48	43.58	stucco on wire mesh over wood
Demolish and Install	S.Y.	6.60	48.25	2.48	57.33	framing.
Clean	S.Y.		2.81		2.81	
Paint	S.Y.	2.05	4.81		6.86	
Minimum Charge	Job		190		190	
Thin Coat						
Veneer						
Demolish	S.Y.		5.75		5.75	Includes materials and labor to install
Install	S.Y.	.96	4.50	.39	5.85	thin coat veneer plaster.
Demolish and Install	S.Y.	.96	10.25	.39	11.60	
Clean	S.Y.		2.81		2.81	
Paint	S.Y.	2.05	4.81		6.86	
Minimum Charge	Job		190		190	

Walls / Ceilings

Wall Plaster

	Unit	Material	Labor	Equip.	Total	Specification
Lath						
Metal						
Demolish	S.Y.		1.35		1.35	Includes material and labor to install
Install	S.Y.	4.91	4.77		9.68	wire lath on steel framed walls.
Demolish and Install	S.Y.	4.91	6.12		11.03	
Clean	S.Y.		2.81		2.81	
Minimum Charge	Job		84		84	
Gypsum						
Demolish	S.Y.		1.35		1.35	Includes material and labor to install
Install	S.Y.	4.95	4.47		9.42	gypsum lath on walls.
Demolish and Install	S.Y.	4.95	5.82		10.77	
Clean	S.Y.		2.81		2.81	
Minimum Charge	Job		190		190	
Repairs						
Fill in Small Cracks						
Install	L.F.	.08	3.04		3.12	Includes labor and material to fill-in
Minimum Charge	Job		190		190	hairline cracks up to 1/8" with filler, sand and prep for paint.
Repair up to 1" Crack						
Install	L.F.	.31	3.80		4.11	Includes labor and material to fill-in
Minimum Charge	Job		190		190	hairline cracks up to 1" with filler, sand and prep for paint.
Patch Small Hole (2")						
Install	Ea.	.08	4		4.08	Includes labor and material to patch
Minimum Charge	Job		190		190	up to a 2" diameter hole in plaster.
Patch Large Hole						
Install	Ea.	9.15	127		136.15	Includes labor and material to patch
Minimum Charge	Job		190		190	up to a 20" diameter hole in plaster.
Patch Plaster Section						
Install	Ea.	2.13	3.80		5.93	Includes labor and material to repair
Minimum Charge	Job		190		190	section of plaster wall, area less than 20 S.F.
Repair Inside Corner						
Install	L.F.	3.64	9.50		13.14	Includes material and labor to patch
Minimum Charge	Job		190		190	plaster corners.

Ceiling Plaster

	Unit	Material	Labor	Equip.	Total	Specification
Interior						
Cement						
Demolish	S.Y.		5.75		5.75	Includes material and labor to install
Install	S.Y.	14.15	24.50	2.12	40.77	cement plaster on interior wood
Demolish and Install	S.Y.	14.15	30.25	2.12	46.52	framing with metal lath.
Clean	S.Y.		2.95		2.95	
Paint	S.Y.	2.05	4.81		6.86	
Minimum Charge	Job		190		190	
Acoustic						
Demolish	S.Y.		5.75		5.75	Includes material and labor to install
Install	S.Y.	11.15	29	2.11	42.26	acoustical cement plaster on interior
Demolish and Install	S.Y.	11.15	34.75	2.11	48.01	wood framing with metal lath.
Clean	S.Y.		2.95		2.95	
Paint	S.Y.	2.05	4.81		6.86	
Minimum Charge	Job		190		190	

Walls / Ceilings

Ceiling Plaster		Unit	Material	Labor	Equip.	Total	Specification
Gypsum							
	Demolish	S.Y.		5.75		5.75	Includes material and labor to install
	Install	S.Y.	9.65	25	1.80	36.45	gypsum wall or ceiling plaster on
	Demolish and Install	S.Y.	9.65	30.75	1.80	42.20	interior wood framing with metal lath.
	Clean	S.Y.		2.95		2.95	
	Paint	S.Y.	2.05	4.81		6.86	
	Minimum Charge	Job		190		190	
Stucco							
	Demolish	S.Y.		13.75		13.75	Includes material and labor to install
	Install	S.Y.	6.60	34.50	2.48	43.58	stucco on wire mesh over wood
	Demolish and Install	S.Y.	6.60	48.25	2.48	57.33	framing.
	Clean	S.Y.		2.95		2.95	
	Paint	S.Y.	2.05	4.81		6.86	
	Minimum Charge	Job		190		190	
Thin Coat							
Veneer							
	Demolish	S.Y.		5.75		5.75	Includes materials and labor to install
	Install	S.Y.	.96	4.50	.39	5.85	thin coat veneer plaster.
	Demolish and Install	S.Y.	.96	10.25	.39	11.60	
	Clean	S.Y.		2.95		2.95	
	Paint	S.Y.	2.05	4.81		6.86	
	Minimum Charge	Job		190		190	
Lath							
Metal							
	Demolish	S.Y.		1.35		1.35	Includes material and labor to install
	Install	S.Y.	4.91	5.95		10.86	wire lath on steel framed ceilings.
	Demolish and Install	S.Y.	4.91	7.30		12.21	
	Minimum Charge	Job		190		190	
Gypsum							
	Demolish	S.Y.		1.35		1.35	Includes material and labor to install
	Install	S.Y.	4.95	3.72		8.67	gypsum lath on ceilings.
	Demolish and Install	S.Y.	4.95	5.07		10.02	
	Minimum Charge	Job		190		190	
Repairs							
Fill in Small Cracks							
	Install	S.F.	.08	.69		.77	Includes labor and material to fill in
	Minimum Charge	Job		380		380	hairline cracks up to 1/8″ with filler,
							sand and prep for paint.

Gypsum Wallboard		Unit	Material	Labor	Equip.	Total	Specification
Finished							
1/4″							
	Demolish	S.F.		.34		.34	Cost includes material and labor to
	Install	S.F.	.24	1.28		1.52	install 1/4″ drywall, including joint
	Demolish and Install	S.F.	.24	1.62		1.86	taping, premixed compound and
	Clean	S.F.		.28		.28	screws.
	Paint	S.F.	.26	.32		.58	
	Minimum Charge	Job		211		211	
3/8″							
	Demolish	S.F.		.34		.34	Cost includes material and labor to
	Install	S.F.	.23	1.29		1.52	install 3/8″ drywall including joint
	Demolish and Install	S.F.	.23	1.63		1.86	tape, premixed compound and
	Clean	S.F.		.28		.28	screws.
	Paint	S.F.	.26	.32		.58	
	Minimum Charge	Job		211		211	

Walls / Ceilings

Gypsum Wallboard		Unit	Material	Labor	Equip.	Total	Specification
1/2"							
	Demolish	S.F.		.34		.34	Cost includes material and labor to
	Install	S.F.	.24	1.30		1.54	install 1/2" drywall including joint
	Demolish and Install	S.F.	.24	1.64		1.88	tape, premixed compound and
	Clean	S.F.		.28		.28	screws.
	Paint	S.F.	.26	.32		.58	
	Minimum Charge	Job		211		211	
1/2" Water Resistant							
	Demolish	S.F.		.34		.34	Cost includes material and labor to
	Install	S.F.	.29	1.11		1.40	install 1/2" WR drywall including joint
	Demolish and Install	S.F.	.29	1.45		1.74	tape, premixed compound and
	Clean	S.F.		.28		.28	screws.
	Paint	S.F.	.26	.32		.58	
	Minimum Charge	Job		211		211	
1/2" Fire-rated							
	Demolish	S.F.		.34		.34	Cost includes material and labor to
	Install	S.F.	.26	1.30		1.56	install 1/2" FR drywall including
	Demolish and Install	S.F.	.26	1.64		1.90	screws. Taping and finishing not
	Clean	S.F.		.28		.28	included.
	Paint	S.F.	.26	.32		.58	
	Minimum Charge	Job		211		211	
5/8"							
	Demolish	S.F.		.34		.34	Cost includes material and labor to
	Install	S.F.	.29	1.26		1.55	install 5/8" drywall including joint
	Demolish and Install	S.F.	.29	1.60		1.89	tape, premixed compound and
	Clean	S.F.		.28		.28	screws.
	Paint	S.F.	.26	.32		.58	
	Minimum Charge	Job		211		211	
5/8" Fire-rated							
	Demolish	S.F.		.34		.34	Cost includes material and labor to
	Install	S.F.	.31	1.48		1.79	install 5/8" FR drywall, including joint
	Demolish and Install	S.F.	.31	1.82		2.13	taping, premixed compound and
	Clean	S.F.		.28		.28	screws.
	Paint	S.F.	.26	.32		.58	
	Minimum Charge	Job		211		211	
2-hour fire-rated							
	Demolish	S.F.		.86		.86	Cost includes material and labor to
	Install	S.F.	.65	2.94		3.59	install 2 layers of 5/8" type X drywall
	Demolish and Install	S.F.	.65	3.80		4.45	including joint taping, premixed
	Clean	S.F.		.28		.28	compound and screws.
	Paint	S.F.	.26	.32		.58	
	Minimum Charge	Job		211		211	
Unfinished							
1/4"							
	Demolish	S.F.		.34		.34	Cost includes material and labor to
	Install	S.F.	.20	.44		.64	install 1/4" drywall including screws.
	Demolish and Install	S.F.	.20	.78		.98	Taping and finishing not included.
	Clean	S.F.		.28		.28	
	Minimum Charge	Job		211		211	
3/8"							
	Demolish	S.F.		.34		.34	Cost includes material and labor to
	Install	S.F.	.19	.45		.64	install 3/8" drywall including screws.
	Demolish and Install	S.F.	.19	.79		.98	Tape, finish or texture is not included.
	Clean	S.F.		.28		.28	
	Minimum Charge	Job		211		211	

Walls / Ceilings

Gypsum Wallboard		Unit	Material	Labor	Equip.	Total	Specification
1/2″							
	Demolish	S.F.		.34		.34	Cost includes material and labor to
	Install	S.F.	.20	.45		.65	install 1/2″ drywall including screws.
	Demolish and Install	S.F.	.20	.79		.99	Taping and finishing not included.
	Clean	S.F.		.28		.28	
	Minimum Charge	Job		211		211	
1/2″ Water Resistant							
	Demolish	S.F.		.34		.34	Cost includes material and labor to
	Install	S.F.	.25	.45		.70	install 1/2″ WR drywall including
	Demolish and Install	S.F.	.25	.79		1.04	screws. Taping and finishing is not
	Clean	S.F.		.28		.28	included.
	Minimum Charge	Job		211		211	
1/2″ Fire-rated							
	Demolish	S.F.		.34		.34	Cost includes labor and material to
	Install	S.F.	.20	.31		.51	install 1/2″ FR drywall including
	Demolish and Install	S.F.	.20	.65		.85	screws. Tape, finish or texture is not
	Clean	S.F.		.28		.28	included.
	Minimum Charge	Job		211		211	
5/8″							
	Demolish	S.F.		.34		.34	Cost includes material and labor to
	Install	S.F.	.25	.50		.75	install 5/8″ drywall including screws.
	Demolish and Install	S.F.	.25	.84		1.09	Taping and finishing not included.
	Clean	S.F.		.28		.28	
	Minimum Charge	Job		211		211	
5/8″ Fire-rated							
	Demolish	S.F.		.34		.34	Cost includes material and labor to
	Install	S.F.	.26	.50		.76	install 5/8″ FR drywall including
	Demolish and Install	S.F.	.26	.84		1.10	screws. Taping and finishing not
	Clean	S.F.		.28		.28	included.
	Minimum Charge	Job		211		211	
2-hour Fire-rated							
	Demolish	S.F.		.86		.86	Cost includes material and labor to
	Install	S.F.	.65	2.94		3.59	install 2 layers of 5/8″ type X drywall
	Demolish and Install	S.F.	.65	3.80		4.45	including joint taping, premixed
	Clean	S.F.		.28		.28	compound and screws.
	Minimum Charge	Job		211		211	
Accessories							
Tape & Finish							
	Install	S.F.	.06	.42		.48	Includes labor and material for taping,
	Minimum Charge	Job		211		211	sanding and finish.
Texture							
Sprayed							
	Install	S.F.	.04	.45		.49	Includes labor and material to install
	Minimum Charge	Job		211		211	by spray, texture finish.
Trowel							
	Install	S.F.		.25		.25	Includes labor and material to apply
	Minimum Charge	Job		211		211	hand troweled texture.
Patch Hole in Gypsum Wallboard							
	Install	S.F.	.43	56.50		56.93	Includes labor and material to patch a
	Minimum Charge	Job		211		211	one square foot area of damaged
							drywall.
Furring Strips							
	Demolish	S.F.		.09		.09	Cost includes material and labor to
	Install	S.F.	.21	.94		1.15	install 1″ x 2″ furring strips, 16″ O.C.
	Demolish and Install	S.F.	.21	1.03		1.24	per S.F. of surface area to be covered.
	Minimum Charge	Job		106		106	

175

Walls / Ceilings

Ceramic Wall Tile	Unit	Material	Labor	Equip.	Total	Specification
Thin Set						
2″ x 2″						
Demolish	S.F.		1.15		1.15	Cost includes material and labor to
Install	S.F.	7.10	6.20		13.30	install 2″ x 2″ ceramic tile in organic
Demolish and Install	S.F.	7.10	7.35		14.45	adhesive including cutting and grout.
Clean	S.F.		.59		.59	
Minimum Charge	Job		186		186	
4-1/4″ x 4-1/4″						
Demolish	S.F.		1.15		1.15	Cost includes material and labor to
Install	S.F.	2.43	3.51		5.94	install 4-1/4″ x 4-1/4″ ceramic tile in
Demolish and Install	S.F.	2.43	4.66		7.09	organic adhesive including grout.
Clean	S.F.		.59		.59	
Minimum Charge	Job		186		186	
6″ x 6″						
Demolish	S.F.		1.15		1.15	Cost includes material and labor to
Install	S.F.	3.31	3.81		7.12	install 6″ x 6″ ceramic tile in organic
Demolish and Install	S.F.	3.31	4.96		8.27	adhesive including and grout.
Clean	S.F.		.59		.59	
Minimum Charge	Job		186		186	
8″ x 8″						
Demolish	S.F.		1.15		1.15	Cost includes material and labor to
Install	S.F.	4.63	3.92		8.55	install 8″ x 8″ ceramic tile in organic
Demolish and Install	S.F.	4.63	5.07		9.70	adhesive including cutting and grout.
Clean	S.F.		.59		.59	
Minimum Charge	Job		186		186	
Thick Set						
2″ x 2″						
Demolish	S.F.		1.38		1.38	Cost includes material and labor to
Install	S.F.	6.95	7.45		14.40	install 2″ x 2″ ceramic tile in mortar
Demolish and Install	S.F.	6.95	8.83		15.78	bed including cutting and grout.
Clean	S.F.		.59		.59	
Minimum Charge	Job		186		186	
4-1/4″ x 4-1/4″						
Demolish	S.F.		1.38		1.38	Cost includes material and labor to
Install	S.F.	8.15	7.45		15.60	install 4-1/4″ x 4-1/4″ ceramic tile in
Demolish and Install	S.F.	8.15	8.83		16.98	mortar bed including cutting and
Clean	S.F.		.59		.59	grout.
Minimum Charge	Job		186		186	
6″ x 6″						
Demolish	S.F.		1.38		1.38	Cost includes material and labor to
Install	S.F.	10.55	6.75		17.30	install 6″ x 6″ ceramic tile in mortar
Demolish and Install	S.F.	10.55	8.13		18.68	bed including cutting and grout.
Clean	S.F.		.59		.59	
Minimum Charge	Job		186		186	
8″ x 8″						
Demolish	S.F.		1.38		1.38	Cost includes material and labor to
Install	S.F.	12.10	6.20		18.30	install 8″ x 8″ ceramic tile in mortar
Demolish and Install	S.F.	12.10	7.58		19.68	bed including cutting and grout.
Clean	S.F.		.59		.59	
Minimum Charge	Job		186		186	
Re-grout						
Install	S.F.	.15	3.72		3.87	Includes material and labor to regrout
Minimum Charge	Job		186		186	ceramic tile walls.

Wall Tile		Unit	Material	Labor	Equip.	Total	Specification
Mirror							
Custom Cut Wall							
	Demolish	S.F.		.66		.66	Includes material and labor to install
	Install	S.F.	2.40	4.17		6.57	distortion-free float glass mirror, 1/8"
	Demolish and Install	S.F.	2.40	4.83		7.23	to 3/16" thick with cut and polished
	Clean	S.F.	.02	.18		.20	edges.
	Minimum Charge	Job		203		203	
Beveled Wall							
	Demolish	S.F.		.66		.66	Includes material and labor to install
	Install	S.F.	6.60	7.40		14	distortion-free float glass mirror, 1/8"
	Demolish and Install	S.F.	6.60	8.06		14.66	to 3/16" thick with beveled edges.
	Clean	S.F.	.02	.18		.20	
	Minimum Charge	Job		203		203	
Smoked Wall							
	Demolish	S.F.		.66		.66	Includes material and labor to install
	Install	S.F.	8.90	7.40		16.30	distortion-free smoked float glass
	Demolish and Install	S.F.	8.90	8.06		16.96	mirror, 1/8" to 3/16" thick with cut
	Clean	S.F.	.02	.18		.20	and polished edges.
	Minimum Charge	Job		203		203	
Marble							
Good Grade							
	Demolish	S.F.		2.15		2.15	Cost includes material and labor to
	Install	S.F.	9.90	16.15		26.05	install 3/8" x 12" x 12" marble tile in
	Demolish and Install	S.F.	9.90	18.30		28.20	mortar bed with grout.
	Clean	S.F.		.59		.59	
	Minimum Charge	Job		186		186	
Better Grade							
	Demolish	S.F.		2.15		2.15	Cost includes material and labor to
	Install	S.F.	11	16.15		27.15	install 3/8" x 12" x 12" marble, in
	Demolish and Install	S.F.	11	18.30		29.30	mortar bed with grout.
	Clean	S.F.		.59		.59	
	Minimum Charge	Job		186		186	
Premium Grade							
	Demolish	S.F.		2.15		2.15	Cost includes material and labor to
	Install	S.F.	21	16.15		37.15	install 3/8" x 12" x 12" marble, in
	Demolish and Install	S.F.	21	18.30		39.30	mortar bed with grout.
	Clean	S.F.		.59		.59	
	Minimum Charge	Job		186		186	
Glass Block							
	Demolish	S.F.		1.72		1.72	Cost includes material and labor to
	Install	S.F.	30.50	16.60		47.10	install 3-7/8" thick, 6" x 6" smooth
	Demolish and Install	S.F.	30.50	18.32		48.82	face glass block including mortar and
	Clean	S.F.		.59		.59	cleaning.
	Minimum Charge	Job		375		375	
Cork							
	Demolish	S.F.		.27		.27	Cost includes material and labor to
	Install	S.F.	6.10	1.51		7.61	install 3/16" thick cork tile and
	Demolish and Install	S.F.	6.10	1.78		7.88	installation.
	Clean	S.F.		.59		.59	
	Minimum Charge	Job		181		181	
Re-grout							
	Install	S.F.	.15	3.72		3.87	Includes material and labor to regrout
	Minimum Charge	Job		186		186	ceramic tile walls.

Walls / Ceilings

Wall Paneling	Unit	Material	Labor	Equip.	Total	Specification
Plywood						
Good Grade						
Demolish	S.F.		.34		.34	Cost includes material and labor to
Install	S.F.	1	1.69		2.69	install 4' x 8' x 1/4" paneling,
Demolish and Install	S.F.	1	2.03		3.03	including nails and adhesive.
Clean	S.F.		.28		.28	
Paint	S.F.	.31	.69		1	
Minimum Charge	Job		211		211	
Better Grade						
Demolish	S.F.		.34		.34	Cost includes material and labor to
Install	S.F.	1.51	2.01		3.52	install 4' x 8' x 1/4" plywood wall
Demolish and Install	S.F.	1.51	2.35		3.86	panel, including nails and adhesive.
Clean	S.F.		.28		.28	
Paint	S.F.	.31	.69		1	
Minimum Charge	Job		211		211	
Premium Grade						
Demolish	S.F.		.34		.34	Cost includes material and labor to
Install	S.F.	2.21	2.42		4.63	install 4' x 8' x 1/4" plywood wall
Demolish and Install	S.F.	2.21	2.76		4.97	panel, including nails and adhesive.
Clean	S.F.		.28		.28	
Paint	S.F.	.31	.69		1	
Minimum Charge	Job		211		211	
Hardboard						
Embossed Paneling						
Demolish	S.F.		.34		.34	Includes material and labor to install
Install	S.F.	.76	1.69		2.45	hardboard embossed paneling
Demolish and Install	S.F.	.76	2.03		2.79	including adhesives.
Clean	S.F.		.28		.28	
Minimum Charge	Job		211		211	
Plastic Molding						
Demolish	L.F.		.34		.34	Includes material and labor to install
Install	L.F.	.75	1.06		1.81	plastic trim for paneling.
Demolish and Install	L.F.	.75	1.40		2.15	
Clean	L.F.	.02	.18		.20	
Minimum Charge	Job		211		211	
Hardboard						
Embossed Brick						
Demolish	S.F.		.34		.34	Cost includes material and labor to
Install	S.F.	2.34	1.88		4.22	install 4' x 8' non-ceramic mineral
Demolish and Install	S.F.	2.34	2.22		4.56	brick-like veneer wall panels including
Clean	S.F.		.28		.28	adhesive.
Paint	S.F.	.23	.54		.77	
Minimum Charge	Job		211		211	
Embossed Ceramic Tile						
Demolish	S.F.		.34		.34	Includes material and labor to install
Install	S.F.	.76	1.69		2.45	hardboard embossed paneling
Demolish and Install	S.F.	.76	2.03		2.79	including adhesives.
Clean	S.F.		.28		.28	
Paint	S.F.	.23	.54		.77	
Minimum Charge	Job		211		211	
Pegboard						
Demolish	S.F.		.34		.34	Cost includes material and labor to
Install	S.F.	.46	1.69		2.15	install 4' x 8' x 1/4" pegboard panel,
Demolish and Install	S.F.	.46	2.03		2.49	including adhesive.
Clean	S.F.		.28		.28	
Paint	S.F.	.23	.54		.77	
Minimum Charge	Job		211		211	

Wall Paneling	Unit	Material	Labor	Equip.	Total	Specification
Furring Strips						
Demolish	S.F.		.09		.09	Cost includes material and labor to
Install	S.F.	.21	.94		1.15	install 1″ x 2″ furring strips, 16″ O.C.
Demolish and Install	S.F.	.21	1.03		1.24	per S.F. of surface area to be covered.
Minimum Charge	Job		211		211	
Molding						
Unfinished Base						
Demolish	L.F.		.57		.57	Cost includes material and labor to
Install	L.F.	.69	1.28		1.97	install 1/2″ x 2-1/2″ pine or birch
Demolish and Install	L.F.	.69	1.85		2.54	base molding.
Reinstall	L.F.		1.02		1.02	
Clean	L.F.	.02	.18		.20	
Paint	L.F.	.06	.39		.45	
Minimum Charge	Job		211		211	
Unfinished Corner						
Demolish	L.F.		.57		.57	Includes material and labor to install
Install	L.F.	2.35	1.76		4.11	unfinished pine corner molding.
Demolish and Install	L.F.	2.35	2.33		4.68	
Reinstall	L.F.		1.41		1.41	
Clean	Ea.	.81	28.50		29.31	
Paint	L.F.	.14	.58		.72	
Minimum Charge	Job		211		211	
Unfinished Cove						
Demolish	L.F.		.57		.57	Cost includes material and labor to
Install	L.F.	2.20	1.57		3.77	install 3/4″ x 3-1/2″ unfinished pine
Demolish and Install	L.F.	2.20	2.14		4.34	cove molding.
Reinstall	L.F.		1.25		1.25	
Clean	Ea.	.81	28.50		29.31	
Paint	L.F.	.14	.58		.72	
Minimum Charge	Job		211		211	
Unfinished Crown						
Demolish	L.F.		.69		.69	Includes material and labor to install
Install	L.F.	2.95	1.69		4.64	unfinished pine crown molding.
Demolish and Install	L.F.	2.95	2.38		5.33	
Reinstall	L.F.		1.35		1.35	
Clean	L.F.	.03	.24		.27	
Paint	L.F.	.10	.39		.49	
Minimum Charge	Job		211		211	
Prefinished Base						
Demolish	L.F.		.57		.57	Cost includes material and labor to
Install	L.F.	1.41	1.36		2.77	install 9/16″ x 3-5/16″ prefinished
Demolish and Install	L.F.	1.41	1.93		3.34	base molding.
Reinstall	L.F.		1.08		1.08	
Clean	L.F.	.02	.18		.20	
Paint	L.F.	.06	.39		.45	
Minimum Charge	Job		211		211	
Prefinished Corner						
Demolish	L.F.		.57		.57	Includes material and labor to install
Install	L.F.	2.64	1.76		4.40	inside or outside prefinished corner
Demolish and Install	L.F.	2.64	2.33		4.97	molding.
Reinstall	L.F.		1.41		1.41	
Clean	L.F.		.12		.12	
Paint	L.F.	.14	.58		.72	
Minimum Charge	Job		211		211	

Walls / Ceilings

Wall Paneling

Wall Paneling		Unit	Material	Labor	Equip.	Total	Specification
Prefinished Crown							
	Demolish	L.F.		.69		.69	Cost includes material and labor to
	Install	L.F.	1.65	1.69		3.34	install 3/4" x 3-13/16" prefinished
	Demolish and Install	L.F.	1.65	2.38		4.03	crown molding.
	Reinstall	L.F.		1.35		1.35	
	Clean	L.F.	.03	.24		.27	
	Paint	L.F.	.10	.39		.49	
	Minimum Charge	Job		211		211	

Wallcovering

Wallcovering		Unit	Material	Labor	Equip.	Total	Specification
Surface Prep							
Patch Walls							
	Install	S.F.	.11	3.80		3.91	Includes labor and material to repair
	Minimum Charge	Job		181		181	section of plaster wall, area less than 20 S.F.
Sizing							
	Install	S.F.	.08	.18		.26	Includes labor and materials to apply
	Minimum Charge	Job		181		181	wall sizing.
Good Grade							
Walls							
	Demolish	S.F.		.67		.67	Includes material and labor to install
	Install	S.F.	1.02	.75		1.77	pre-pasted, vinyl-coated wallcovering.
	Demolish and Install	S.F.	1.02	1.42		2.44	
	Clean	S.F.	.01	.07		.08	
	Minimum Charge	Job		181		181	
Ceiling							
	Demolish	S.F.		.67		.67	Includes material and labor to install
	Install	S.F.	1.02	.75		1.77	pre-pasted, vinyl-coated wallcovering.
	Demolish and Install	S.F.	1.02	1.42		2.44	
	Clean	S.F.	.02	.18		.20	
	Minimum Charge	Job		181		181	
Better Grade							
Walls							
	Demolish	S.F.		.67		.67	Includes material and labor to install
	Install	S.F.	1.14	.75		1.89	pre-pasted, vinyl-coated wallcovering.
	Demolish and Install	S.F.	1.14	1.42		2.56	
	Clean	S.F.	.01	.07		.08	
	Minimum Charge	Job		181		181	
Ceiling							
	Demolish	S.F.		.67		.67	Includes material and labor to install
	Install	S.F.	1.14	.75		1.89	pre-pasted, vinyl-coated wallcovering.
	Demolish and Install	S.F.	1.14	1.42		2.56	
	Clean	S.F.	.02	.18		.20	
	Minimum Charge	Job		181		181	
Premium Grade							
Walls							
	Demolish	S.F.		.84		.84	Includes material and labor to install
	Install	S.F.	1.23	.75		1.98	pre-pasted, vinyl-coated wallcovering.
	Demolish and Install	S.F.	1.23	1.59		2.82	
	Clean	S.F.	.01	.07		.08	
	Minimum Charge	Job		181		181	
Ceiling							
	Demolish	S.F.		.84		.84	Includes material and labor to install
	Install	S.F.	1.23	.75		1.98	pre-pasted, vinyl-coated wallcovering.
	Demolish and Install	S.F.	1.23	1.59		2.82	
	Clean	S.F.	.02	.18		.20	
	Minimum Charge	Job		181		181	

Walls / Ceilings

Wallcovering		Unit	Material	Labor	Equip.	Total	Specification
Mural							
Pre-printed							
	Demolish	Ea.		108		108	Includes material and labor to install
	Install	Ea.	276	272		548	wallcovering with a pre-printed mural.
	Demolish and Install	Ea.	276	380		656	
	Clean	Ea.		7.75		7.75	
	Minimum Charge	Job		181		181	
Hand-painted							
	Demolish	Ea.		108		108	Includes material and labor to install
	Install	Ea.	535	550		1085	wallcovering with a hand-painted
	Demolish and Install	Ea.	535	658		1193	mural.
	Clean	Ea.		7.75		7.75	
	Minimum Charge	Job		181		181	
Vinyl-coated							
Good Grade							
	Demolish	Ea.		19.10		19.10	Includes material and labor to install
	Install	Ea.	16.95	25.50		42.45	pre-pasted, vinyl-coated wallcovering.
	Demolish and Install	Ea.	16.95	44.60		61.55	Based upon 30 S.F. rolls.
	Clean	Ea.		7.75		7.75	
	Minimum Charge	Job		181		181	
Better Grade							
	Demolish	Ea.		19.10		19.10	Includes material and labor to install
	Install	Ea.	26.50	25.50		52	pre-pasted, vinyl-coated wallcovering.
	Demolish and Install	Ea.	26.50	44.60		71.10	Based upon 30 S.F. rolls.
	Clean	Ea.		7.75		7.75	
	Minimum Charge	Job		181		181	
Premium Grade							
	Demolish	Ea.		19.10		19.10	Includes material and labor to install
	Install	Ea.	31.50	25.50		57	pre-pasted, vinyl-coated wallcovering.
	Demolish and Install	Ea.	31.50	44.60		76.10	Based upon 30 S.F. rolls.
	Clean	Ea.		7.75		7.75	
	Minimum Charge	Job		181		181	
Blank Underliner							
	Demolish	Ea.		19.10		19.10	Includes material and labor to install
	Install	Ea.	9.90	26		35.90	blank stock liner paper or bridging
	Demolish and Install	Ea.	9.90	45.10		55	paper. Based upon 30 S.F. rolls.
	Clean	Ea.		7.75		7.75	
	Minimum Charge	Job		181		181	
Solid Vinyl							
Good Grade							
	Demolish	Roll		22.50		22.50	Includes material and labor to install
	Install	Ea.	24	25.50		49.50	pre-pasted, vinyl wallcovering. Based
	Demolish and Install	Ea.	24	48		72	upon 30 S.F. rolls.
	Clean	Ea.		7.75		7.75	
	Minimum Charge	Job		181		181	
Better Grade							
	Demolish	Roll		22.50		22.50	Includes material and labor to install
	Install	Ea.	31.50	25.50		57	pre-pasted, vinyl wallcovering. Based
	Demolish and Install	Ea.	31.50	48		79.50	upon 30 S.F. rolls.
	Clean	Ea.		7.75		7.75	
	Minimum Charge	Job		181		181	
Premium Grade							
	Demolish	Roll		22.50		22.50	Includes material and labor to install
	Install	Ea.	50.50	25.50		76	pre-pasted, vinyl wallcovering. Based
	Demolish and Install	Ea.	50.50	48		98.50	upon 30 S.F. rolls.
	Clean	Ea.		7.75		7.75	
	Minimum Charge	Job		181		181	

Walls / Ceilings

Wallcovering		Unit	Material	Labor	Equip.	Total	Specification
Blank Underliner							
	Demolish	Ea.		19.10		19.10	Includes material and labor to install
	Install	Ea.	9.90	26		35.90	blank stock liner paper or bridging
	Demolish and Install	Ea.	9.90	45.10		55	paper. Based upon 30 S.F. rolls.
	Clean	Ea.		7.75		7.75	
	Minimum Charge	Job		181		181	
Grasscloth / String							
Good Grade							
	Demolish	Ea.		19.10		19.10	Includes material and labor to install
	Install	Ea.	53.50	33		86.50	grasscloth or string wallcovering.
	Demolish and Install	Ea.	53.50	52.10		105.60	Based upon 30 S.F. rolls.
	Clean	Ea.		10.30		10.30	
	Minimum Charge	Job		181		181	
Better Grade							
	Demolish	Ea.		19.10		19.10	Includes material and labor to install
	Install	Ea.	60.50	33		93.50	grasscloth or string wallcovering.
	Demolish and Install	Ea.	60.50	52.10		112.60	Based upon 30 S.F. rolls.
	Clean	Ea.		10.30		10.30	
	Minimum Charge	Job		181		181	
Premium Grade							
	Demolish	Ea.		19.10		19.10	Includes material and labor to install
	Install	Ea.	67.50	33		100.50	grasscloth or string wallcovering.
	Demolish and Install	Ea.	67.50	52.10		119.60	Based upon 30 S.F. rolls.
	Clean	Ea.		10.30		10.30	
	Minimum Charge	Job		181		181	
Blank Underliner							
	Demolish	Ea.		19.10		19.10	Includes material and labor to install
	Install	Ea.	9.90	26		35.90	blank stock liner paper or bridging
	Demolish and Install	Ea.	9.90	45.10		55	paper. Based upon 30 S.F. rolls.
	Clean	Ea.		7.75		7.75	
	Minimum Charge	Job		181		181	
Designer							
Premium Grade							
	Demolish	S.F.		.84		.84	Includes material and labor to install
	Install	Ea.	113	38		151	paperback vinyl, untrimmed and
	Demolish and Install	Ea.	113	38.84		151.84	hand-painted. Based upon 30 S.F.
	Clean	Ea.		7.75		7.75	rolls.
	Minimum Charge	Job		181		181	
Blank Underliner							
	Demolish	Ea.		19.10		19.10	Includes material and labor to install
	Install	Ea.	9.90	26		35.90	blank stock liner paper or bridging
	Demolish and Install	Ea.	9.90	45.10		55	paper. Based upon 30 S.F. rolls.
	Clean	Ea.		7.75		7.75	
	Minimum Charge	Job		181		181	
Foil							
	Demolish	Ea.		19.10		19.10	Includes material and labor to install
	Install	Ea.	28.50	33		61.50	foil coated wallcovering. Based upon
	Demolish and Install	Ea.	28.50	52.10		80.60	30 S.F. rolls.
	Clean	Ea.		7.75		7.75	
	Minimum Charge	Job		181		181	
Wallcovering Border							
Good Grade							
	Demolish	L.F.		1.25		1.25	Includes material and labor install
	Install	L.F.	1.65	1.39		3.04	vinyl-coated border 4" to 6" wide.
	Demolish and Install	L.F.	1.65	2.64		4.29	
	Clean	L.F.		.12		.12	
	Minimum Charge	Job		181		181	

Walls / Ceilings

Wallcovering

Wallcovering		Unit	Material	Labor	Equip.	Total	Specification
Better Grade							
	Demolish	L.F.		1.25		1.25	Includes material and labor install
	Install	L.F.	2.97	1.39		4.36	vinyl-coated border 4" to 6" wide.
	Demolish and Install	L.F.	2.97	2.64		5.61	
	Clean	L.F.		.12		.12	
	Minimum Charge	Job		181		181	
Premium Grade							
	Demolish	L.F.		1.25		1.25	Includes material and labor install
	Install	L.F.	3.69	1.39		5.08	vinyl-coated border 4" to 6" wide.
	Demolish and Install	L.F.	3.69	2.64		6.33	
	Clean	L.F.		.12		.12	
	Minimum Charge	Job		181		181	

Acoustical Ceiling

Acoustical Ceiling		Unit	Material	Labor	Equip.	Total	Specification
Complete System							
2' x 2' Square Edge							
	Demolish	S.F.		.57		.57	Includes material and labor to install
	Install	S.F.	1.87	1.30		3.17	T-bar grid acoustical ceiling system,
	Demolish and Install	S.F.	1.87	1.87		3.74	random pinhole 5/8" thick square
	Reinstall	S.F.		1.04		1.04	edge tile.
	Clean	S.F.	.03	.23		.26	
	Minimum Charge	Job		211		211	
2' x 2' Reveal Edge							
	Demolish	S.F.		.57		.57	Includes material and labor to install
	Install	S.F.	2.86	1.69		4.55	T-bar grid acoustical ceiling system,
	Demolish and Install	S.F.	2.86	2.26		5.12	random pinhole 5/8" thick reveal
	Reinstall	S.F.		1.35		1.35	edge tile.
	Clean	S.F.	.03	.23		.26	
	Minimum Charge	Job		211		211	
2' x 2' Fire-rated							
	Demolish	S.F.		.57		.57	Includes material and labor to install
	Install	S.F.	3.21	1.30		4.51	T-bar grid FR acoustical ceiling system,
	Demolish and Install	S.F.	3.21	1.87		5.08	random pinhole 5/8" thick square
	Reinstall	S.F.		1.04		1.04	edge tile.
	Clean	S.F.	.03	.23		.26	
	Minimum Charge	Job		211		211	
2' x 4' Square Edge							
	Demolish	S.F.		.57		.57	Cost includes material and labor to
	Install	S.F.	1.57	1.11		2.68	install 2' x 4' T-bar grid acoustical
	Demolish and Install	S.F.	1.57	1.68		3.25	ceiling system, random pinhole 5/8"
	Reinstall	S.F.		.89		.89	thick square edge tile.
	Clean	S.F.	.03	.23		.26	
	Minimum Charge	Job		211		211	
2' x 4' Reveal Edge							
	Demolish	S.F.		.57		.57	Includes material and labor to install
	Install	S.F.	2.86	1.54		4.40	T-bar grid acoustical ceiling system,
	Demolish and Install	S.F.	2.86	2.11		4.97	random pinhole 5/8" thick reveal
	Reinstall	S.F.		1.23		1.23	edge tile.
	Clean	S.F.	.03	.23		.26	
	Minimum Charge	Job		211		211	
2' x 4' Fire-rated							
	Demolish	S.F.		.57		.57	Includes material and labor to install
	Install	S.F.	3.21	1.30		4.51	T-bar grid FR acoustical ceiling system,
	Demolish and Install	S.F.	3.21	1.87		5.08	random pinhole 5/8" thick square
	Reinstall	S.F.		1.04		1.04	edge tile.
	Clean	S.F.	.03	.23		.26	
	Minimum Charge	Job		211		211	

Acoustical Ceiling	Unit	Material	Labor	Equip.	Total	Specification
T-bar Grid						
2' x 2'						
Demolish	S.F.		.67		.67	Includes material and labor to install
Install	S.F.	.95	.65		1.60	T-bar grid acoustical ceiling
Demolish and Install	S.F.	.95	1.32		2.27	suspension system.
Reinstall	S.F.		.52		.52	
Clean	S.F.	.03	.23		.26	
Minimum Charge	Job		190		190	
2' x 4'						
Demolish	S.F.		.67		.67	Includes material and labor to install
Install	S.F.	.74	.53		1.27	T-bar grid acoustical ceiling
Demolish and Install	S.F.	.74	1.20		1.94	suspension system.
Reinstall	S.F.		.42		.42	
Clean	S.F.	.03	.23		.26	
Minimum Charge	Job		190		190	
Tile Panels						
2' x 2' Square Edge						
Demolish	S.F.		.19		.19	Includes material and labor to install
Install	S.F.	.84	.68		1.52	square edge panels 5/8" thick with
Demolish and Install	S.F.	.84	.87		1.71	pinhole pattern. T-bar grid not
Reinstall	S.F.		.54		.54	included.
Clean	S.F.	.03	.23		.26	
Minimum Charge	Job		211		211	
2' x 2' Reveal Edge						
Demolish	S.F.		.19		.19	Includes material and labor to install
Install	S.F.	1.74	.90		2.64	reveal edge panels 5/8" thick with
Demolish and Install	S.F.	1.74	1.09		2.83	pinhole pattern. T-bar grid not
Reinstall	S.F.		.72		.72	included.
Clean	S.F.	.03	.23		.26	
Minimum Charge	Job		211		211	
2' x 2' Fire-rated						
Demolish	S.F.		.19		.19	Includes material and labor to install
Install	S.F.	1.27	.63		1.90	FR square edge panels 5/8" thick with
Demolish and Install	S.F.	1.27	.82		2.09	pinhole pattern. T-bar grid not
Reinstall	S.F.		.50		.50	included.
Clean	S.F.	.03	.23		.26	
Minimum Charge	Job		211		211	
2' x 4' Square Edge						
Demolish	S.F.		.19		.19	Includes material and labor to install
Install	S.F.	.84	.68		1.52	square edge panels 5/8" thick with
Demolish and Install	S.F.	.84	.87		1.71	pinhole pattern. T-bar grid not
Reinstall	S.F.		.54		.54	included.
Clean	S.F.	.03	.23		.26	
Minimum Charge	Job		211		211	
2' x 4' Reveal Edge						
Demolish	S.F.		.19		.19	Includes material and labor to install
Install	S.F.	1.74	.90		2.64	reveal edge panels 5/8" thick with
Demolish and Install	S.F.	1.74	1.09		2.83	pinhole pattern. T-bar grid not
Reinstall	S.F.		.72		.72	included.
Clean	S.F.	.03	.23		.26	
Minimum Charge	Job		211		211	
2' x 4' Fire-rated						
Demolish	S.F.		.19		.19	Includes material and labor to install
Install	S.F.	1.27	.63		1.90	FR square edge panels 5/8" thick with
Demolish and Install	S.F.	1.27	.82		2.09	pinhole pattern. T-bar grid not
Reinstall	S.F.		.50		.50	included.
Clean	S.F.	.03	.23		.26	
Minimum Charge	Job		211		211	

Walls / Ceilings

Acoustical Ceiling		Unit	Material	Labor	Equip.	Total	Specification
Bleach Out Stains							
	Install	S.F.	.14	.31		.45	Includes labor and material to apply
	Minimum Charge	Job		211		211	laundry type chlorine bleach to remove light staining of acoustic tile material.
Adhesive Tile							
Good Grade							
	Demolish	S.F.		.76		.76	Cost includes material and labor to
	Install	S.F.	1.16	1.41		2.57	install 12" x 12" tile with embossed
	Demolish and Install	S.F.	1.16	2.17		3.33	texture pattern. Furring strips not
	Clean	S.F.	.02	.18		.20	included.
	Minimum Charge	Job		211		211	
Better Grade							
	Demolish	S.F.		.76		.76	Cost includes material and labor to
	Install	S.F.	1.65	1.41		3.06	install 12" x 12" tile with embossed
	Demolish and Install	S.F.	1.65	2.17		3.82	texture pattern. Furring strips not
	Clean	S.F.	.02	.18		.20	included.
	Minimum Charge	Job		211		211	
Premium Grade							
	Demolish	S.F.		.76		.76	Cost includes material and labor to
	Install	S.F.	5.45	1.41		6.86	install 12" x 12" tile with embossed
	Demolish and Install	S.F.	5.45	2.17		7.62	texture pattern. Furring strips not
	Clean	S.F.	.02	.18		.20	included.
	Minimum Charge	Job		211		211	
Bleach Out Stains							
	Install	S.F.	.14	.31		.45	Includes labor and material to apply
	Minimum Charge	Job		211		211	laundry type chlorine bleach to remove light staining of acoustic tile material.
Stapled Tile							
Good Grade							
	Demolish	S.F.		.46		.46	Cost includes material and labor to
	Install	S.F.	1.16	1.41		2.57	install 12" x 12" tile with embossed
	Demolish and Install	S.F.	1.16	1.87		3.03	texture pattern. Furring strips not
	Clean	S.F.	.02	.18		.20	included.
	Minimum Charge	Job		211		211	
Better Grade							
	Demolish	S.F.		.46		.46	Cost includes material and labor to
	Install	S.F.	1.65	1.41		3.06	install 12" x 12" tile with embossed
	Demolish and Install	S.F.	1.65	1.87		3.52	texture pattern. Furring strips not
	Clean	S.F.	.02	.18		.20	included.
	Minimum Charge	Job		211		211	
Premium Grade							
	Demolish	S.F.		.46		.46	Cost includes material and labor to
	Install	S.F.	5.45	1.41		6.86	install 12" x 12" tile with embossed
	Demolish and Install	S.F.	5.45	1.87		7.32	texture pattern. Furring strips not
	Clean	S.F.	.02	.18		.20	included.
	Minimum Charge	Job		211		211	
Bleach Out Stains							
	Install	S.F.	.14	.31		.45	Includes labor and material to apply
	Minimum Charge	Job		211		211	laundry type chlorine bleach to remove light staining of acoustic tile material.

Walls / Ceilings

Acoustical Ceiling	Unit	Material	Labor	Equip.	Total	Specification
Blown						
Scrape Off Existing						
Install	S.F.		.64		.64	Includes labor and material to remove
Minimum Charge	Job		211		211	existing blown acoustical ceiling material.
Seal Ceiling Gypsum Wallboard						
Install	S.F.	.07	.31		.38	Includes labor and material to seal
Minimum Charge	Job		180		180	existing drywall with shellac-based material.
Reblow Existing						
Install	S.F.	.04	.45		.49	Includes labor and material to install
Minimum Charge	Job		510	72	582	by spray, texture finish.
Bleach Out Stains						
Install	S.F.	.14	.31		.45	Includes labor and material to apply
Minimum Charge	Job		211		211	laundry type chlorine bleach to remove light staining of acoustic tile material.
New Blown						
Install	S.F.	.04	.45		.49	Includes labor and material to install
Minimum Charge	Job		180		180	by spray, texture finish.
Apply Glitter To						
Install	S.F.	.22	.16		.38	Includes labor and material to apply
Minimum Charge	Job		180		180	glitter material to existing or new drywall.
Furring Strips						
Demolish	S.F.		.09		.09	Cost includes material and labor to
Install	S.F.	.21	.94		1.15	install 1" x 2" furring strips, 16" O.C.
Demolish and Install	S.F.	.21	1.03		1.24	per S.F. of surface area to be covered.
Minimum Charge	Job		106		106	

Finished Stair

Bi-Fold Door

Raised Panel Door

Hollow Core Door	Unit	Material	Labor	Equip.	Total	Specification
Pre-hung Wood						
1' 6" x 6' 8"						
Demolish	Ea.		39		39	Cost includes material and labor to
Install	Ea.	151	23.50		174.50	install 1' 6" x 6' 8" pre-hung 1-3/8"
Demolish and Install	Ea.	151	62.50		213.50	lauan door with split pine jamb
Reinstall	Ea.		23.49		23.49	including casing, stops, and hinges.
Clean	Ea.	3.27	12.90		16.17	
Paint	Ea.	10.10	51.50		61.60	
Minimum Charge	Job		211		211	
2' x 6' 8"						
Demolish	Ea.		39		39	Cost includes material and labor to
Install	Ea.	165	25		190	install 2' x 6' 8" pre-hung 1-3/8"
Demolish and Install	Ea.	165	64		229	lauan door with split pine jamb
Reinstall	Ea.		24.87		24.87	including casing, stops, and hinges.
Clean	Ea.	3.27	12.90		16.17	
Paint	Ea.	10.10	51.50		61.60	
Minimum Charge	Job		211		211	
2' 4" x 6' 8"						
Demolish	Ea.		39		39	Cost includes material and labor to
Install	Ea.	169	25		194	install 2' 4" x 6' 8" pre-hung 1-3/8"
Demolish and Install	Ea.	169	64		233	lauan door with split pine jamb
Reinstall	Ea.		24.87		24.87	including casing, stops, and hinges.
Clean	Ea.	3.27	12.90		16.17	
Paint	Ea.	10.10	51.50		61.60	
Minimum Charge	Job		211		211	
2' 6" x 6' 8"						
Demolish	Ea.		39		39	Cost includes material and labor to
Install	Ea.	174	25		199	install 2' 6" x 6' 8" pre-hung 1-3/8"
Demolish and Install	Ea.	174	64		238	lauan door with split pine jamb
Reinstall	Ea.		24.87		24.87	including casing, stops, and hinges.
Clean	Ea.	3.27	12.90		16.17	
Paint	Ea.	10.10	51.50		61.60	
Minimum Charge	Job		211		211	
2' 8" x 6' 8"						
Demolish	Ea.		39		39	Cost includes material and labor to
Install	Ea.	168	25		193	install 2' 8" x 6' 8" pre-hung 1-3/8"
Demolish and Install	Ea.	168	64		232	lauan door with split pine jamb,
Reinstall	Ea.		24.87		24.87	including casing, stop and hinges.
Clean	Ea.	3.27	12.90		16.17	
Paint	Ea.	10.10	51.50		61.60	
Minimum Charge	Job		211		211	

Hollow Core Door

	Unit	Material	Labor	Equip.	Total	Specification
3' x 6' 8"						
Demolish	Ea.		39		39	Cost includes material and labor to
Install	Ea.	174	26.50		200.50	install 3' x 6' 8" pre-hung 1-3/8"
Demolish and Install	Ea.	174	65.50		239.50	lauan door with split pine jamb
Reinstall	Ea.		26.43		26.43	including casing, stops, and hinges.
Clean	Ea.	3.27	12.90		16.17	
Paint	Ea.	10.10	51.50		61.60	
Minimum Charge	Job		211		211	
Pocket Type						
Demolish	Ea.		58.50		58.50	Cost includes material and labor to
Install	Ea.	44	47		91	install 2' 8" x 6' 8" lauan door.
Demolish and Install	Ea.	44	105.50		149.50	
Reinstall	Ea.		46.98		46.98	
Clean	Ea.	3.27	12.90		16.17	
Paint	Ea.	10.10	51.50		61.60	
Minimum Charge	Job		211		211	
Stain						
Install	Ea.	6.80	30		36.80	Includes labor and material to stain
Minimum Charge	Job		180		180	single door and trim on both sides.
Pre-hung Masonite						
1' 6" x 6' 8"						
Demolish	Ea.		39		39	Cost includes material and labor to
Install	Ea.	211	23.50		234.50	install 1' 6" x 6' 8" pre-hung 1-3/8"
Demolish and Install	Ea.	211	62.50		273.50	masonite door with split pine jamb
Reinstall	Ea.		23.49		23.49	including casing, stops, and hinges.
Clean	Ea.	3.27	12.90		16.17	
Paint	Ea.	10.10	51.50		61.60	
Minimum Charge	Job		211		211	
2' x 6' 8"						
Demolish	Ea.		39		39	Cost includes material and labor to
Install	Ea.	220	25		245	install 2' x 6' 8" pre-hung 1-3/8"
Demolish and Install	Ea.	220	64		284	masonite door with split pine jamb
Reinstall	Ea.		24.87		24.87	including casing, stops, and hinges.
Clean	Ea.	3.27	12.90		16.17	
Paint	Ea.	10.10	51.50		61.60	
Minimum Charge	Job		211		211	
2' 4" x 6' 8"						
Demolish	Ea.		39		39	Cost includes material and labor to
Install	Ea.	222	25		247	install 2' 4" x 6' 8" pre-hung 1-3/8"
Demolish and Install	Ea.	222	64		286	masonite door with split pine jamb
Reinstall	Ea.		24.87		24.87	including casing, stops, and hinges.
Clean	Ea.	3.27	12.90		16.17	
Paint	Ea.	10.10	51.50		61.60	
Minimum Charge	Job		211		211	
2' 8" x 6' 8"						
Demolish	Ea.		39		39	Cost includes material and labor to
Install	Ea.	226	25		251	install 2' 8" x 6' 8" pre-hung 1-3/8"
Demolish and Install	Ea.	226	64		290	masonite door with split pine jamb
Reinstall	Ea.		24.87		24.87	including casing, stops, and hinges.
Clean	Ea.	3.27	12.90		16.17	
Paint	Ea.	10.10	51.50		61.60	
Minimum Charge	Job		211		211	
3' x 6' 8"						
Demolish	Ea.		39		39	Cost includes material and labor to
Install	Ea.	230	26.50		256.50	install 3' x 6' 8" pre-hung 1-3/8"
Demolish and Install	Ea.	230	65.50		295.50	masonite door with split pine jamb
Reinstall	Ea.		26.43		26.43	including casing, stops, and hinges.
Clean	Ea.	3.27	12.90		16.17	
Paint	Ea.	10.10	51.50		61.60	
Minimum Charge	Job		211		211	

Finish Carpentry

Hollow Core Door

Hollow Core Door		Unit	Material	Labor	Equip.	Total	Specification
Pocket Type							
	Demolish	Ea.		58.50		58.50	Cost includes material and labor to
	Install	Ea.	44	47		91	install 2' 8" x 6' 8" lauan door.
	Demolish and Install	Ea.	44	105.50		149.50	
	Reinstall	Ea.		46.98		46.98	
	Clean	Ea.	3.27	12.90		16.17	
	Paint	Ea.	10.10	51.50		61.60	
	Minimum Charge	Job		211		211	
Stain							
	Install	Ea.	6.80	30		36.80	Includes labor and material to stain
	Minimum Charge	Job		180		180	single door and trim on both sides.

Hollow Core Door Only

1' 6" x 6' 8"		Unit	Material	Labor	Equip.	Total	Specification
	Demolish	Ea.		8.40		8.40	Cost includes material and labor to
	Install	Ea.	47.50	22.50		70	install 1' 6" x 6' 8" lauan hollow core
	Demolish and Install	Ea.	47.50	30.90		78.40	door slab.
	Reinstall	Ea.		22.25		22.25	
	Clean	Ea.	1.36	9.70		11.06	
	Paint	Ea.	4.86	45		49.86	
	Minimum Charge	Job		211		211	
2' x 6' 8'							
	Demolish	Ea.		8.40		8.40	Cost includes material and labor to
	Install	Ea.	50.50	23.50		74	install 2' x 6' 8" lauan hollow core
	Demolish and Install	Ea.	50.50	31.90		82.40	door slab.
	Reinstall	Ea.		23.49		23.49	
	Clean	Ea.	1.36	9.70		11.06	
	Paint	Ea.	4.86	45		49.86	
	Minimum Charge	Job		211		211	
2' 4" x 6' 8'							
	Demolish	Ea.		8.40		8.40	Cost includes material and labor to
	Install	Ea.	54.50	23.50		78	install 2' 4" x 6' 8" lauan hollow core
	Demolish and Install	Ea.	54.50	31.90		86.40	door slab.
	Reinstall	Ea.		23.49		23.49	
	Clean	Ea.	1.36	9.70		11.06	
	Paint	Ea.	4.86	45		49.86	
	Minimum Charge	Job		211		211	
2' 8" x 6' 8'							
	Demolish	Ea.		8.40		8.40	Cost includes material and labor to
	Install	Ea.	56	23.50		79.50	install 2' 8" x 6' 8" lauan hollow core
	Demolish and Install	Ea.	56	31.90		87.90	door slab.
	Reinstall	Ea.		23.49		23.49	
	Clean	Ea.	1.36	9.70		11.06	
	Paint	Ea.	4.86	45		49.86	
	Minimum Charge	Job		211		211	
3' x 6' 8'							
	Demolish	Ea.		8.40		8.40	Cost includes material and labor to
	Install	Ea.	59	25		84	install 3' x 6' 8" lauan hollow core
	Demolish and Install	Ea.	59	33.40		92.40	door slab.
	Reinstall	Ea.		24.87		24.87	
	Clean	Ea.	1.36	9.70		11.06	
	Paint	Ea.	4.86	45		49.86	
	Minimum Charge	Job		211		211	
Pocket Type							
	Demolish	Ea.		8.40		8.40	Cost includes material and labor to
	Install	Ea.	44	47		91	install 2' 8" x 6' 8" lauan door.
	Demolish and Install	Ea.	44	55.40		99.40	
	Reinstall	Ea.		46.98		46.98	
	Clean	Ea.	1.36	9.70		11.06	
	Paint	Ea.	4.86	45		49.86	
	Minimum Charge	Job		211		211	

Finish Carpentry

Hollow Core Door		Unit	Material	Labor	Equip.	Total	Specification
Stain							
	Install	Ea.	3.76	40		43.76	Includes labor and materials to stain a
	Minimum Charge	Job		180		180	flush style door on both sides by
							brush.
Casing Trim							
Single Width							
	Demolish	Ea.		7.45		7.45	Cost includes material and labor to
	Install	Opng.	20.50	35		63.50	install 11/16" x 2-1/2" pine ranch
	Demolish and Install	Opng.	28.50	42.45		70.95	style casing for one side of a standard
	Reinstall	Opng.		28.19		28.19	door opening.
	Clean	Opng.	.46	3.10		3.56	
	Paint	Ea.	5.20	7.20		12.40	
	Minimum Charge	Job		211		211	
Double Width							
	Demolish	Ea.		8.85		8.85	Cost includes material and labor to
	Install	Opng.	33.50	42.50		76	install 11/16" x 2-1/2" pine ranch
	Demolish and Install	Opng.	33.50	51.35		84.85	style door casing for one side of a
	Reinstall	Opng.		33.82		33.82	double door opening.
	Clean	Opng.	.54	6.20		6.74	
	Paint	Ea.	5.20	8.20		13.40	
	Minimum Charge	Job		211		211	
Hardware							
Doorknob w / Lock							
	Demolish	Ea.		16.80		16.80	Includes material and labor to install
	Install	Ea.	45	26.50		71.50	privacy lockset.
	Demolish and Install	Ea.	45	43.30		88.30	
	Reinstall	Ea.		21.14		21.14	
	Clean	Ea.	.33	9.70		10.03	
	Minimum Charge	Job		211		211	
Doorknob							
	Demolish	Ea.		16.80		16.80	Includes material and labor to install
	Install	Ea.	29.50	26.50		56	residential passage lockset, keyless
	Demolish and Install	Ea.	29.50	43.30		72.80	bored type.
	Reinstall	Ea.		21.14		21.14	
	Clean	Ea.	.33	9.70		10.03	
	Minimum Charge	Job		211		211	
Deadbolt							
	Demolish	Ea.		14.70		14.70	Includes material and labor to install a
	Install	Ea.	70	30		100	deadbolt.
	Demolish and Install	Ea.	70	44.70		114.70	
	Reinstall	Ea.		24.16		24.16	
	Clean	Ea.	.33	9.70		10.03	
	Minimum Charge	Job		211		211	
Lever Handle							
	Demolish	Ea.		16.80		16.80	Includes material and labor to install
	Install	Ea.	40	42.50		82.50	residential passage lockset, keyless
	Demolish and Install	Ea.	40	59.30		99.30	bored type with lever handle.
	Reinstall	Ea.		33.82		33.82	
	Clean	Ea.	.33	9.70		10.03	
	Minimum Charge	Job		211		211	
Closer							
	Demolish	Ea.		8.80		8.80	Includes material and labor to install
	Install	Ea.	145	65		210	pneumatic light duty closer for interior
	Demolish and Install	Ea.	145	73.80		218.80	type doors.
	Reinstall	Ea.		52.04		52.04	
	Clean	Ea.	.17	12.90		13.07	
	Minimum Charge	Job		211		211	

Finish Carpentry

Hollow Core Door		Unit	Material	Labor	Equip.	Total	Specification
Push Plate							
	Demolish	Ea.		13.05		13.05	Includes material and labor to install
	Install	Ea.	29.50	35		64.50	bronze push-pull plate.
	Demolish and Install	Ea.	29.50	48.05		77.55	
	Reinstall	Ea.		28.19		28.19	
	Clean	Ea.	.33	9.70		10.03	
	Minimum Charge	Job		211		211	
Kickplate							
	Demolish	Ea.		13.05		13.05	Includes material and labor to install
	Install	Ea.	23.50	28		51.50	aluminum kickplate, 10" x 28".
	Demolish and Install	Ea.	23.50	41.05		64.55	
	Reinstall	Ea.		22.55		22.55	
	Clean	Ea.	.33	9.70		10.03	
	Minimum Charge	Job		211		211	
Exit Sign							
	Demolish	Ea.		23		23	Includes material and labor to install a
	Install	Ea.	118	57.50		175.50	wall mounted interior electric exit sign.
	Demolish and Install	Ea.	118	80.50		198.50	
	Reinstall	Ea.		45.92		45.92	
	Clean	Ea.	.02	7.10		7.12	
	Minimum Charge	Job		211		211	
Jamb							
	Demolish	Ea.		29.50		29.50	Includes material and labor to install
	Install	Ea.	157	11.45		168.45	flat pine jamb with square cut heads
	Demolish and Install	Ea.	157	40.95		197.95	and rabbeted sides for 6' 8" high and
	Clean	Ea.	.10	7.75		7.85	3-9/16" door including trim sets for
	Paint	Ea.	6.75	24		30.75	both sides.
	Minimum Charge	Job		211		211	
Shave & Refit							
	Install	Ea.		35		35	Includes labor to shave and rework
	Minimum Charge	Job		211		211	door to fit opening at the job site.

Solid Core Door		Unit	Material	Labor	Equip.	Total	Specification
Pre-hung Wood							
2' 4" x 6' 8"							
	Demolish	Ea.		39		39	Cost includes material and labor to
	Install	Ea.	250	25		275	install 2' 4" x 6' 8" pre-hung 1-3/4"
	Demolish and Install	Ea.	250	64		314	door, 4-1/2" split jamb, including
	Reinstall	Ea.		24.87		24.87	casing and stop, hinges, aluminum sill,
	Clean	Ea.	3.27	12.90		16.17	weatherstripping.
	Paint	Ea.	10.95	51.50		62.45	
	Minimum Charge	Job		211		211	
2' 6" x 6' 8"							
	Demolish	Ea.		39		39	Cost includes material and labor to
	Install	Ea.	248	25		273	install 2' 6" x 6' 8" pre-hung 1-3/4"
	Demolish and Install	Ea.	248	64		312	door, 4-1/2" split jamb, including
	Reinstall	Ea.		24.87		24.87	casing and stop, hinges, aluminum sill,
	Clean	Ea.	3.27	12.90		16.17	weatherstripping.
	Paint	Ea.	10.95	51.50		62.45	
	Minimum Charge	Job		211		211	

Finish Carpentry

Solid Core Door

Solid Core Door	Unit	Material	Labor	Equip.	Total	Specification
2′ 8″ x 6′ 8″						
Demolish	Ea.		39		39	Cost includes material and labor to
Install	Ea.	243	25		268	install 2′ 8″ x 6′ 8″ pre-hung 1-3/4″
Demolish and Install	Ea.	243	64		307	door, 4-1/2″ split jamb, including
Reinstall	Ea.		24.87		24.87	casing and stop, hinges, aluminum sill,
Clean	Ea.	3.27	12.90		16.17	weatherstripping.
Paint	Ea.	10.95	51.50		62.45	
Minimum Charge	Job		211		211	
3′ x 6′ 8″						
Demolish	Ea.		39		39	Cost includes material and labor to
Install	Ea.	252	26.50		278.50	install 3′ x 6′ 8″ pre-hung 1-3/4″
Demolish and Install	Ea.	252	65.50		317.50	door, including casing and stop,
Reinstall	Ea.		26.43		26.43	hinges, jamb, aluminum sill,
Clean	Ea.	3.27	12.90		16.17	weatherstripped.
Paint	Ea.	10.95	51.50		62.45	
Minimum Charge	Job		211		211	
3′ 6″ x 6′ 8″						
Demolish	Ea.		39		39	Cost includes material and labor to
Install	Ea.	395	28		423	install 3′ 6″ x 6′ 8″ pre-hung 1-3/4″
Demolish and Install	Ea.	395	67		462	door, 4-1/2″ split jamb, including
Reinstall	Ea.		28.19		28.19	casing and stop, hinges, aluminum sill,
Clean	Ea.	3.27	12.90		16.17	weatherstripping.
Paint	Ea.	10.95	51.50		62.45	
Minimum Charge	Job		211		211	
Raised Panel						
Demolish	Ea.		39		39	Cost includes material and labor to
Install	Ea.	485	23.50		508.50	install 3′ x 6′ 8″ pre-hung 1-3/4″
Demolish and Install	Ea.	485	62.50		547.50	door, jamb, including casing and
Reinstall	Ea.		23.49		23.49	stop, hinges, aluminum sill,
Clean	Ea.	3.27	12.90		16.17	weatherstripping.
Paint	Ea.	10.95	120		130.95	
Minimum Charge	Job		211		211	
Stain						
Install	Ea.	6.80	30		36.80	Includes labor and material to stain
Minimum Charge	Job		180		180	single door and trim on both sides.

Bifold Door

Bifold Door	Unit	Material	Labor	Equip.	Total	Specification
Wood, Single						
3′						
Demolish	Ea.		29.50		29.50	Cost includes material and labor to
Install	Ea.	78.50	65		143.50	install 3′ wide solid wood bi-fold door
Demolish and Install	Ea.	78.50	94.50		173	not including jamb, trim, track, and
Reinstall	Ea.		65.05		65.05	hardware.
Clean	Ea.	3.27	12.90		16.17	
Paint	Ea.	10.10	51.50		61.60	
Minimum Charge	Job		211		211	
5′						
Demolish	Ea.		29.50		29.50	Includes material and labor to install
Install	Ea.	116	77		193	pair of 2′ 6″ wide solid wood bi-fold
Demolish and Install	Ea.	116	106.50		222.50	doors not including jamb, trim, track,
Reinstall	Ea.		76.87		76.87	and hardware.
Clean	Ea.	1.43	22		23.43	
Paint	Ea.	33.50	90		123.50	
Minimum Charge	Job		211		211	

Finish Carpentry

Bifold Door		Unit	Material	Labor	Equip.	Total	Specification
6'							
	Demolish	Ea.		29.50		29.50	Includes material and labor to install
	Install	Ea.	141	84.50		225.50	pair of 3' wide solid wood bi-fold
	Demolish and Install	Ea.	141	114		255	doors not including jamb, trim, track,
	Reinstall	Ea.		84.56		84.56	and hardware.
	Clean	Ea.	1.63	26		27.63	
	Paint	Ea.	33.50	90		123.50	
	Minimum Charge	Job		211		211	
8'							
	Demolish	Ea.		29.50		29.50	Includes material and labor to install
	Install	Ea.	215	94		309	pair of 4' wide solid wood bi-fold
	Demolish and Install	Ea.	215	123.50		338.50	doors not including jamb, trim, track,
	Reinstall	Ea.		93.96		93.96	and hardware.
	Clean	Ea.	2.24	31		33.24	
	Paint	Ea.	17.60	90		107.60	
	Minimum Charge	Job		211		211	
Mirrored							
6'							
	Demolish	Ea.		29.50		29.50	Includes material and labor to install
	Install	Ea.	450	47		497	pair of 3' wide mirrored bi-fold doors
	Demolish and Install	Ea.	450	76.50		526.50	including jamb, trim, track, and
	Reinstall	Ea.		46.98		46.98	hardware.
	Clean	Ea.	1.63	12.90		14.53	
	Minimum Charge	Job		211		211	
8'							
	Demolish	Ea.		29.50		29.50	Includes material and labor to install
	Install	Ea.	710	70.50		780.50	pair of 4' wide mirrored bi-fold doors
	Demolish and Install	Ea.	710	100		810	including jamb, trim, track, and
	Reinstall	Ea.		70.47		70.47	hardware.
	Clean	Ea.	2.24	15.50		17.74	
	Minimum Charge	Job		211		211	
Louvered							
3'							
	Demolish	Ea.		29.50		29.50	Cost includes material and labor to
	Install	Ea.	227	65		292	install 3' wide louvered bi-fold door
	Demolish and Install	Ea.	227	94.50		321.50	not including jamb, trim, track, and
	Reinstall	Ea.		65.05		65.05	hardware.
	Clean	Ea.	3.27	12.90		16.17	
	Paint	Ea.	10.10	51.50		61.60	
	Minimum Charge	Job		211		211	
5'							
	Demolish	Ea.		29.50		29.50	Includes material and labor to install
	Install	Ea.	286	77		363	pair of 2' 6" wide louvered wood
	Demolish and Install	Ea.	286	106.50		392.50	bi-fold doors not including jamb, trim,
	Reinstall	Ea.		76.87		76.87	track, and hardware.
	Clean	Ea.	1.43	22		23.43	
	Paint	Ea.	33.50	90		123.50	
	Minimum Charge	Job		211		211	
6'							
	Demolish	Ea.		29.50		29.50	Includes material and labor to install
	Install	Ea.	315	84.50		399.50	pair of 3' wide louvered wood bi-fold
	Demolish and Install	Ea.	315	114		429	doors not including jamb, trim, track,
	Reinstall	Ea.		84.56		84.56	and hardware.
	Clean	Ea.	1.63	26		27.63	
	Paint	Ea.	33.50	90		123.50	
	Minimum Charge	Job		211		211	

Finish Carpentry

Bifold Door		Unit	Material	Labor	Equip.	Total	Specification
8'							
	Demolish	Ea.		29.50		29.50	Includes material and labor to install
	Install	Ea.	490	84.50		574.50	pair of 4' wide solid wood bi-fold
	Demolish and Install	Ea.	490	114		604	doors not including jamb, trim, track,
	Reinstall	Ea.		84.56		84.56	and hardware.
	Clean	Ea.	7.75	22		29.75	
	Paint	Ea.	17.60	90		107.60	
	Minimum Charge	Job		211		211	
Accordion Type							
	Demolish	S.F.		1.01		1.01	Includes material and labor to install
	Install	Ea.	43.50	84.50		128	vinyl folding accordion closet doors,
	Demolish and Install	Ea.	43.50	85.51		129.01	including track and frame.
	Reinstall	Ea.		84.56		84.56	
	Clean	S.F.		.38		.38	
	Minimum Charge	Job		211		211	
Casing Trim							
Single Width							
	Demolish	Ea.		7.45		7.45	Cost includes material and labor to
	Install	Opng.	28.50	35		63.50	install 11/16" x 2-1/2" pine ranch
	Demolish and Install	Opng.	28.50	42.45		70.95	style casing for one side of a standard
	Reinstall	Opng.		28.19		28.19	door opening.
	Clean	Opng.	.46	3.10		3.56	
	Paint	Ea.	5.20	7.20		12.40	
	Minimum Charge	Job		211		211	
Double Width							
	Demolish	Ea.		8.85		8.85	Cost includes material and labor to
	Install	Opng.	33.50	42.50		76	install 11/16" x 2-1/2" pine ranch
	Demolish and Install	Opng.	33.50	51.35		84.85	style door casing for one side of a
	Reinstall	Opng.		33.82		33.82	double door opening.
	Clean	Opng.	.54	6.20		6.74	
	Paint	Ea.	5.20	8.20		13.40	
	Minimum Charge	Job		211		211	
Jamb							
	Demolish	Ea.		29.50		29.50	Includes material and labor to install
	Install	Ea.	157	11.45		168.45	flat pine jamb with square cut heads
	Demolish and Install	Ea.	157	40.95		197.95	and rabbeted sides for 6' 8" high and
	Clean	Ea.	.10	7.75		7.85	3-9/16" door including trim sets for
	Paint	Ea.	6.75	24		30.75	both sides.
	Minimum Charge	Job		106		106	
Stain							
	Install	Ea.	6.80	30		36.80	Includes labor and material to stain
	Minimum Charge	Job		180		180	single door and trim on both sides.
Shave & Refit							
	Install	Ea.		35		35	Includes labor to shave and rework
	Minimum Charge	Job		106		106	door to fit opening at the job site.

Finish Carpentry

Louver Door

	Unit	Material	Labor	Equip.	Total	Specification
Full						
2' x 6' 8"						
Demolish	Ea.		8.40		8.40	Cost includes material and labor to
Install	Ea.	260	42.50		302.50	install 2' x 6' 8" pine full louver door,
Demolish and Install	Ea.	260	50.90		310.90	jamb, hinges.
Reinstall	Ea.		42.28		42.28	
Clean	Ea.	6.05	38.50		44.55	
Paint	Ea.	10.10	51.50		61.60	
Minimum Charge	Job		211		211	
2' 8" x 6' 8"						
Demolish	Ea.		8.40		8.40	Cost includes material and labor to
Install	Ea.	293	42.50		335.50	install 2' 8" x 6' 8" pine full louver
Demolish and Install	Ea.	293	50.90		343.90	door, jamb, hinges.
Reinstall	Ea.		42.28		42.28	
Clean	Ea.	6.05	38.50		44.55	
Paint	Ea.	10.10	51.50		61.60	
Minimum Charge	Job		211		211	
3' x 6' 8"						
Demolish	Ea.		8.40		8.40	Cost includes material and labor to
Install	Ea.	305	44.50		349.50	install 3' x 6' 8" pine full louver door,
Demolish and Install	Ea.	305	52.90		357.90	jamb, hinges.
Reinstall	Ea.		44.51		44.51	
Clean	Ea.	6.05	38.50		44.55	
Paint	Ea.	10.10	51.50		61.60	
Minimum Charge	Job		211		211	
Half						
2' x 6' 8"						
Demolish	Ea.		8.40		8.40	Cost includes material and labor to
Install	Ea.	380	42.50		422.50	install 2' x 6' 8" pine half louver door,
Demolish and Install	Ea.	380	50.90		430.90	jamb, hinges.
Reinstall	Ea.		42.28		42.28	
Clean	Ea.	6.05	38.50		44.55	
Paint	Ea.	10.10	51.50		61.60	
Minimum Charge	Job		211		211	
2' 8" x 6' 8"						
Demolish	Ea.		8.40		8.40	Cost includes material and labor to
Install	Ea.	405	42.50		447.50	install 2' 8" x 6' 8" pine half louver
Demolish and Install	Ea.	405	50.90		455.90	door, jamb, hinges.
Reinstall	Ea.		42.28		42.28	
Clean	Ea.	6.05	38.50		44.55	
Paint	Ea.	10.10	51.50		61.60	
Minimum Charge	Job		211		211	
3' x 6' 8"						
Demolish	Ea.		8.40		8.40	Cost includes material and labor to
Install	Ea.	420	44.50		464.50	install 3' x 6' 8" pine half louver door,
Demolish and Install	Ea.	420	52.90		472.90	jamb, hinges.
Reinstall	Ea.		44.51		44.51	
Clean	Ea.	6.05	38.50		44.55	
Paint	Ea.	10.10	51.50		61.60	
Minimum Charge	Job		211		211	
Jamb						
Demolish	Ea.		29.50		29.50	Includes material and labor to install
Install	Ea.	157	11.45		168.45	flat pine jamb with square cut heads
Demolish and Install	Ea.	157	40.95		197.95	and rabbeted sides for 6' 8" high and
Clean	Ea.	.10	7.75		7.85	3-9/16" door including trim sets for
Paint	Ea.	6.75	24		30.75	both sides.
Minimum Charge	Job		106		106	

Finish Carpentry

Louver Door	Unit	Material	Labor	Equip.	Total	Specification
Stain						
Install	Ea.	6.80	30		36.80	Includes labor and material to stain
Minimum Charge	Job		180		180	single door and trim on both sides.
Shave & Refit						
Install	Ea.		35		35	Includes labor to shave and rework
Minimum Charge	Job		180		180	door to fit opening at the job site.

Bypass Sliding Door	Unit	Material	Labor	Equip.	Total	Specification
Wood						
5'						
Demolish	Ea.		39		39	Includes material and labor to install
Install	Opng.	206	77		283	pair of 2' 6" wide lauan hollow core
Demolish and Install	Opng.	206	116		322	wood bi-pass doors with jamb, track
Reinstall	Opng.		76.87		76.87	hardware.
Clean	Ea.	1.43	22		23.43	
Paint	Ea.	33.50	90		123.50	
Minimum Charge	Job		211		211	
6'						
Demolish	Ea.		39		39	Includes material and labor to install
Install	Opng.	230	84.50		314.50	pair of 3' wide lauan hollow core
Demolish and Install	Opng.	230	123.50		353.50	wood bi-pass doors with jamb, track
Reinstall	Opng.		84.56		84.56	and hardware.
Clean	Ea.	1.63	26		27.63	
Paint	Ea.	33.50	90		123.50	
Minimum Charge	Job		211		211	
8'						
Demolish	Ea.		39		39	Includes material and labor to install
Install	Opng.	345	106		451	pair of 4' wide lauan hollow core
Demolish and Install	Opng.	345	145		490	wood bi-pass doors with jamb, track
Reinstall	Opng.		105.70		105.70	and hardware.
Clean	Ea.	2.24	31		33.24	
Paint	Ea.	17.60	90		107.60	
Minimum Charge	Job		211		211	
Mirrored						
6'						
Demolish	Ea.		39		39	Includes material and labor to install
Install	Opng.	290	84.50		374.50	pair of 3' wide mirrored bi-pass doors
Demolish and Install	Opng.	290	123.50		413.50	with jamb, trim, track hardware.
Reinstall	Opng.		84.56		84.56	
Clean	Ea.	1.63	12.90		14.53	
Minimum Charge	Job		211		211	
8'						
Demolish	Ea.		39		39	Includes material and labor to install
Install	Opng.	475	94		569	pair of 4' wide mirrored bi-pass doors
Demolish and Install	Opng.	475	133		608	with jamb, trim, track hardware.
Reinstall	Opng.		93.96		93.96	
Clean	Ea.	2.24	15.50		17.74	
Minimum Charge	Job		211		211	

Finish Carpentry

Bypass Sliding Door

Bypass Sliding Door	Unit	Material	Labor	Equip.	Total	Specification
Casing Trim						
Single Width						
Demolish	Ea.		7.45		7.45	Cost includes material and labor to
Install	Opng.	28.50	35		63.50	install 11/16" x 2-1/2" pine ranch
Demolish and Install	Opng.	28.50	42.45		70.95	style casing for one side of a standard
Reinstall	Opng.		28.19		28.19	door opening.
Clean	Opng.	.46	3.10		3.56	
Paint	Ea.	5.20	7.20		12.40	
Minimum Charge	Job		211		211	
Double Width						
Demolish	Ea.		8.85		8.85	Cost includes material and labor to
Install	Opng.	33.50	42.50		76	install 11/16" x 2-1/2" pine ranch
Demolish and Install	Opng.	33.50	51.35		84.85	style door casing for one side of a
Reinstall	Opng.		33.82		33.82	double door opening.
Clean	Opng.	.54	6.20		6.74	
Paint	Ea.	5.20	8.20		13.40	
Minimum Charge	Job		211		211	
Jamb						
Demolish	Ea.		29.50		29.50	Includes material and labor to install
Install	Ea.	157	11.45		168.45	flat pine jamb with square cut heads
Demolish and Install	Ea.	157	40.95		197.95	and rabbeted sides for 6' 8" high and
Clean	Ea.	.10	7.75		7.85	3-9/16" door including trim sets for
Paint	Ea.	6.75	24		30.75	both sides.
Minimum Charge	Job		106		106	
Stain						
Install	Ea.	6.80	30		36.80	Includes labor and material to stain
Minimum Charge	Job		180		180	single door and trim on both sides.
Shave & Refit						
Install	Ea.		35		35	Includes labor to shave and rework
Minimum Charge	Job		106		106	door to fit opening at the job site.

Base Molding

Base Molding	Unit	Material	Labor	Equip.	Total	Specification
One Piece						
1-5/8"						
Demolish	L.F.		.57		.57	Includes material and labor to install
Install	L.F.	.48	1.25		1.73	pine or birch base molding.
Demolish and Install	L.F.	.48	1.82		2.30	
Reinstall	L.F.		1		1	
Clean	L.F.	.02	.18		.20	
Paint	L.F.	.06	.39		.45	
Minimum Charge	Job		211		211	
2-1/2"						
Demolish	L.F.		.57		.57	Cost includes material and labor to
Install	L.F.	.69	1.28		1.97	install 1/2" x 2-1/2" pine or birch
Demolish and Install	L.F.	.69	1.85		2.54	base molding.
Reinstall	L.F.		1.02		1.02	
Clean	L.F.	.02	.18		.20	
Paint	L.F.	.06	.39		.45	
Minimum Charge	Job		211		211	

Finish Carpentry

Base Molding	Unit	Material	Labor	Equip.	Total	Specification
3-1/2″						
Demolish	L.F.		.57		.57	Cost includes material and labor to
Install	L.F.	1.10	1.36		2.46	install 9/16″ x 3-1/2″ pine or birch
Demolish and Install	L.F.	1.10	1.93		3.03	base molding.
Reinstall	L.F.		1.08		1.08	
Clean	L.F.	.02	.18		.20	
Paint	L.F.	.09	.43		.52	
Minimum Charge	Job		211		211	
6″						
Demolish	L.F.		.57		.57	Cost includes material and labor to
Install	L.F.	.75	1.36		2.11	install 1″ x 6″ pine or birch base
Demolish and Install	L.F.	.75	1.93		2.68	molding.
Reinstall	L.F.		1.08		1.08	
Clean	L.F.	.03	.26		.29	
Paint	L.F.	.10	.54		.64	
Minimum Charge	Job		211		211	
8″						
Demolish	L.F.		.69		.69	Cost includes material and labor to
Install	L.F.	1	1.36		2.36	install 1″ x 8″ pine or birch base
Demolish and Install	L.F.	1	2.05		3.05	molding.
Reinstall	L.F.		1.08		1.08	
Clean	L.F.	.03	.31		.34	
Paint	L.F.	.11	.60		.71	
Minimum Charge	Job		211		211	
Premium / Custom Grade						
Demolish	L.F.		.69		.69	Includes material and labor to install
Install	L.F.	2.89	4.07		6.96	custom 2 or 3 piece oak or other
Demolish and Install	L.F.	2.89	4.76		7.65	hardwood, stain grade trim.
Reinstall	L.F.		3.25		3.25	
Clean	L.F.	.03	.31		.34	
Paint	L.F.	.12	.60		.72	
Minimum Charge	Job		211		211	
One Piece Oak						
1-5/8″						
Demolish	L.F.		.57		.57	Cost includes material and labor to
Install	L.F.	1.05	1.25		2.30	install 1/2″ x 1-5/8″ oak base
Demolish and Install	L.F.	1.05	1.82		2.87	molding.
Reinstall	L.F.		1		1	
Clean	L.F.	.02	.18		.20	
Paint	L.F.	.12	.53		.65	
Minimum Charge	Job		211		211	
2-1/2″						
Demolish	L.F.		.57		.57	Cost includes material and labor to
Install	L.F.	1.47	1.28		2.75	install 1/2″ x 2-1/2″ oak base
Demolish and Install	L.F.	1.47	1.85		3.32	molding.
Reinstall	L.F.		1.02		1.02	
Clean	L.F.	.02	.18		.20	
Paint	L.F.	.12	.55		.67	
Minimum Charge	Job		211		211	
3-1/2″						
Demolish	L.F.		.57		.57	Cost includes material and labor to
Install	L.F.	1.78	1.30		3.08	install 1/2″ x 3-1/2″ oak base
Demolish and Install	L.F.	1.78	1.87		3.65	molding.
Reinstall	L.F.		1.04		1.04	
Clean	L.F.	.02	.18		.20	
Paint	L.F.	.13	.56		.69	
Minimum Charge	Job		211		211	

Finish Carpentry

Base Molding

Base Molding		Unit	Material	Labor	Equip.	Total	Specification
6"							
	Demolish	L.F.		.57		.57	Cost includes material and labor to
	Install	L.F.	1.72	1.36		3.08	install 1" x 6" oak base molding.
	Demolish and Install	L.F.	1.72	1.93		3.65	
	Reinstall	L.F.		1.08		1.08	
	Clean	L.F.	.03	.31		.34	
	Paint	L.F.	.13	.58		.71	
	Minimum Charge	Job		211		211	
8"							
	Demolish	L.F.		.69		.69	Cost includes material and labor to
	Install	L.F.	2.29	1.36		3.65	install 1" x 8" oak base molding.
	Demolish and Install	L.F.	2.29	2.05		4.34	
	Reinstall	L.F.		1.08		1.08	
	Clean	L.F.	.03	.31		.34	
	Paint	L.F.	.13	.60		.73	
	Minimum Charge	Job		211		211	

Vinyl Base Molding

		Unit	Material	Labor	Equip.	Total	Specification
2-1/2"							
	Demolish	L.F.		.57		.57	Cost includes material and labor to
	Install	L.F.	1.50	1.18		2.68	install 2-1/2" vinyl or rubber cove
	Demolish and Install	L.F.	1.50	1.75		3.25	base including adhesive.
	Reinstall	L.F.		1.18		1.18	
	Clean	L.F.		.28		.28	
	Minimum Charge	Job		192		192	
4"							
	Demolish	L.F.		.57		.57	Cost includes material and labor to
	Install	Ea.	1.87	1.18		3.05	install 4" vinyl or rubber cove base
	Demolish and Install	Ea.	1.87	1.75		3.62	including adhesive.
	Reinstall	Ea.		1.18		1.18	
	Clean	L.F.		.28		.28	
	Minimum Charge	Job		192		192	
6"							
	Demolish	L.F.		.57		.57	Cost includes material and labor to
	Install	L.F.	1.42	1.18		2.60	install 6" vinyl or rubber cove base
	Demolish and Install	L.F.	1.42	1.75		3.17	including adhesive.
	Reinstall	L.F.		1.18		1.18	
	Clean	L.F.		.28		.28	
	Minimum Charge	Job		192		192	

Prefinished

		Unit	Material	Labor	Equip.	Total	Specification
	Demolish	L.F.		.57		.57	Cost includes material and labor to
	Install	L.F.	1.41	1.36		2.77	install 9/16" x 3-5/16" prefinished
	Demolish and Install	L.F.	1.41	1.93		3.34	base molding.
	Reinstall	L.F.		1.08		1.08	
	Clean	L.F.	.02	.18		.20	
	Paint	L.F.	.09	.43		.52	
	Minimum Charge	Job		211		211	

Shoe

		Unit	Material	Labor	Equip.	Total	Specification
	Demolish	L.F.		.57		.57	Includes material and labor to install
	Install	L.F.	.48	1.25		1.73	pine or birch base molding.
	Demolish and Install	L.F.	.48	1.82		2.30	
	Reinstall	L.F.		1		1	
	Clean	L.F.	.02	.18		.20	
	Paint	L.F.	.14	.58		.72	
	Minimum Charge	Job		211		211	

Finish Carpentry

Base Molding	Unit	Material	Labor	Equip.	Total	Specification
Carpeted						
Demolish	L.F.		.23		.23	Includes material and labor to install
Install	L.F.	1.49	2.25		3.74	carpet baseboard with adhesive.
Demolish and Install	L.F.	1.49	2.48		3.97	
Reinstall	L.F.		1.80		1.80	
Clean	L.F.		.12		.12	
Minimum Charge	Job		186		186	
Ceramic Tile						
Demolish	L.F.		1.25		1.25	Includes material and labor to install
Install	L.F.	3.98	5.20		9.18	thin set ceramic tile cove base
Demolish and Install	L.F.	3.98	6.45		10.43	including grout.
Clean	L.F.		.59		.59	
Minimum Charge	Job		186		186	
Quarry Tile						
Demolish	L.F.		1.25		1.25	Includes material and labor to install
Install	L.F.	5.90	6.05		11.95	quarry tile base including grout.
Demolish and Install	L.F.	5.90	7.30		13.20	
Clean	L.F.		.59		.59	
Minimum Charge	Job		186		186	
Caulk and Renail						
Install	L.F.	.01	.66		.67	Includes labor and material to renail
Minimum Charge	Job		211		211	base and caulk where necessary.

Molding	Unit	Material	Labor	Equip.	Total	Specification
Oak Base						
Demolish	L.F.		.57		.57	Cost includes material and labor to
Install	L.F.	1.47	1.28		2.75	install 1/2" x 2-1/2" oak base
Demolish and Install	L.F.	1.47	1.85		3.32	molding.
Reinstall	L.F.		1.02		1.02	
Clean	L.F.	.02	.18		.20	
Paint	L.F.	.12	.55		.67	
Minimum Charge	Job		211		211	
Shoe						
Demolish	L.F.		.57		.57	Includes material and labor to install
Install	L.F.	.48	1.25		1.73	pine or birch base molding.
Demolish and Install	L.F.	.48	1.82		2.30	
Reinstall	L.F.		1		1	
Clean	L.F.	.02	.18		.20	
Paint	L.F.	.14	.58		.72	
Minimum Charge	Job		211		211	
Quarter Round						
Demolish	L.F.		.57		.57	Includes material and labor to install
Install	L.F.	.77	1.66		2.43	pine base quarter round molding.
Demolish and Install	L.F.	.77	2.23		3	
Reinstall	L.F.		1.33		1.33	
Clean	L.F.	.02	.18		.20	
Paint	L.F.	.12	.90		1.02	
Minimum Charge	Job		211		211	
Chair Rail						
Demolish	L.F.		.57		.57	Cost includes material and labor to
Install	L.F.	1.85	1.57		3.42	install 5/8" x 2-1/2" oak chair rail
Demolish and Install	L.F.	1.85	2.14		3.99	molding.
Reinstall	L.F.		1.25		1.25	
Clean	L.F.	.02	.18		.20	
Paint	L.F.	.12	.90		1.02	
Minimum Charge	Job		211		211	

Finish Carpentry

Molding

Molding		Unit	Material	Labor	Equip.	Total	Specification
Crown							
	Demolish	L.F.		.57		.57	Includes material and labor to install
	Install	L.F.	2.95	1.69		4.64	unfinished pine crown molding.
	Demolish and Install	L.F.	2.95	2.26		5.21	
	Reinstall	L.F.		1.35		1.35	
	Clean	L.F.	.03	.24		.27	
	Paint	L.F.	.12	.90		1.02	
	Minimum Charge	Job		211		211	
Cove							
	Demolish	L.F.		.57		.57	Cost includes material and labor to
	Install	L.F.	2.40	1.66		4.06	install 1/2" x 2-3/4" pine cove
	Demolish and Install	L.F.	2.40	2.23		4.63	molding.
	Reinstall	L.F.		1.33		1.33	
	Clean	L.F.	.02	.18		.20	
	Paint	L.F.	.12	.90		1.02	
	Minimum Charge	Job		211		211	
Corner							
	Demolish	L.F.		.57		.57	Includes material and labor to install
	Install	L.F.	1.89	.66		2.55	inside or outside pine corner molding.
	Demolish and Install	L.F.	1.89	1.23		3.12	
	Reinstall	L.F.		.53		.53	
	Clean	L.F.	.02	.18		.20	
	Paint	L.F.	.12	.90		1.02	
	Minimum Charge	Job		211		211	
Picture							
	Demolish	L.F.		.57		.57	Cost includes material and labor to
	Install	L.F.	1.68	1.76		3.44	install 9/16" x 2-1/2" pine casing.
	Demolish and Install	L.F.	1.68	2.33		4.01	
	Reinstall	L.F.		1.41		1.41	
	Clean	L.F.	.02	.18		.20	
	Paint	L.F.	.12	.90		1.02	
	Minimum Charge	Job		211		211	
Single Cased Opening							
	Demolish	Ea.		29.50		29.50	Includes material and labor to install
	Install	Ea.	26	26.50		52.50	flat pine jamb with square cut heads
	Demolish and Install	Ea.	26	56		82	and rabbeted sides for 6' 8" high
	Clean	Ea.	.23	6.45		6.68	opening including trim sets for both
	Paint	Ea.	1.47	9		10.47	sides.
	Minimum Charge	Job		211		211	
Double Cased Opening							
	Demolish	Ea.		29.50		29.50	Includes material and labor to install
	Install	Ea.	51	30		81	flat pine jamb with square cut heads
	Demolish and Install	Ea.	51	59.50		110.50	and rabbeted sides for 6' 8" high
	Clean	Ea.	.26	7.55		7.81	opening including trim sets for both
	Paint	Ea.	1.72	10.60		12.32	sides.
	Minimum Charge	Job		211		211	

Window Trim Set

Window Trim Set		Unit	Material	Labor	Equip.	Total	Specification
Single							
	Demolish	Ea.		7.65		7.65	Cost includes material and labor to
	Install	Opng.	48.50	42.50		91	install 11/16" x 2-1/2" pine ranch
	Demolish and Install	Opng.	48.50	50.15		98.65	style window casing.
	Reinstall	Opng.		33.82		33.82	
	Clean	Ea.	.33	4.42		4.75	
	Paint	Ea.	1.39	8.60		9.99	
	Minimum Charge	Job		211		211	

Finish Carpentry

Window Trim Set

	Unit	Material	Labor	Equip.	Total	Specification
Double						
Demolish	Ea.		8.60		8.60	Cost includes material and labor to
Install	Ea.	62.50	53		115.50	install 11/16" x 2-1/2" pine ranch
Demolish and Install	Ea.	62.50	61.60		124.10	style trim for casing one side of a
Reinstall	Ea.		42.28		42.28	double window opening.
Clean	Ea.	.58	5.15		5.73	
Paint	Ea.	2.32	14.45		16.77	
Minimum Charge	Job		211		211	
Triple						
Demolish	Ea.		9.85		9.85	Cost includes material and labor to
Install	Opng.	69	70.50		139.50	install 11/16" x 2-1/2" pine ranch
Demolish and Install	Opng.	69	80.35		149.35	style trim for casing one side of a
Reinstall	Opng.		56.37		56.37	triple window opening.
Clean	Ea.	.81	6.20		7.01	
Paint	Ea.	3.26	20		23.26	
Minimum Charge	Job		211		211	
Window Casing Per L.F.						
Demolish	L.F.		.57		.57	Cost includes material and labor to
Install	L.F.	1.69	1.69		3.38	install 11/16" x 2-1/2" pine ranch
Demolish and Install	L.F.	1.69	2.26		3.95	style trim for casing.
Reinstall	L.F.		1.35		1.35	
Clean	L.F.	.02	.18		.20	
Paint	L.F.	.12	.90		1.02	
Minimum Charge	Job		211		211	

Window Sill

	Unit	Material	Labor	Equip.	Total	Specification
Wood						
Demolish	L.F.		.57		.57	Includes material and labor to install
Install	L.F.	2.92	2.11		5.03	flat, wood window stool.
Demolish and Install	L.F.	2.92	2.68		5.60	
Reinstall	L.F.		1.69		1.69	
Clean	L.F.	.02	.18		.20	
Paint	L.F.	.12	.90		1.02	
Minimum Charge	Job		211		211	
Marble						
Demolish	L.F.		.86		.86	Includes material and labor to install
Install	L.F.	11.50	8.80		20.30	marble window sills, 6" x 3/4" thick.
Demolish and Install	L.F.	11.50	9.66		21.16	
Reinstall	L.F.		7.04		7.04	
Clean	L.F.	.02	.18		.20	
Minimum Charge	Job		186		186	

Shelving

	Unit	Material	Labor	Equip.	Total	Specification
Pine						
18"						
Demolish	L.F.		.98		.98	Cost includes material and labor to
Install	L.F.	7	4.45		11.45	install 1" thick custom shelving board
Demolish and Install	L.F.	7	5.43		12.43	18" wide.
Reinstall	L.F.		3.56		3.56	
Clean	L.F.	.04	.52		.56	
Paint	L.F.	1.49	.57		2.06	
Minimum Charge	Job		211		211	

Finish Carpentry

Shelving

Shelving		Unit	Material	Labor	Equip.	Total	Specification
24"							
	Demolish	L.F.		.98		.98	Cost includes material and labor to
	Install	L.F.	9.35	4.97		14.32	install 1" thick custom shelving board.
	Demolish and Install	L.F.	9.35	5.95		15.30	
	Reinstall	L.F.		3.98		3.98	
	Clean	L.F.	.07	.77		.84	
	Paint	L.F.	2.02	.76		2.78	
	Minimum Charge	Job		211		211	
Particle Board							
	Demolish	L.F.		.91		.91	Cost includes material and labor to
	Install	L.F.	1.29	2.23		3.52	install 3/4" thick particle shelving
	Demolish and Install	L.F.	1.29	3.14		4.43	board 18" wide.
	Reinstall	L.F.		1.78		1.78	
	Clean	L.F.	.04	.52		.56	
	Paint	S.F.	.14	.84		.98	
	Minimum Charge	Job		211		211	
Plywood							
	Demolish	L.F.		.91		.91	Cost includes material and labor to
	Install	L.F.	.85	6.50		7.35	install 3/4" thick plywood shelving
	Demolish and Install	L.F.	.85	7.41		8.26	board.
	Reinstall	L.F.		5.20		5.20	
	Clean	L.F.	.04	.52		.56	
	Paint	S.F.	.14	.84		.98	
	Minimum Charge	Job		211		211	
Hardwood (oak)							
	Demolish	L.F.		.91		.91	Includes material and labor to install
	Install	L.F.	5.20	2.92		8.12	oak shelving, 1" x 12", incl. cleats and
	Demolish and Install	L.F.	5.20	3.83		9.03	bracing.
	Reinstall	L.F.		2.33		2.33	
	Clean	L.F.	.04	.52		.56	
	Paint	S.F.	.14	.84		.98	
	Minimum Charge	Job		211		211	
Custom Bookcase							
	Demolish	L.F.		11.45		11.45	Includes material and labor to install
	Install	L.F.	36	42.50		78.50	custom modular bookcase unit with
	Demolish and Install	L.F.	36	53.95		89.95	clear pine faced frames, shelves 12"
	Reinstall	L.F.		33.82		33.82	O.C., 7' high, 8" deep.
	Clean	L.F.	1.31	11.05		12.36	
	Paint	L.F.	1.60	28		29.60	
	Minimum Charge	Job		211		211	
Glass							
	Demolish	L.F.		.91		.91	Includes material and labor to install
	Install	L.F.	33	2.92		35.92	glass shelving, 1" x 12", incl. cleats
	Demolish and Install	L.F.	33	3.83		36.83	and bracing.
	Reinstall	L.F.		2.33		2.33	
	Clean	L.F.	.06	.48		.54	
	Minimum Charge	Job		203		203	

Closet Shelving

Closet Shelving		Unit	Material	Labor	Equip.	Total	Specification
Closet Shelf and Rod							
	Demolish	L.F.		.91		.91	Cost includes material and labor to
	Install	L.F.	3.10	7.05		10.15	install 1" thick custom shelving board
	Demolish and Install	L.F.	3.10	7.96		11.06	18" wide and 1" diameter clothes pole
	Reinstall	L.F.		5.64		5.64	with brackets 3' O.C.
	Clean	L.F.	.03	.39		.42	
	Paint	L.F.	.23	.90		1.13	
	Minimum Charge	Job		211		211	

Finish Carpentry

Closet Shelving

Closet Shelving	Unit	Material	Labor	Equip.	Total	Specification
Clothing Rod						
Demolish	L.F.		.60		.60	Includes material and labor to install
Install	L.F.	2.60	2.11		4.71	fir closet pole, 1-5/8" diameter.
Demolish and Install	L.F.	2.60	2.71		5.31	
Reinstall	L.F.		1.69		1.69	
Clean	L.F.	.03	.53		.56	
Paint	L.F.	.21	.98		1.19	
Minimum Charge	Job		211		211	

Stair Assemblies

Stair Assemblies	Unit	Material	Labor	Equip.	Total	Specification
Straight Hardwood						
Demolish	Ea.		410		410	Includes material and labor to install
Install	Flight	1250	282		1532	factory cut and assembled straight
Demolish and Install	Flight	1250	692		1942	closed box stairs with oak treads and
Clean	Flight	3.27	26.50		29.77	prefinished stair rail with balusters.
Paint	Flight	21	48		69	
Minimum Charge	Job		211		211	
Clear Oak Tread						
Demolish	Ea.		10.70		10.70	Includes material and labor to install
Install	Ea.	82.50	23.50		106	1-1/4" thick clear oak treads per riser.
Demolish and Install	Ea.	82.50	34.20		116.70	
Clean	Ea.	.10	2.67		2.77	
Paint	Ea.	.69	2.67		3.36	
Minimum Charge	Job		211		211	
Landing						
Demolish	S.F.		3.58		3.58	Includes material and labor to install
Install	S.F.	10.30	23.50		33.80	clear oak landing.
Demolish and Install	S.F.	10.30	27.08		37.38	
Clean	S.F.		.39		.39	
Paint	S.F.	.44	.89		1.33	
Minimum Charge	Job		211		211	
Refinish / Stain						
Install	S.F.	1.44	2.38		3.82	Includes labor and material to sand
Minimum Charge	Job		180		180	and finish (3 passes) with two coats of urethane on a new floor.
Spiral Hardwood						
Demolish	Ea.		410		410	Cost includes material and labor to
Install	Flight	4000	565		4565	install 4' - 6' diameter spiral stairs with
Demolish and Install	Flight	4000	975		4975	oak treads, factory cut and assembled
Clean	Flight	3.27	26.50		29.77	with double handrails.
Paint	Flight	21	48		69	
Minimum Charge	Job		211		211	

Stairs

Stairs	Unit	Material	Labor	Equip.	Total	Specification
Disappearing						
Demolish	Ea.		21.50		21.50	Includes material and labor to install a
Install	Ea.	275	121		396	folding staircase.
Demolish and Install	Ea.	275	142.50		417.50	
Reinstall	Ea.		96.64		96.64	
Clean	Ea.	.50	19.35		19.85	
Minimum Charge	Job		211		211	

Finish Carpentry

Stair Components

	Unit	Material	Labor	Equip.	Total	Specification
Handrail w / Balusters						Includes material and labor to install
Demolish	L.F.		2.87		2.87	prefinished assembled stair rail with
Install	L.F.	42	8.80		50.80	brackets, turned balusters and newel.
Demolish and Install	L.F.	42	11.67		53.67	
Clean	L.F.	.65	2.14		2.79	
Paint	L.F.	2.29	4.24		6.53	
Minimum Charge	Job		211		211	
Bannister						Includes material and labor to install
Demolish	L.F.		2.87		2.87	built-up oak railings.
Install	L.F.	38	7.05		45.05	
Demolish and Install	L.F.	38	9.92		47.92	
Clean	L.F.	.65	1.94		2.59	
Paint	L.F.	2.29	3.28		5.57	
Minimum Charge	Job		211		211	

Cabinets and Countertops

Wall and Base Cabinets

Countertop

Wall Cabinets	Unit	Material	Labor	Equip.	Total	Specification
Good Grade Laminated						
12″ Wide, 1 Door						
Demolish	Ea.		8.60		8.60	Includes material and labor to install
Install	Ea.	161	38.50		199.50	good modular unit with melamine
Demolish and Install	Ea.	161	47.10		208.10	laminated to particle board, in
Reinstall	Ea.		30.75		30.75	textured colors or wood grain print
Clean	Ea.	.50	5.95		6.45	finish, hinges and pulls.
Paint	Ea.	.68	12.90		13.58	
Minimum Charge	Job		211		211	
15″ Wide, 1 Door						
Demolish	Ea.		10.75		10.75	Includes material and labor to install
Install	Ea.	180	39.50		219.50	good modular unit with melamine
Demolish and Install	Ea.	180	50.25		230.25	laminated to particle board, in
Reinstall	Ea.		31.61		31.61	textured colors or wood grain print
Clean	Ea.	.62	7.35		7.97	finish, hinges and pulls.
Paint	Ea.	.86	16.40		17.26	
Minimum Charge	Job		211		211	
18″ Wide, 1 Door						
Demolish	Ea.		13		13	Includes material and labor to install
Install	Ea.	198	40.50		238.50	good modular unit with melamine
Demolish and Install	Ea.	198	53.50		251.50	laminated to particle board, in
Reinstall	Ea.		32.37		32.37	textured colors or wood grain print
Clean	Ea.	.74	8.85		9.59	finish, hinges and pulls.
Paint	Ea.	1.02	19		20.02	
Minimum Charge	Job		211		211	
21″ Wide, 1 Door						
Demolish	Ea.		14.95		14.95	Includes material and labor to install
Install	Ea.	208	41		249	good modular unit with melamine
Demolish and Install	Ea.	208	55.95		263.95	laminated to particle board, in
Reinstall	Ea.		32.84		32.84	textured colors or wood grain print
Clean	Ea.	.86	10.30		11.16	finish, hinges and pulls.
Paint	Ea.	1.20	22.50		23.70	
Minimum Charge	Job		211		211	
24″ w / Blind Corner						
Demolish	Ea.		17.20		17.20	Includes material and labor to install
Install	Ea.	231	41.50		272.50	good modular unit with melamine
Demolish and Install	Ea.	231	58.70		289.70	laminated to particle board, in
Reinstall	Ea.		33.32		33.32	textured colors or wood grain print
Clean	Ea.	.98	11.90		12.88	finish, hinges and pulls.
Paint	Ea.	1.36	26		27.36	
Minimum Charge	Job		211		211	

207

Cabinets and Countertops

Wall Cabinets	Unit	Material	Labor	Equip.	Total	Specification
27″ Wide, 2 Door						
Demolish	Ea.		19.10		19.10	Includes material and labor to install
Install	Ea.	260	42.50		302.50	good modular unit with melamine
Demolish and Install	Ea.	260	61.60		321.60	laminated to particle board, in
Reinstall	Ea.		34.17		34.17	textured colors or wood grain print
Clean	Ea.	1.10	13.45		14.55	finish, hinges and pulls.
Paint	Ea.	1.55	30		31.55	
Minimum Charge	Job		211		211	
30″ Wide, 2 Door						
Demolish	Ea.		21.50		21.50	Includes material and labor to install
Install	Ea.	270	44		314	good modular unit with melamine
Demolish and Install	Ea.	270	65.50		335.50	laminated to particle board, in
Reinstall	Ea.		35.05		35.05	textured colors or wood grain print
Clean	Ea.	1.22	14.75		15.97	finish, hinges and pulls.
Paint	Ea.	1.72	33		34.72	
Minimum Charge	Job		211		211	
36″ Wide, 2 Door						
Demolish	Ea.		25.50		25.50	Includes material and labor to install
Install	Ea.	305	45		350	good modular unit with melamine
Demolish and Install	Ea.	305	70.50		375.50	laminated to particle board, in
Reinstall	Ea.		35.98		35.98	textured colors or wood grain print
Clean	Ea.	1.47	18.20		19.67	finish, hinges and pulls.
Paint	Ea.	2.08	40		42.08	
Minimum Charge	Job		211		211	
48″ Wide, 2 Door						
Demolish	Ea.		34.50		34.50	Includes material and labor to install
Install	Ea.	375	46		421	good modular unit with melamine
Demolish and Install	Ea.	375	80.50		455.50	laminated to particle board, in
Reinstall	Ea.		36.77		36.77	textured colors or wood grain print
Clean	Ea.	1.96	24		25.96	finish, hinges and pulls.
Paint	Ea.	2.75	51.50		54.25	
Minimum Charge	Job		211		211	
Above-Appliance						
Demolish	Ea.		21.50		21.50	Includes material and labor to install
Install	Ea.	183	34		217	good modular unit with melamine
Demolish and Install	Ea.	183	55.50		238.50	laminated to particle board, in
Reinstall	Ea.		27.28		27.28	textured colors or wood grain print
Clean	Ea.	1.22	14.75		15.97	finish, hinges and pulls.
Paint	Ea.	1.72	33		34.72	
Minimum Charge	Job		211		211	
Better Grade Wood						
12″ Wide, 1 Door						
Demolish	Ea.		8.60		8.60	Includes material and labor to install
Install	Ea.	215	38.50		253.50	custom modular unit with solid
Demolish and Install	Ea.	215	47.10		262.10	hardwood faced frames, hardwood
Reinstall	Ea.		30.75		30.75	door frames and drawer fronts,
Clean	Ea.	.50	5.95		6.45	hardwood veneer on raised door
Paint	Ea.	.68	13.90		14.58	panels, hinges and pulls.
Minimum Charge	Job		211		211	
15″ Wide, 1 Door						
Demolish	Ea.		10.75		10.75	Includes material and labor to install
Install	Ea.	240	39.50		279.50	custom modular unit with solid
Demolish and Install	Ea.	240	50.25		290.25	hardwood faced frames, hardwood
Reinstall	Ea.		31.61		31.61	door frames and drawer fronts,
Clean	Ea.	.62	7.35		7.97	hardwood veneer on raised door
Paint	Ea.	.86	17.20		18.06	panels, hinges and pulls.
Minimum Charge	Job		211		211	

Cabinets and Countertops

Wall Cabinets	Unit	Material	Labor	Equip.	Total	Specification
18″ Wide, 1 Door						
Demolish	Ea.		13		13	Includes material and labor to install
Install	Ea.	264	40.50		304.50	custom modular unit with solid
Demolish and Install	Ea.	264	53.50		317.50	hardwood faced frames, hardwood
Reinstall	Ea.		32.37		32.37	door frames and drawer fronts,
Clean	Ea.	.74	8.85		9.59	hardwood veneer on raised door
Paint	Ea.	1.02	21		22.02	panels, hinges and pulls.
Minimum Charge	Job		211		211	
21″ Wide, 1 Door						
Demolish	Ea.		14.95		14.95	Includes material and labor to install
Install	Ea.	277	41		318	custom modular unit with solid
Demolish and Install	Ea.	277	55.95		332.95	hardwood faced frames, hardwood
Reinstall	Ea.		32.84		32.84	door frames and drawer fronts,
Clean	Ea.	.86	10.30		11.16	hardwood veneer on raised door
Paint	Ea.	1.20	24		25.20	panels, hinges and pulls.
Minimum Charge	Job		211		211	
24″ w / Blind Corner						
Demolish	Ea.		17.20		17.20	Includes material and labor to install
Install	Ea.	310	41.50		351.50	custom modular unit with solid
Demolish and Install	Ea.	310	58.70		368.70	hardwood faced frames, hardwood
Reinstall	Ea.		33.32		33.32	door frames and drawer fronts,
Clean	Ea.	.98	11.90		12.88	hardwood veneer on raised door
Paint	Ea.	1.36	28		29.36	panels, hinges and pulls.
Minimum Charge	Job		211		211	
27″ Wide, 2 Door						
Demolish	Ea.		19.10		19.10	Includes material and labor to install
Install	Ea.	345	42.50		387.50	custom modular unit with solid
Demolish and Install	Ea.	345	61.60		406.60	hardwood faced frames, hardwood
Reinstall	Ea.		34.17		34.17	door frames and drawer fronts,
Clean	Ea.	1.10	13.45		14.55	hardwood veneer on raised door
Paint	Ea.	1.55	30		31.55	panels, hinges and pulls.
Minimum Charge	Job		211		211	
30″ Wide, 2 Door						
Demolish	Ea.		21.50		21.50	Includes material and labor to install
Install	Ea.	360	44		404	custom modular unit with solid
Demolish and Install	Ea.	360	65.50		425.50	hardwood faced frames, hardwood
Reinstall	Ea.		35.05		35.05	door frames and drawer fronts,
Clean	Ea.	1.22	14.75		15.97	hardwood veneer on raised door
Paint	Ea.	1.72	36		37.72	panels, hinges and pulls.
Minimum Charge	Job		211		211	
36″ Wide, 2 Door						
Demolish	Ea.		25.50		25.50	Includes material and labor to install
Install	Ea.	405	45		450	custom modular unit with solid
Demolish and Install	Ea.	405	70.50		475.50	hardwood faced frames, hardwood
Reinstall	Ea.		35.98		35.98	door frames and drawer fronts,
Clean	Ea.	1.47	18.20		19.67	hardwood veneer on raised door
Paint	Ea.	2.08	40		42.08	panels, hinges and pulls.
Minimum Charge	Job		211		211	
48″ Wide, 2 Door						
Demolish	Ea.		34.50		34.50	Includes material and labor to install
Install	Ea.	500	46		546	custom modular unit with solid
Demolish and Install	Ea.	500	80.50		580.50	hardwood faced frames, hardwood
Reinstall	Ea.		36.77		36.77	door frames and drawer fronts,
Clean	Ea.	1.96	24		25.96	hardwood veneer on raised door
Paint	Ea.	2.75	51.50		54.25	panels, hinges and pulls.
Minimum Charge	Job		211		211	

209

Cabinets and Countertops

Wall Cabinets	Unit	Material	Labor	Equip.	Total	Specification
Above-appliance						
Demolish	Ea.		21.50		21.50	Includes material and labor to install
Install	Ea.	243	34		277	custom modular unit with solid
Demolish and Install	Ea.	243	55.50		298.50	hardwood faced frames, hardwood
Reinstall	Ea.		27.28		27.28	door frames and drawer fronts,
Clean	Ea.	1.22	14.75		15.97	hardwood veneer on raised door
Paint	Ea.	1.72	36		37.72	panels, hinges and pulls.
Minimum Charge	Job		211		211	
Premium Grade Wood						
12″ Wide, 1 Door						
Demolish	Ea.		8.60		8.60	Includes material and labor to install
Install	Ea.	257	38.50		295.50	premium modular unit with solid
Demolish and Install	Ea.	257	47.10		304.10	hardwood faced frames, hardwood
Reinstall	Ea.		30.75		30.75	door frames and drawer fronts,
Clean	Ea.	.50	5.95		6.45	hardwood veneer on raised door
Paint	Ea.	.68	15.05		15.73	panels, hinges and pulls.
Minimum Charge	Job		211		211	
15″ Wide, 1 Door						
Demolish	Ea.		10.75		10.75	Includes material and labor to install
Install	Ea.	320	39.50		359.50	premium modular unit with solid
Demolish and Install	Ea.	320	50.25		370.25	hardwood faced frames, hardwood
Reinstall	Ea.		31.61		31.61	door frames and drawer fronts,
Clean	Ea.	.62	7.35		7.97	hardwood veneer on raised door
Paint	Ea.	.86	19		19.86	panels, hinges and pulls.
Minimum Charge	Job		211		211	
18″ Wide, 1 Door						
Demolish	Ea.		13		13	Includes material and labor to install
Install	Ea.	315	40.50		355.50	premium modular unit with solid
Demolish and Install	Ea.	315	53.50		368.50	hardwood faced frames, hardwood
Reinstall	Ea.		32.37		32.37	door frames and drawer fronts,
Clean	Ea.	.74	8.85		9.59	hardwood veneer on raised door
Paint	Ea.	1.02	22.50		23.52	panels, hinges and pulls.
Minimum Charge	Job		211		211	
21″ Wide, 1 Door						
Demolish	Ea.		14.95		14.95	Includes material and labor to install
Install	Ea.	335	41		376	premium modular unit with solid
Demolish and Install	Ea.	335	55.95		390.95	hardwood faced frames, hardwood
Reinstall	Ea.		32.84		32.84	door frames and drawer fronts,
Clean	Ea.	.86	10.30		11.16	hardwood veneer on raised door
Paint	Ea.	1.20	26		27.20	panels, hinges and pulls.
Minimum Charge	Job		211		211	
24″ w / Blind Corner						
Demolish	Ea.		17.20		17.20	Includes material and labor to install
Install	Ea.	370	41.50		411.50	premium modular unit with solid
Demolish and Install	Ea.	370	58.70		428.70	hardwood faced frames, hardwood
Reinstall	Ea.		33.32		33.32	door frames and drawer fronts,
Clean	Ea.	.98	11.90		12.88	hardwood veneer on raised door
Paint	Ea.	1.36	30		31.36	panels, hinges and pulls.
Minimum Charge	Job		211		211	
27″ Wide, 2 Door						
Demolish	Ea.		19.10		19.10	Includes material and labor to install
Install	Ea.	345	42.50		387.50	premium modular unit with solid
Demolish and Install	Ea.	345	61.60		406.60	hardwood faced frames, hardwood
Reinstall	Ea.		34.17		34.17	door frames and drawer fronts,
Clean	Ea.	1.10	13.45		14.55	hardwood veneer on raised door
Paint	Ea.	1.55	33		34.55	panels, hinges and pulls.
Minimum Charge	Job		211		211	

Cabinets and Countertops

Wall Cabinets	Unit	Material	Labor	Equip.	Total	Specification
30″ Wide, 2 Door						
Demolish	Ea.		21.50		21.50	Includes material and labor to install
Install	Ea.	430	44		474	premium modular unit with solid
Demolish and Install	Ea.	430	65.50		495.50	hardwood faced frames, hardwood
Reinstall	Ea.		35.05		35.05	door frames and drawer fronts,
Clean	Ea.	1.22	14.75		15.97	hardwood veneer on raised door
Paint	Ea.	1.72	36		37.72	panels, hinges and pulls.
Minimum Charge	Job		211		211	
36″ Wide, 2 Door						
Demolish	Ea.		25.50		25.50	Includes material and labor to install
Install	Ea.	490	45		535	premium modular unit with solid
Demolish and Install	Ea.	490	70.50		560.50	hardwood faced frames, hardwood
Reinstall	Ea.		35.98		35.98	door frames and drawer fronts,
Clean	Ea.	1.47	18.20		19.67	hardwood veneer on raised door
Paint	Ea.	2.08	45		47.08	panels, hinges and pulls.
Minimum Charge	Job		211		211	
48″ Wide, 2 Door						
Demolish	Ea.		34.50		34.50	Includes material and labor to install
Install	Ea.	600	46		646	premium modular unit with solid
Demolish and Install	Ea.	600	80.50		680.50	hardwood faced frames, hardwood
Reinstall	Ea.		36.77		36.77	door frames and drawer fronts,
Clean	Ea.	1.96	24		25.96	hardwood veneer on raised door
Paint	Ea.	2.75	60		62.75	panels, hinges and pulls.
Minimum Charge	Job		211		211	
Above-appliance						
Demolish	Ea.		21.50		21.50	Includes material and labor to install
Install	Ea.	292	34		326	premium modular unit with solid
Demolish and Install	Ea.	292	55.50		347.50	hardwood faced frames, hardwood
Reinstall	Ea.		27.28		27.28	door frames and drawer fronts,
Clean	Ea.	1.22	14.75		15.97	hardwood veneer on raised door
Paint	Ea.	1.72	36		37.72	panels, hinges, pulls.
Minimum Charge	Job		211		211	
Premium Hardwood						
12″ Wide, 1 Door						
Demolish	Ea.		8.60		8.60	Includes material and labor to install
Install	Ea.	281	38.50		319.50	premium hardwood modular unit with
Demolish and Install	Ea.	281	47.10		328.10	solid hardwood faced frames, drawer
Reinstall	Ea.		30.75		30.75	fronts, and door panels, steel drawer
Clean	Ea.	.50	5.95		6.45	guides, hinges, pulls, laminated
Paint	Ea.	.68	16.40		17.08	interior.
Minimum Charge	Job		211		211	
15″ Wide, 1 Door						
Demolish	Ea.		10.75		10.75	Includes material and labor to install
Install	Ea.	420	39.50		459.50	premium hardwood modular unit with
Demolish and Install	Ea.	420	50.25		470.25	solid hardwood faced frames, drawer
Reinstall	Ea.		31.61		31.61	fronts, and door panels, steel drawer
Clean	Ea.	.62	7.35		7.97	guides, hinges, pulls, laminated
Paint	Ea.	.86	20		20.86	interior.
Minimum Charge	Job		211		211	
18″ Wide, 1 Door						
Demolish	Ea.		13		13	Includes material and labor to install
Install	Ea.	345	40.50		385.50	premium hardwood modular unit with
Demolish and Install	Ea.	345	53.50		398.50	solid hardwood faced frames, drawer
Reinstall	Ea.		32.37		32.37	fronts, and door panels, steel drawer
Clean	Ea.	.74	8.85		9.59	guides, hinges, pulls, laminated
Paint	Ea.	1.02	24		25.02	interior.
Minimum Charge	Job		211		211	

Cabinets and Countertops

Wall Cabinets	Unit	Material	Labor	Equip.	Total	Specification
21″ Wide, 1 Door						
Demolish	Ea.		14.95		14.95	Includes material and labor to install
Install	Ea.	365	41		406	premium hardwood modular unit with
Demolish and Install	Ea.	365	55.95		420.95	solid hardwood faced frames, drawer
Reinstall	Ea.		32.84		32.84	fronts, and door panels, steel drawer
Clean	Ea.	.86	10.30		11.16	guides, hinges, pulls, laminated
Paint	Ea.	1.20	28		29.20	interior.
Minimum Charge	Job		211		211	
24″ w / Blind Corner						
Demolish	Ea.		17.20		17.20	Includes material and labor to install
Install	Ea.	400	41.50		441.50	premium hardwood modular unit with
Demolish and Install	Ea.	400	58.70		458.70	solid hardwood faced frames, drawer
Reinstall	Ea.		33.32		33.32	fronts, and door panels, steel drawer
Clean	Ea.	.98	11.90		12.88	guides, hinges, pulls, laminated
Paint	Ea.	1.36	33		34.36	interior.
Minimum Charge	Job		211		211	
27″ Wide, 2 Door						
Demolish	Ea.		19.10		19.10	Includes material and labor to install
Install	Ea.	455	42.50		497.50	premium hardwood modular unit with
Demolish and Install	Ea.	455	61.60		516.60	solid hardwood faced frames, drawer
Reinstall	Ea.		34.17		34.17	fronts, and door panels, steel drawer
Clean	Ea.	1.10	13.45		14.55	guides, hinges, pulls, laminated
Paint	Ea.	1.55	36		37.55	interior.
Minimum Charge	Job		211		211	
30″ Wide, 2 Door						
Demolish	Ea.		21.50		21.50	Includes material and labor to install
Install	Ea.	470	44		514	premium hardwood modular unit with
Demolish and Install	Ea.	470	65.50		535.50	solid hardwood faced frames, drawer
Reinstall	Ea.		35.05		35.05	fronts, and door panels, steel drawer
Clean	Ea.	1.22	14.75		15.97	guides, hinges, pulls, laminated
Paint	Ea.	1.72	40		41.72	interior.
Minimum Charge	Job		211		211	
36″ Wide, 2 Door						
Demolish	Ea.		25.50		25.50	Includes material and labor to install
Install	Ea.	535	45		580	premium hardwood modular unit with
Demolish and Install	Ea.	535	70.50		605.50	solid hardwood faced frames, drawer
Reinstall	Ea.		35.98		35.98	fronts, and door panels, steel drawer
Clean	Ea.	1.47	18.20		19.67	guides, hinges, pulls, laminated
Paint	Ea.	2.08	51.50		53.58	interior.
Minimum Charge	Job		211		211	
48″ Wide, 2 Door						
Demolish	Ea.		34.50		34.50	Includes material and labor to install
Install	Ea.	655	46		701	premium hardwood modular unit with
Demolish and Install	Ea.	655	80.50		735.50	solid hardwood faced frames, drawer
Reinstall	Ea.		36.77		36.77	fronts, and door panels, steel drawer
Clean	Ea.	1.96	24		25.96	guides, hinges, pulls, laminated
Paint	Ea.	2.75	60		62.75	interior.
Minimum Charge	Job		211		211	
Above-appliance						
Demolish	Ea.		21.50		21.50	Includes material and labor to install
Install	Ea.	320	34		354	premium hardwood modular unit with
Demolish and Install	Ea.	320	55.50		375.50	solid hardwood faced frames, drawer
Reinstall	Ea.		27.28		27.28	fronts, and door panels, steel drawer
Clean	Ea.	1.22	14.75		15.97	guides, hinges, pulls, laminated
Paint	Ea.	1.72	40		41.72	interior.
Minimum Charge	Job		211		211	

Cabinets and Countertops

Wall Cabinets	Unit	Material	Labor	Equip.	Total	Specification
Medicine Cabinet						
Good Grade						
Demolish	Ea.		14.35		14.35	Cost includes material and labor to
Install	Ea.	165	30		195	install 14" x 18" recessed medicine
Demolish and Install	Ea.	165	44.35		209.35	cabinet with stainless steel frame,
Reinstall	Ea.		24.16		24.16	mirror. Light not included.
Clean	Ea.	.02	14.20		14.22	
Paint	Ea.	2.02	22.50		24.52	
Minimum Charge	Job		211		211	
Good Grade Lighted						
Demolish	Ea.		14.35		14.35	Cost includes material and labor to
Install	Ea.	420	35		455	install 14" x 18" recessed medicine
Demolish and Install	Ea.	420	49.35		469.35	cabinet with stainless steel frame,
Reinstall	Ea.		28.19		28.19	mirror and light.
Clean	Ea.	.02	14.20		14.22	
Paint	Ea.	2.02	22.50		24.52	
Minimum Charge	Job		211		211	
Better Grade						
Demolish	Ea.		14.35		14.35	Includes material and labor to install
Install	Ea.	275	30		305	recessed medicine cabinet with
Demolish and Install	Ea.	275	44.35		319.35	polished chrome frame and mirror.
Reinstall	Ea.		24.16		24.16	Light not included.
Clean	Ea.	.02	14.20		14.22	
Paint	Ea.	2.02	22.50		24.52	
Minimum Charge	Job		211		211	
Better Grade Lighted						
Demolish	Ea.		14.35		14.35	Includes material and labor to install
Install	Ea.	530	35		565	recessed medicine cabinet with
Demolish and Install	Ea.	530	49.35		579.35	polished chrome frame, mirror and
Reinstall	Ea.		28.19		28.19	light.
Clean	Ea.	.02	14.20		14.22	
Paint	Ea.	2.02	22.50		24.52	
Minimum Charge	Job		211		211	
Premium Grade						
Demolish	Ea.		14.35		14.35	Cost includes material and labor to
Install	Ea.	330	30		360	install 14" x 24" recessed medicine
Demolish and Install	Ea.	330	44.35		374.35	cabinet with beveled mirror, frameless
Reinstall	Ea.		24.16		24.16	swing door. Light not included.
Clean	Ea.	.02	14.20		14.22	
Paint	Ea.	2.02	22.50		24.52	
Minimum Charge	Job		211		211	
Premium Grade Light						
Demolish	Ea.		14.35		14.35	Cost includes material and labor to
Install	Ea.	580	35		615	install 36" x 30" surface mounted
Demolish and Install	Ea.	580	49.35		629.35	medicine cabinet with beveled tri-view
Reinstall	Ea.		28.19		28.19	mirror, frameless swing door and light.
Clean	Ea.	.02	14.20		14.22	
Paint	Ea.	2.02	22.50		24.52	
Minimum Charge	Job		211		211	
Bath Mirror						
Demolish	Ea.		10.75		10.75	Cost includes material and labor to
Install	Ea.	48	21		69	install 18" x 24" surface mount mirror
Demolish and Install	Ea.	48	31.75		79.75	with stainless steel frame.
Reinstall	Ea.		16.91		16.91	
Clean	Ea.	.02	.94		.96	
Minimum Charge	Job		211		211	

Cabinets and Countertops

Wall Cabinets		Unit	Material	Labor	Equip.	Total	Specification
Garage							
	Demolish	L.F.		8.60		8.60	Includes material and labor to install
	Install	L.F.	87.50	21		108.50	good modular unit with melamine
	Demolish and Install	L.F.	87.50	29.60		117.10	laminated to particle board, in
	Reinstall	L.F.		16.91		16.91	textured colors or wood grain print
	Clean	L.F.	.50	5.95		6.45	finish, hinges and pulls.
	Paint	L.F.	.68	12.90		13.58	
	Minimum Charge	Job		211		211	
Strip and Refinish							
	Install	L.F.	4.61	45		49.61	Includes labor and material to strip,
	Minimum Charge	Job		180		180	prep and refinish exterior of cabinets.

Tall Cabinets		Unit	Material	Labor	Equip.	Total	Specification
Good Grade Laminated							
	Demolish	L.F.		15.30		15.30	Includes material and labor to install
	Install	L.F.	217	47		264	custom modular unit with solid
	Demolish and Install	L.F.	217	62.30		279.30	hardwood faced frames, hardwood
	Reinstall	L.F.		37.58		37.58	door frames and drawer fronts,
	Clean	L.F.	1.31	11.05		12.36	hardwood veneer on raised door
	Paint	L.F.	1.60	26		27.60	panels, hinges and pulls.
	Minimum Charge	Job		211		211	
Better Grade Wood							
	Demolish	L.F.		15.30		15.30	Includes material and labor to install
	Install	L.F.	310	47		357	custom modular unit with solid
	Demolish and Install	L.F.	310	62.30		372.30	hardwood faced frames, hardwood
	Reinstall	L.F.		37.58		37.58	door frames and drawer fronts,
	Clean	L.F.	1.31	11.05		12.36	hardwood veneer on raised door
	Paint	L.F.	1.60	28		29.60	panels, hinges and pulls.
	Minimum Charge	Job		211		211	
Premium Grade Wood							
	Demolish	L.F.		15.30		15.30	Includes material and labor to install
	Install	L.F.	445	47		492	premium modular unit with solid
	Demolish and Install	L.F.	445	62.30		507.30	hardwood faced frames, hardwood
	Reinstall	L.F.		37.58		37.58	door frames and drawer fronts,
	Clean	L.F.	1.31	11.05		12.36	hardwood veneer on raised door
	Paint	L.F.	1.60	30		31.60	panels, hinges and pulls.
	Minimum Charge	Job		211		211	
Premium Hardwood							
	Demolish	L.F.		15.30		15.30	Includes material and labor to install
	Install	L.F.	620	47		667	premium hardwood modular unit with
	Demolish and Install	L.F.	620	62.30		682.30	solid hardwood faced frames, drawer
	Reinstall	L.F.		37.58		37.58	fronts, and door panels, steel drawer
	Clean	L.F.	1.31	11.05		12.36	guides, hinges and pulls, laminated
	Paint	L.F.	1.60	33		34.60	interior.
	Minimum Charge	Job		211		211	

Cabinets and Countertops

Tall Cabinets	Unit	Material	Labor	Equip.	Total	Specification
Single Oven, 27" Wide						
Good Grade Laminated						
Demolish	Ea.		34.50		34.50	Includes material and labor to install
Install	Ea.	755	106		861	good modular unit with melamine
Demolish and Install	Ea.	755	140.50		895.50	laminated to particle board, in
Reinstall	Ea.		84.56		84.56	textured colors or wood grain print
Clean	Ea.	4.08	19.35		23.43	finish, hinges and pulls.
Paint	Ea.	2.68	57.50		60.18	
Minimum Charge	Job		211		211	
Better Grade Veneer						
Demolish	Ea.		34.50		34.50	Includes material and labor to install
Install	Ea.	1325	106		1431	custom modular unit with solid
Demolish and Install	Ea.	1325	140.50		1465.50	hardwood faced frames, hardwood
Reinstall	Ea.		84.56		84.56	door frames and drawer fronts,
Clean	Ea.	4.08	19.35		23.43	hardwood veneer on raised door
Paint	Ea.	2.68	62		64.68	panels, hinges and pulls.
Minimum Charge	Job		211		211	
Premium Grade Veneer						
Demolish	Ea.		34.50		34.50	Includes material and labor to install
Install	Ea.	1200	106		1306	premium modular unit with solid
Demolish and Install	Ea.	1200	140.50		1340.50	hardwood faced frames, hardwood
Reinstall	Ea.		84.56		84.56	door frames and drawer fronts,
Clean	Ea.	4.08	19.35		23.43	hardwood veneer on raised door
Paint	Ea.	2.68	68		70.68	panels, hinges and pulls.
Minimum Charge	Job		211		211	
Premium Hardwood						
Demolish	Ea.		34.50		34.50	Includes material and labor to install
Install	Ea.	1325	106		1431	premium hardwood modular unit with
Demolish and Install	Ea.	1325	140.50		1465.50	solid hardwood faced frames, drawer
Reinstall	Ea.		84.56		84.56	fronts, and door panels, steel drawer
Clean	Ea.	4.08	19.35		23.43	guides, hinges and pulls, laminated
Paint	Ea.	2.68	73.50		76.18	interior.
Minimum Charge	Job		211		211	
Single Oven, 30" Wide						
Good Grade Laminated						
Demolish	Ea.		34.50		34.50	Includes material and labor to install
Install	Ea.	930	106		1036	good modular unit with melamine
Demolish and Install	Ea.	930	140.50		1070.50	laminated to particle board, in
Reinstall	Ea.		84.56		84.56	textured colors or wood grain print
Clean	Ea.	4.08	19.35		23.43	finish, hinges and pulls.
Paint	Ea.	2.68	57.50		60.18	
Minimum Charge	Job		211		211	
Better Grade Veneer						
Demolish	Ea.		34.50		34.50	Includes material and labor to install
Install	Ea.	1025	106		1131	custom modular unit with solid
Demolish and Install	Ea.	1025	140.50		1165.50	hardwood faced frames, hardwood
Reinstall	Ea.		84.56		84.56	door frames and drawer fronts,
Clean	Ea.	4.08	19.35		23.43	hardwood veneer on raised door
Paint	Ea.	2.68	57.50		60.18	panels, hinges and pulls.
Minimum Charge	Job		211		211	

Cabinets and Countertops

Tall Cabinets

	Unit	Material	Labor	Equip.	Total	Specification
Premium Grade Veneer						
Demolish	Ea.		34.50		34.50	Includes material and labor to install
Install	Ea.	1175	106		1281	premium modular unit with solid
Demolish and Install	Ea.	1175	140.50		1315.50	hardwood faced frames, hardwood
Reinstall	Ea.		84.56		84.56	door frames and drawer fronts,
Clean	Ea.	4.08	19.35		23.43	hardwood veneer on raised door
Paint	Ea.	2.68	57.50		60.18	panels, hinges and pulls.
Minimum Charge	Job		211		211	
Premium Hardwood						
Demolish	Ea.		34.50		34.50	Includes material and labor to install
Install	Ea.	1275	106		1381	premium hardwood modular unit with
Demolish and Install	Ea.	1275	140.50		1415.50	solid hardwood faced frames, drawer
Reinstall	Ea.		84.56		84.56	fronts, and door panels, steel drawer
Clean	Ea.	4.08	19.35		23.43	guides, hinges and pulls, laminated
Paint	Ea.	2.68	57.50		60.18	interior.
Minimum Charge	Job		211		211	

Utility 18" W x 12" D

	Unit	Material	Labor	Equip.	Total	Specification
Good Grade Laminated						
Demolish	Ea.		34.50		34.50	Includes material and labor to install
Install	Ea.	625	84.50		709.50	good modular unit with melamine
Demolish and Install	Ea.	625	119		744	laminated to particle board, in
Reinstall	Ea.		67.65		67.65	textured colors or wood grain print
Clean	Ea.	.96	9.40		10.36	finish, hinges and pulls.
Paint	Ea.	1.02	22.50		23.52	
Minimum Charge	Job		211		211	
Better Grade Veneer						
Demolish	Ea.		34.50		34.50	Includes material and labor to install
Install	Ea.	835	84.50		919.50	custom modular unit with solid
Demolish and Install	Ea.	835	119		954	hardwood faced frames, hardwood
Reinstall	Ea.		67.65		67.65	door frames and drawer fronts,
Clean	Ea.	.96	9.40		10.36	hardwood veneer on raised door
Paint	Ea.	1.02	24		25.02	panels, hinges and pulls.
Minimum Charge	Job		211		211	
Premium Grade Veneer						
Demolish	Ea.		34.50		34.50	Includes material and labor to install
Install	Ea.	1000	84.50		1084.50	premium modular unit with solid
Demolish and Install	Ea.	1000	119		1119	hardwood faced frames, hardwood
Reinstall	Ea.		67.65		67.65	door frames and drawer fronts,
Clean	Ea.	.96	9.40		10.36	hardwood veneer on raised door
Paint	Ea.	1.02	28		29.02	panels, hinges and pulls.
Minimum Charge	Job		211		211	
Premium Hardwood						
Demolish	Ea.		34.50		34.50	Includes material and labor to install
Install	Ea.	1100	84.50		1184.50	premium hardwood modular unit with
Demolish and Install	Ea.	1100	119		1219	solid hardwood faced frames, drawer
Reinstall	Ea.		67.65		67.65	fronts, and door panels, steel drawer
Clean	Ea.	.96	9.40		10.36	guides, hinges and pulls, laminated
Paint	Ea.	1.02	30		31.02	interior.
Minimum Charge	Job		211		211	

Cabinets and Countertops

Tall Cabinets	Unit	Material	Labor	Equip.	Total	Specification
Utility 24″ W x 24″ D						
Good Grade Laminated						
Demolish	Ea.		34.50		34.50	Includes material and labor to install
Install	Ea.	610	106		716	good modular unit with melamine
Demolish and Install	Ea.	610	140.50		750.50	laminated to particle board, in
Reinstall	Ea.		84.56		84.56	textured colors or wood grain print
Clean	Ea.	1.25	12.40		13.65	finish, hinges and pulls.
Paint	Ea.	1.36	30		31.36	
Minimum Charge	Job		211		211	
Better Grade Veneer						
Demolish	Ea.		34.50		34.50	Includes material and labor to install
Install	Ea.	815	106		921	custom modular unit with solid
Demolish and Install	Ea.	815	140.50		955.50	hardwood faced frames, hardwood
Reinstall	Ea.		84.56		84.56	door frames and drawer fronts,
Clean	Ea.	1.25	12.40		13.65	hardwood veneer on raised door
Paint	Ea.	1.36	33		34.36	panels, hinges and pulls.
Minimum Charge	Job		211		211	
Premium Grade Veneer						
Demolish	Ea.		34.50		34.50	Includes material and labor to install
Install	Ea.	980	106		1086	premium modular unit with solid
Demolish and Install	Ea.	980	140.50		1120.50	hardwood faced frames, hardwood
Reinstall	Ea.		84.56		84.56	door frames and drawer fronts,
Clean	Ea.	1.25	12.40		13.65	hardwood veneer on raised door
Paint	Ea.	1.36	36		37.36	panels, hinges and pulls.
Minimum Charge	Job		211		211	
Premium Hardwood						
Demolish	Ea.		34.50		34.50	Includes material and labor to install
Install	Ea.	1075	106		1181	premium hardwood modular unit with
Demolish and Install	Ea.	1075	140.50		1215.50	solid hardwood faced frames, drawer
Reinstall	Ea.		84.56		84.56	fronts, and door panels, steel drawer
Clean	Ea.	1.25	12.40		13.65	guides, hinges and pulls, laminated
Paint	Ea.	1.36	40		41.36	interior.
Minimum Charge	Job		211		211	
Plain Shelves						
Install	Ea.	28	23.50		51.50	Includes labor and material to install
Rotating Shelves						tall cabinet shelving, per set.
Install	Ea.	88	21		109	Includes labor and material to install
Strip and Refinish						lazy susan shelving for wall cabinet.
Install	L.F.	9.20	72		81.20	Includes labor and material to strip,
Minimum Charge	Job		180		180	prep and refinish exterior of cabinets.
Stain						
Install	L.F.	.26	3.28		3.54	Includes labor and material to replace
Minimum Charge	Job		180		180	normal prep and stain on doors and
						exterior cabinets.

Cabinets and Countertops

Base Cabinets	Unit	Material	Labor	Equip.	Total	Specification
Good Grade Laminated						
12" w, 1 Door, 1 Drawer						
Demolish	Ea.		8.60		8.60	Includes material and labor to install
Install	Ea.	198	34		232	good modular unit with melamine
Demolish and Install	Ea.	198	42.60		240.60	laminate, textured colors or wood
Reinstall	Ea.		27.28		27.28	grain print finish including hinges and
Clean	Ea.	.65	6.20		6.85	pulls.
Paint	Ea.	.68	15.05		15.73	
Minimum Charge	Job		211		211	
15" w, 1 Door, 1 Drawer						
Demolish	Ea.		10.75		10.75	Includes material and labor to install
Install	Ea.	207	35		242	good modular unit with melamine
Demolish and Install	Ea.	207	45.75		252.75	laminate, textured colors or wood
Reinstall	Ea.		28.19		28.19	grain print finish including hinges and
Clean	Ea.	.81	7.75		8.56	pulls.
Paint	Ea.	.86	19		19.86	
Minimum Charge	Job		211		211	
18" w, 1 Door, 1 Drawer						
Demolish	Ea.		12.75		12.75	Includes material and labor to install
Install	Ea.	224	36.50		260.50	good modular unit with melamine
Demolish and Install	Ea.	224	49.25		273.25	laminate, textured colors or wood
Reinstall	Ea.		29.03		29.03	grain print finish including hinges and
Clean	Ea.	.96	9.40		10.36	pulls.
Paint	Ea.	1.02	22.50		23.52	
Minimum Charge	Job		211		211	
21" w, 1 Door, 1 Drawer						
Demolish	Ea.		14.95		14.95	Includes material and labor to install
Install	Ea.	235	37.50		272.50	good modular unit with melamine
Demolish and Install	Ea.	235	52.45		287.45	laminate, textured colors or wood
Reinstall	Ea.		29.80		29.80	grain print finish including hinges and
Clean	Ea.	1.17	11.05		12.22	pulls.
Paint	Ea.	1.20	26		27.20	
Minimum Charge	Job		211		211	
24" w, 1 Door, 1 Drawer						
Demolish	Ea.		17.20		17.20	Includes material and labor to install
Install	Ea.	271	38		309	good modular unit with melamine
Demolish and Install	Ea.	271	55.20		326.20	laminate, textured colors or wood
Reinstall	Ea.		30.34		30.34	grain print finish including hinges and
Clean	Ea.	1.25	12.40		13.65	pulls.
Paint	Ea.	1.36	30		31.36	
Minimum Charge	Job		211		211	
30" Wide Sink Base						
Demolish	Ea.		21.50		21.50	Includes material and labor to install
Install	Ea.	271	39.50		310.50	good modular unit with melamine
Demolish and Install	Ea.	271	61		332	laminate, textured colors or wood
Reinstall	Ea.		31.61		31.61	grain print finish including hinges and
Clean	Ea.	1.63	15.50		17.13	pulls.
Paint	Ea.	1.72	36		37.72	
Minimum Charge	Job		211		211	
36" Wide Sink Base						
Demolish	Ea.		26.50		26.50	Includes material and labor to install
Install	Ea.	305	41.50		346.50	good modular unit with melamine
Demolish and Install	Ea.	305	68		373	laminate, textured colors or wood
Reinstall	Ea.		33.32		33.32	grain print finish including hinges and
Clean	Ea.	1.82	18.20		20.02	pulls.
Paint	Ea.	2.08	45		47.08	
Minimum Charge	Job		211		211	

Cabinets and Countertops

Base Cabinets	Unit	Material	Labor	Equip.	Total	Specification
36" Blind Corner						
Demolish	Ea.		26.50		26.50	Includes material and labor to install
Install	Ea.	470	47		517	good modular unit with melamine
Demolish and Install	Ea.	470	73.50		543.50	laminate, textured colors or wood
Reinstall	Ea.		37.58		37.58	grain print finish including hinges and
Clean	Ea.	1.82	18.20		20.02	pulls.
Paint	Ea.	2.08	45		47.08	
Minimum Charge	Job		211		211	
42" w, 2 Door, 2 Drawer						
Demolish	Ea.		31.50		31.50	Includes material and labor to install
Install	Ea.	370	42.50		412.50	good modular unit with melamine
Demolish and Install	Ea.	370	74		444	laminate, textured colors or wood
Reinstall	Ea.		34.17		34.17	grain print finish including hinges and
Clean	Ea.	2.33	22		24.33	pulls.
Paint	Ea.	2.38	51.50		53.88	
Minimum Charge	Job		211		211	
Clean Interior						
Clean	L.F.	.65	4.69		5.34	Includes labor and materials to clean
Minimum Charge	Job		155		155	the interior of a 27" cabinet.
Better Grade Wood						
12" w, 1 Door, 1 Drawer						
Demolish	Ea.		8.60		8.60	Includes material and labor to install
Install	Ea.	264	34		298	custom modular unit including solid
Demolish and Install	Ea.	264	42.60		306.60	hardwood faced frames, hardwood
Reinstall	Ea.		27.28		27.28	door frames and drawer fronts,
Clean	Ea.	.65	6.20		6.85	hardwood veneer on raised door
Paint	Ea.	.68	16.40		17.08	panels, hinges and pulls.
Minimum Charge	Job		211		211	
15" w, 1 Door, 1 Drawer						
Demolish	Ea.		10.75		10.75	Includes material and labor to install
Install	Ea.	275	35		310	custom modular unit including solid
Demolish and Install	Ea.	275	45.75		320.75	hardwood faced frames, hardwood
Reinstall	Ea.		28.19		28.19	door frames and drawer fronts,
Clean	Ea.	.81	7.75		8.56	hardwood veneer on raised door
Paint	Ea.	.86	21		21.86	panels, hinges and pulls.
Minimum Charge	Job		211		211	
18" w, 1 Door, 1 Drawer						
Demolish	Ea.		12.75		12.75	Includes material and labor to install
Install	Ea.	299	36.50		335.50	custom modular unit including solid
Demolish and Install	Ea.	299	49.25		348.25	hardwood faced frames, hardwood
Reinstall	Ea.		29.03		29.03	door frames and drawer fronts,
Clean	Ea.	.96	9.40		10.36	hardwood veneer on raised door
Paint	Ea.	1.02	24		25.02	panels, hinges and pulls.
Minimum Charge	Job		211		211	
21" w, 1 Door, 1 Drawer						
Demolish	Ea.		14.95		14.95	Includes material and labor to install
Install	Ea.	315	37.50		352.50	custom modular unit including solid
Demolish and Install	Ea.	315	52.45		367.45	hardwood faced frames, hardwood
Reinstall	Ea.		29.80		29.80	door frames and drawer fronts,
Clean	Ea.	1.17	11.05		12.22	hardwood veneer on raised door
Paint	Ea.	1.20	28		29.20	panels, hinges and pulls.
Minimum Charge	Job		211		211	

Cabinets and Countertops

Base Cabinets	Unit	Material	Labor	Equip.	Total	Specification
24" w, 1 Door, 1 Drawer						
Demolish	Ea.		17.20		17.20	Includes material and labor to install
Install	Ea.	365	38		403	custom modular unit including solid
Demolish and Install	Ea.	365	55.20		420.20	hardwood faced frames, hardwood
Reinstall	Ea.		30.34		30.34	door frames and drawer fronts,
Clean	Ea.	1.25	12.40		13.65	hardwood veneer on raised door
Paint	Ea.	1.36	33		34.36	panels, hinges and pulls.
Minimum Charge	Job		211		211	
30" Wide Sink Base						
Demolish	Ea.		21.50		21.50	Includes material and labor to install
Install	Ea.	360	39.50		399.50	custom modular unit including solid
Demolish and Install	Ea.	360	61		421	hardwood faced frames, hardwood
Reinstall	Ea.		31.61		31.61	door frames and drawer fronts,
Clean	Ea.	1.63	15.50		17.13	hardwood veneer on raised door
Paint	Ea.	1.72	40		41.72	panels, hinges and pulls.
Minimum Charge	Job		211		211	
36" Wide Sink Base						
Demolish	Ea.		26.50		26.50	Includes material and labor to install
Install	Ea.	405	41.50		446.50	custom modular unit including solid
Demolish and Install	Ea.	405	68		473	hardwood faced frames, hardwood
Reinstall	Ea.		33.32		33.32	door frames and drawer fronts,
Clean	Ea.	1.82	18.20		20.02	hardwood veneer on raised door
Paint	Ea.	2.08	51.50		53.58	panels, hinges and pulls.
Minimum Charge	Job		211		211	
36" Blind Corner						
Demolish	Ea.		26.50		26.50	Includes material and labor to install
Install	Ea.	620	47		667	custom modular unit including solid
Demolish and Install	Ea.	620	73.50		693.50	hardwood faced frames, hardwood
Reinstall	Ea.		37.58		37.58	door frames and drawer fronts,
Clean	Ea.	1.82	18.20		20.02	hardwood veneer on raised door
Paint	Ea.	2.08	51.50		53.58	panels, hinges and pulls.
Minimum Charge	Job		211		211	
42" w, 2 Door, 2 Drawer						
Demolish	Ea.		31.50		31.50	Includes material and labor to install
Install	Ea.	495	42.50		537.50	modular unit including solid hardwood
Demolish and Install	Ea.	495	74		569	faced frames, hardwood door frames
Reinstall	Ea.		34.17		34.17	and drawer fronts, hardwood veneer
Clean	Ea.	2.33	22		24.33	on raised door panels, hinges and
Paint	Ea.	2.38	60		62.38	pulls.
Minimum Charge	Job		211		211	
Clean Interior						
Clean	L.F.	.65	4.69		5.34	Includes labor and materials to clean
Minimum Charge	Job		155		155	the interior of a 27" cabinet.
Premium Grade Wood						
12" w, 1 Door, 1 Drawer						
Demolish	Ea.		8.60		8.60	Includes material and labor to install
Install	Ea.	315	17.05		332.05	premium hardwood modular unit with
Demolish and Install	Ea.	315	25.65		340.65	solid hardwood faced frames, drawer
Reinstall	Ea.		13.64		13.64	fronts, and door panels, steel drawer
Clean	Ea.	.65	6.20		6.85	guides, hinges, pulls and laminated
Paint	Ea.	.68	18.05		18.73	interior.
Minimum Charge	Job		211		211	

Cabinets and Countertops

Base Cabinets	Unit	Material	Labor	Equip.	Total	Specification
15″ w, 1 Door, 1 Drawer						
Demolish	Ea.		10.75		10.75	Includes material and labor to install
Install	Ea.	330	35		365	premium modular unit including solid
Demolish and Install	Ea.	330	45.75		375.75	hardwood faced frames, hardwood
Reinstall	Ea.		28.19		28.19	door frames and drawer fronts,
Clean	Ea.	.81	7.75		8.56	hardwood veneer on raised door
Paint	Ea.	.86	22.50		23.36	panels, hinges and pulls.
Minimum Charge	Job		211		211	
18″ w, 1 Door, 1 Drawer						
Demolish	Ea.		12.75		12.75	Includes material and labor to install
Install	Ea.	360	36.50		396.50	premium modular unit including solid
Demolish and Install	Ea.	360	49.25		409.25	hardwood faced frames, hardwood
Reinstall	Ea.		29.03		29.03	door frames and drawer fronts,
Clean	Ea.	.96	9.40		10.36	hardwood veneer on raised door
Paint	Ea.	1.02	28		29.02	panels, hinges and pulls.
Minimum Charge	Job		211		211	
21″ w, 1 Door, 1 Drawer						
Demolish	Ea.		14.95		14.95	Includes material and labor to install
Install	Ea.	380	37.50		417.50	premium modular unit including solid
Demolish and Install	Ea.	380	52.45		432.45	hardwood faced frames, hardwood
Reinstall	Ea.		29.80		29.80	door frames and drawer fronts,
Clean	Ea.	1.17	11.05		12.22	hardwood veneer on raised door
Paint	Ea.	1.20	30		31.20	panels, hinges and pulls.
Minimum Charge	Job		211		211	
24″ w, 1 Door, 1 Drawer						
Demolish	Ea.		17.20		17.20	Includes material and labor to install
Install	Ea.	435	38		473	premium modular unit including solid
Demolish and Install	Ea.	435	55.20		490.20	hardwood faced frames, hardwood
Reinstall	Ea.		30.34		30.34	door frames and drawer fronts,
Clean	Ea.	1.25	12.40		13.65	hardwood veneer on raised door
Paint	Ea.	1.36	36		37.36	panels, hinges and pulls.
Minimum Charge	Job		211		211	
30″ Wide Sink Base						
Demolish	Ea.		21.50		21.50	Includes material and labor to install
Install	Ea.	435	39.50		474.50	premium modular unit including solid
Demolish and Install	Ea.	435	61		496	hardwood faced frames, hardwood
Reinstall	Ea.		31.61		31.61	door frames and drawer fronts,
Clean	Ea.	1.63	15.50		17.13	hardwood veneer on raised door
Paint	Ea.	1.72	45		46.72	panels, hinges and pulls.
Minimum Charge	Job		211		211	
36″ Wide Sink Base						
Demolish	Ea.		26.50		26.50	Includes material and labor to install
Install	Ea.	485	41.50		526.50	premium modular unit including solid
Demolish and Install	Ea.	485	68		553	hardwood faced frames, hardwood
Reinstall	Ea.		33.32		33.32	door frames and drawer fronts,
Clean	Ea.	1.82	18.20		20.02	hardwood veneer on raised door
Paint	Ea.	2.08	51.50		53.58	panels, hinges and pulls.
Minimum Charge	Job		211		211	
36″ Blind Corner						
Demolish	Ea.		26.50		26.50	Includes material and labor to install
Install	Ea.	750	47		797	premium modular unit including solid
Demolish and Install	Ea.	750	73.50		823.50	hardwood faced frames, hardwood
Reinstall	Ea.		37.58		37.58	door frames and drawer fronts,
Clean	Ea.	1.82	18.20		20.02	hardwood veneer on raised door
Paint	Ea.	2.08	51.50		53.58	panels, hinges and pulls.
Minimum Charge	Job		211		211	

Cabinets and Countertops

Base Cabinets	Unit	Material	Labor	Equip.	Total	Specification
42″ w, 2 Door, 2 Drawer						Includes material and labor to install
Demolish	Ea.		31.50		31.50	premium modular unit including solid
Install	Ea.	655	42.50		697.50	hardwood faced frames, hardwood
Demolish and Install	Ea.	655	74		729	door frames and drawer fronts,
Reinstall	Ea.		34.17		34.17	hardwood veneer on raised door
Clean	Ea.	2.33	22		24.33	panels, hinges and pulls.
Paint	Ea.	2.38	60		62.38	
Minimum Charge	Job		211		211	
Clean Interior						
Clean	L.F.	.65	4.69		5.34	Includes labor and materials to clean
Minimum Charge	Job		155		155	the interior of a 27″ cabinet.
Premium Hardwood						
12″ w, 1 Door, 1 Drawer						Includes material and labor to install
Demolish	Ea.		8.60		8.60	premium hardwood modular unit with
Install	Ea.	345	34		379	solid hardwood faced frames, drawer
Demolish and Install	Ea.	345	42.60		387.60	fronts, and door panels, steel drawer
Reinstall	Ea.		27.28		27.28	guides, hinges, pulls and laminated
Clean	Ea.	.65	6.20		6.85	interior.
Paint	Ea.	.68	20		20.68	
Minimum Charge	Job		211		211	
15″ w, 1 Door, 1 Drawer						Includes material and labor to install
Demolish	Ea.		10.75		10.75	premium hardwood modular unit with
Install	Ea.	365	35		400	solid hardwood faced frames, drawer
Demolish and Install	Ea.	365	45.75		410.75	fronts, and door panels, steel drawer
Reinstall	Ea.		28.19		28.19	guides, hinges, pulls and laminated
Clean	Ea.	.81	7.75		8.56	interior.
Paint	Ea.	.86	26		26.86	
Minimum Charge	Job		211		211	
18″ w, 1 Door, 1 Drawer						Includes material and labor to install
Demolish	Ea.		12.75		12.75	premium hardwood modular unit with
Install	Ea.	390	36.50		426.50	solid hardwood faced frames, drawer
Demolish and Install	Ea.	390	49.25		439.25	fronts, and door panels, steel drawer
Reinstall	Ea.		29.03		29.03	guides, hinges, pulls and laminated
Clean	Ea.	.96	9.40		10.36	interior.
Paint	Ea.	1.02	30		31.02	
Minimum Charge	Job		211		211	
21″ w, 1 Door, 1 Drawer						Includes material and labor to install
Demolish	Ea.		14.95		14.95	premium hardwood modular unit with
Install	Ea.	415	37.50		452.50	solid hardwood faced frames, drawer
Demolish and Install	Ea.	415	52.45		467.45	fronts, and door panels, steel drawer
Reinstall	Ea.		29.80		29.80	guides, hinges, pulls and laminated
Clean	Ea.	1.17	11.05		12.22	interior.
Paint	Ea.	1.20	33		34.20	
Minimum Charge	Job		211		211	
24″ w, 1 Door, 1 Drawer						Includes material and labor to install
Demolish	Ea.		17.20		17.20	premium hardwood modular unit with
Install	Ea.	475	38		513	solid hardwood faced frames, drawer
Demolish and Install	Ea.	475	55.20		530.20	fronts, and door panels, steel drawer
Reinstall	Ea.		30.34		30.34	guides, hinges and pulls, laminated
Clean	Ea.	1.25	12.40		13.65	interior.
Paint	Ea.	1.36	40		41.36	
Minimum Charge	Job		211		211	

Cabinets and Countertops

Base Cabinets		Unit	Material	Labor	Equip.	Total	Specification
30″ Wide Sink Base							Includes material and labor to install
	Demolish	Ea.		21.50		21.50	premium hardwood modular unit with
	Install	Ea.	475	39.50		514.50	solid hardwood faced frames, drawer
	Demolish and Install	Ea.	475	61		536	fronts, and door panels, steel drawer
	Reinstall	Ea.		31.61		31.61	guides, hinges, pulls and laminated
	Clean	Ea.	1.63	15.50		17.13	interior.
	Paint	Ea.	1.72	45		46.72	
	Minimum Charge	Job		211		211	
36″ Wide Sink Base							Includes material and labor to install
	Demolish	Ea.		26.50		26.50	premium hardwood modular unit with
	Install	Ea.	530	41.50		571.50	solid hardwood faced frames, drawer
	Demolish and Install	Ea.	530	68		598	fronts, and door panels, steel drawer
	Reinstall	Ea.		33.32		33.32	guides, hinges, pulls and laminated
	Clean	Ea.	1.82	18.20		20.02	interior.
	Paint	Ea.	2.08	60		62.08	
	Minimum Charge	Job		211		211	
36″ Blind Corner							Includes material and labor to install
	Demolish	Ea.		26.50		26.50	premium hardwood modular unit with
	Install	Ea.	815	47		862	solid hardwood faced frames, drawer
	Demolish and Install	Ea.	815	73.50		888.50	fronts, and door panels, steel drawer
	Reinstall	Ea.		37.58		37.58	guides, hinges, pulls and laminated
	Clean	Ea.	1.82	18.20		20.02	interior.
	Paint	Ea.	2.08	60		62.08	
	Minimum Charge	Job		211		211	
42″ w, 2 Door, 2 Drawer							Includes material and labor to install
	Demolish	Ea.		31.50		31.50	premium hardwood modular unit with
	Install	Ea.	865	42.50		907.50	solid hardwood faced frames, drawer
	Demolish and Install	Ea.	865	74		939	fronts, and door panels, steel drawer
	Reinstall	Ea.		34.17		34.17	guides, hinges, pulls and laminated
	Clean	Ea.	2.33	22		24.33	interior.
	Paint	Ea.	2.38	72		74.38	
	Minimum Charge	Job		211		211	
Clean Interior							Includes labor and materials to clean
	Clean	L.F.	.65	4.69		5.34	the interior of a 27″ cabinet.
	Minimum Charge	Job		155		155	

Island

		Unit	Material	Labor	Equip.	Total	Specification
24″ Wide							Includes material and labor to install
	Demolish	Ea.		17.20		17.20	custom modular unit with solid
	Install	Ea.	590	35		625	hardwood faced frames, hardwood
	Demolish and Install	Ea.	590	52.20		642.20	door frames and drawer fronts,
	Reinstall	Ea.		28.19		28.19	hardwood veneer on raised door
	Clean	Ea.	1.82	18.20		20.02	panels, hinges and pulls.
	Paint	Ea.	2.05	51.50		53.55	
	Minimum Charge	Job		211		211	
30″ Wide							Includes material and labor to install
	Demolish	Ea.		21.50		21.50	custom modular unit with solid
	Install	Ea.	645	44		689	hardwood faced frames, hardwood
	Demolish and Install	Ea.	645	65.50		710.50	door frames and drawer fronts,
	Reinstall	Ea.		35.23		35.23	hardwood veneer on raised door
	Clean	Ea.	2.33	22		24.33	panels, hinges and pulls.
	Paint	Ea.	2.57	60		62.57	
	Minimum Charge	Job		211		211	

Cabinets and Countertops

Base Cabinets	Unit	Material	Labor	Equip.	Total	Specification
36" Wide						
Demolish	Ea.		26.50		26.50	Includes material and labor to install
Install	Ea.	940	53		993	custom modular unit with solid
Demolish and Install	Ea.	940	79.50		1019.50	hardwood faced frames, hardwood
Reinstall	Ea.		42.28		42.28	door frames and drawer fronts,
Clean	Ea.	2.72	28		30.72	hardwood veneer on raised door
Paint	Ea.	3.11	72		75.11	panels, hinges and pulls.
Minimum Charge	Job		211		211	
48" Wide						
Demolish	Ea.		34.50		34.50	Includes material and labor to install
Install	Ea.	1175	70.50		1245.50	custom modular unit with solid
Demolish and Install	Ea.	1175	105		1280	hardwood faced frames, hardwood
Reinstall	Ea.		56.37		56.37	door frames and drawer fronts,
Clean	Ea.	3.27	34.50		37.77	hardwood veneer on raised door
Paint	Ea.	3.56	90		93.56	panels, hinges and pulls.
Minimum Charge	Job		211		211	
60" Wide						
Demolish	Ea.		38		38	Includes material and labor to install
Install	Ea.	1300	88		1388	custom modular unit with solid
Demolish and Install	Ea.	1300	126		1426	hardwood faced frames, hardwood
Reinstall	Ea.		70.47		70.47	door frames and drawer fronts,
Clean	Ea.	4.08	44		48.08	hardwood veneer on raised door
Paint	Ea.	3.56	120		123.56	panels, hinges and pulls.
Minimum Charge	Job		211		211	
Clean Interior						
Clean	L.F.	.65	4.69		5.34	Includes labor and materials to clean
Minimum Charge	Job		155		155	the interior of a 27" cabinet.
Liquor Bar						
Demolish	L.F.		8.60		8.60	Includes material and labor to install
Install	L.F.	173	42.50		215.50	custom modular unit with solid
Demolish and Install	L.F.	173	51.10		224.10	hardwood faced frames, hardwood
Reinstall	L.F.		33.82		33.82	door frames and drawer fronts,
Clean	L.F.	.65	6.45		7.10	hardwood veneer on raised door
Paint	L.F.	5.50	16.40		21.90	panels, hinges and pulls.
Minimum Charge	Job		211		211	
Clean Interior						
Clean	L.F.	.65	4.69		5.34	Includes labor and materials to clean
Minimum Charge	Job		155		155	the interior of a 27" cabinet.
Seal Interior						
Install	L.F.	.48	5.90		6.38	Includes labor and materials to seal a
Minimum Charge	Job		180		180	cabinet interior.
Back Bar						
Demolish	L.F.		8.60		8.60	Includes material and labor to install
Install	L.F.	173	42.50		215.50	custom modular unit with solid
Demolish and Install	L.F.	173	51.10		224.10	hardwood faced frames, hardwood
Reinstall	L.F.		33.82		33.82	door frames and drawer fronts,
Clean	L.F.	.65	6.45		7.10	hardwood veneer on raised door
Paint	L.F.	5.50	16.40		21.90	panels, hinges and pulls.
Minimum Charge	Job		211		211	
Clean Interior						
Clean	L.F.	.65	4.69		5.34	Includes labor and materials to clean
Minimum Charge	Job		155		155	the interior of a 27" cabinet.

Cabinets and Countertops

Base Cabinets		Unit	Material	Labor	Equip.	Total	Specification
Seal Interior							
	Install	L.F.	.48	5.90		6.38	Includes labor and materials to seal a
	Minimum Charge	Job		180		180	cabinet interior.
Storage (garage)							
	Demolish	L.F.		8.60		8.60	Includes material and labor to install
	Install	L.F.	130	42.50		172.50	good modular unit with melamine
	Demolish and Install	L.F.	130	51.10		181.10	laminate, textured colors or wood
	Reinstall	L.F.		33.82		33.82	grain print finish including hinges and
	Clean	L.F.	.65	6.45		7.10	pulls.
	Paint	L.F.	5.50	16.40		21.90	
	Minimum Charge	Job		211		211	
Clean Interior							
	Clean	L.F.	.65	4.69		5.34	Includes labor and materials to clean
	Minimum Charge	Job		155		155	the interior of a 27" cabinet.
Seal Interior							
	Install	L.F.	.48	5.90		6.38	Includes labor and materials to seal a
	Minimum Charge	Job		180		180	cabinet interior.
Built-in Desk							
	Demolish	L.F.		8.60		8.60	Includes material and labor to install
	Install	L.F.	173	42.50		215.50	custom modular unit with solid
	Demolish and Install	L.F.	173	51.10		224.10	hardwood faced frames, hardwood
	Reinstall	L.F.		33.82		33.82	door frames and drawer fronts,
	Clean	L.F.	.65	6.45		7.10	hardwood veneer on raised door
	Paint	L.F.	5.50	16.40		21.90	panels, hinges and pulls.
	Minimum Charge	Job		211		211	
Built-in Bookcase							
	Demolish	L.F.		11.45		11.45	Includes material and labor to install
	Install	L.F.	36	42.50		78.50	custom modular bookcase unit with
	Demolish and Install	L.F.	36	53.95		89.95	clear pine faced frames, shelves 12"
	Reinstall	L.F.		33.82		33.82	O.C., 7' high, 8" deep.
	Clean	L.F.	1.31	11.05		12.36	
	Paint	L.F.	1.60	28		29.60	
	Minimum Charge	Job		211		211	
Strip and Refinish							
	Install	L.F.	4.61	45		49.61	Includes labor and material to strip,
	Minimum Charge	Job		180		180	prep and refinish exterior of cabinets.
Stain							
	Install	L.F.	.59	1.68		2.27	Includes labor and material to stain
	Minimum Charge	Job		211		211	door and exterior of cabinets.

Countertop		Unit	Material	Labor	Equip.	Total	Specification
Laminated							
With Splash							
	Demolish	L.F.		5.75		5.75	Includes material and labor to install
	Install	L.F.	45.50	15.10		60.60	one piece laminated top with 4"
	Demolish and Install	L.F.	45.50	20.85		66.35	backsplash.
	Reinstall	L.F.		12.08		12.08	
	Clean	L.F.	.08	.77		.85	
	Minimum Charge	Job		211		211	

Cabinets and Countertops

Countertop		Unit	Material	Labor	Equip.	Total	Specification
Roll Top							
	Demolish	L.F.		5.75		5.75	Includes material and labor to install
	Install	L.F.	11	14.10		25.10	one piece laminated top with rolled
	Demolish and Install	L.F.	11	19.85		30.85	drip edge (post formed) and
	Reinstall	L.F.		11.27		11.27	backsplash.
	Clean	L.F.	.12	1.17		1.29	
	Minimum Charge	Job		211		211	
Ceramic Tile							
	Demolish	S.F.		2.87		2.87	Cost includes material and labor to
	Install	S.F.	18.15	8.45		26.60	install 4-1/4" x 4-1/4" to 6" x 6"
	Demolish and Install	S.F.	18.15	11.32		29.47	glazed tile set in mortar bed and grout
	Clean	S.F.	.06	.59		.65	on particle board substrate.
	Minimum Charge	Job		186		186	
Cultured Marble							
	Demolish	L.F.		5.75		5.75	Includes material and labor to install
	Install	L.F.	49.50	41.50		91	marble countertop, 24" wide, no
	Demolish and Install	L.F.	49.50	47.25		96.75	backsplash.
	Reinstall	L.F.		33.18		33.18	
	Clean	L.F.	.12	1.17		1.29	
	Minimum Charge	Job		186		186	
With Splash							
	Demolish	L.F.		5.75		5.75	Includes material and labor to install
	Install	L.F.	57	41.50		98.50	marble countertop, 24" wide, with
	Demolish and Install	L.F.	57	47.25		104.25	backsplash.
	Reinstall	L.F.		33.18		33.18	
	Clean	L.F.	.12	1.17		1.29	
	Minimum Charge	Job		186		186	
Quarry Tile							
	Demolish	S.F.		2.87		2.87	Includes material and labor to install
	Install	S.F.	9.30	8.45		17.75	quarry tile countertop with no
	Demolish and Install	S.F.	9.30	11.32		20.62	backsplash.
	Clean	S.F.	.06	.59		.65	
	Minimum Charge	Job		186		186	
With Splash							
	Demolish	S.F.		2.87		2.87	Includes material and labor to install
	Install	S.F.	12.15	9.40		21.55	quarry tile countertop with backsplash.
	Demolish and Install	S.F.	12.15	12.27		24.42	
	Clean	S.F.	.06	.59		.65	
	Minimum Charge	Job		186		186	
Butcher Block							
	Demolish	S.F.		2.87		2.87	Includes material and labor to install
	Install	S.F.	40.50	7.55		48.05	solid laminated maple countertop with
	Demolish and Install	S.F.	40.50	10.42		50.92	no backsplash.
	Reinstall	S.F.		6.04		6.04	
	Clean	S.F.	.03	.24		.27	
	Minimum Charge	Job		211		211	
Solid Surface							
	Demolish	L.F.		5.75		5.75	Includes material and labor to install
	Install	L.F.	59	30		89	solid surface counter top 22" deep.
	Demolish and Install	L.F.	59	35.75		94.75	
	Reinstall	L.F.		24.16		24.16	
	Clean	L.F.	.12	1.17		1.29	
	Minimum Charge	Job		211		211	

Cabinets and Countertops

Vanity Cabinets	Unit	Material	Labor	Equip.	Total	Specification
Good Grade Laminated						
24" Wide, 2 Door						
Demolish	Ea.		17.20		17.20	Includes material and labor to install
Install	Ea.	238	42.50		280.50	modular unit with melamine laminated
Demolish and Install	Ea.	238	59.70		297.70	to particle board, textured colors or
Reinstall	Ea.		33.82		33.82	wood grain print finish, hinges and
Clean	Ea.	1.25	12.40		13.65	pulls.
Paint	Ea.	1.36	30		31.36	
Minimum Charge	Job		211		211	
30" Wide, 2 Door						
Demolish	Ea.		21.50		21.50	Includes material and labor to install
Install	Ea.	284	53		337	modular unit with melamine laminated
Demolish and Install	Ea.	284	74.50		358.50	to particle board, textured colors or
Reinstall	Ea.		42.28		42.28	wood grain print finish, hinges and
Clean	Ea.	1.63	15.50		17.13	pulls.
Paint	Ea.	1.72	36		37.72	
Minimum Charge	Job		211		211	
36" Wide, 2 Door						
Demolish	Ea.		26.50		26.50	Includes material and labor to install
Install	Ea.	275	63.50		338.50	modular unit with melamine laminated
Demolish and Install	Ea.	275	90		365	to particle board, textured colors or
Reinstall	Ea.		50.75		50.75	wood grain print finish, hinges and
Clean	Ea.	1.82	18.20		20.02	pulls.
Paint	Ea.	2.08	45		47.08	
Minimum Charge	Job		211		211	
42" w, 3 Doors / Drawer						
Demolish	Ea.		31.50		31.50	Includes material and labor to install
Install	Ea.	435	68.50		503.50	modular unit with melamine laminated
Demolish and Install	Ea.	435	100		535	to particle board, textured colors or
Reinstall	Ea.		54.64		54.64	wood grain print finish, hinges and
Clean	Ea.	2.33	22		24.33	pulls.
Paint	Ea.	2.38	51.50		53.88	
Minimum Charge	Job		211		211	
48" w, 3 Doors / Drawer						
Demolish	Ea.		34.50		34.50	Includes material and labor to install
Install	Ea.	365	74		439	modular unit with melamine laminated
Demolish and Install	Ea.	365	108.50		473.50	to particle board, textured colors or
Reinstall	Ea.		59.18		59.18	wood grain print finish, hinges and
Clean	Ea.	3.27	34.50		37.77	pulls.
Paint	Ea.	3.56	90		93.56	
Minimum Charge	Job		211		211	
Clean Interior						
Install	L.F.	.65	6.45		7.10	Includes labor and materials to clean
Minimum Charge	Job		155		155	base cabinetry.
Better Grade Wood						
24" Wide, 2 Door						
Demolish	Ea.		17.20		17.20	Includes material and labor to install
Install	Ea.	315	42.50		357.50	custom modular unit with solid
Demolish and Install	Ea.	315	59.70		374.70	hardwood faced frames, hardwood
Reinstall	Ea.		33.82		33.82	door frames and drawer fronts,
Clean	Ea.	1.25	12.40		13.65	hardwood veneer on raised door
Paint	Ea.	1.36	33		34.36	panels, hinges and pulls.
Minimum Charge	Job		211		211	

Cabinets and Countertops

Vanity Cabinets	Unit	Material	Labor	Equip.	Total	Specification
30" Wide, 2 Door						Includes material and labor to install
Demolish	Ea.		21.50		21.50	custom modular unit with solid
Install	Ea.	380	53		433	hardwood faced frames, hardwood
Demolish and Install	Ea.	380	74.50		454.50	door frames and drawer fronts,
Reinstall	Ea.		42.28		42.28	hardwood veneer on raised door
Clean	Ea.	1.63	15.50		17.13	panels, hinges and pulls.
Paint	Ea.	1.72	40		41.72	
Minimum Charge	Job		211		211	
36" Wide, 2 Door						Includes material and labor to install
Demolish	Ea.		26.50		26.50	custom modular unit with solid
Install	Ea.	370	63.50		433.50	hardwood faced frames, hardwood
Demolish and Install	Ea.	370	90		460	door frames and drawer fronts,
Reinstall	Ea.		50.75		50.75	hardwood veneer on raised door
Clean	Ea.	1.82	18.20		20.02	panels, hinges and pulls.
Paint	Ea.	2.08	51.50		53.58	
Minimum Charge	Job		211		211	
42" w, 3 Doors / Drawer						Includes material and labor to install
Demolish	Ea.		31.50		31.50	custom modular unit with solid
Install	Ea.	435	68.50		503.50	hardwood faced frames, hardwood
Demolish and Install	Ea.	435	100		535	door frames and drawer fronts,
Reinstall	Ea.		54.64		54.64	hardwood veneer on raised door
Clean	Ea.	2.33	22		24.33	panels, hinges and pulls.
Paint	Ea.	2.38	60		62.38	
Minimum Charge	Job		211		211	
48" w, 3 Doors / Drawer						Includes material and labor to install
Demolish	Ea.		34.50		34.50	custom modular unit with solid
Install	Ea.	485	74		559	hardwood faced frames, hardwood
Demolish and Install	Ea.	485	108.50		593.50	door frames and drawer fronts,
Reinstall	Ea.		59.18		59.18	hardwood veneer on raised door
Clean	Ea.	3.27	34.50		37.77	panels, hinges and pulls.
Paint	Ea.	3.56	90		93.56	
Minimum Charge	Job		211		211	
Clean Interior						Includes labor and materials to clean
Install	L.F.	.65	6.45		7.10	base cabinetry.
Minimum Charge	Job		155		155	
Premium Grade Wood						
24" Wide, 2 Door						Includes material and labor to install
Demolish	Ea.		17.20		17.20	premium modular unit with solid
Install	Ea.	335	42.50		377.50	hardwood faced frames, hardwood
Demolish and Install	Ea.	335	59.70		394.70	door frames and drawer fronts,
Reinstall	Ea.		33.82		33.82	hardwood veneer on raised door
Clean	Ea.	1.25	12.40		13.65	panels, hinges and pulls.
Paint	Ea.	1.36	36		37.36	
Minimum Charge	Job		211		211	
30" Wide, 2 Door						Includes material and labor to install
Demolish	Ea.		21.50		21.50	premium modular unit with solid
Install	Ea.	470	53		523	hardwood faced frames, hardwood
Demolish and Install	Ea.	470	74.50		544.50	door frames and drawer fronts,
Reinstall	Ea.		42.28		42.28	hardwood veneer on raised door
Clean	Ea.	1.63	15.50		17.13	panels, hinges and pulls.
Paint	Ea.	1.72	45		46.72	
Minimum Charge	Job		211		211	

Cabinets and Countertops

Vanity Cabinets	Unit	Material	Labor	Equip.	Total	Specification
36" Wide, 2 Door						
Demolish	Ea.		26.50		26.50	Includes material and labor to install
Install	Ea.	450	63.50		513.50	premium modular unit with solid
Demolish and Install	Ea.	450	90		540	hardwood faced frames, hardwood
Reinstall	Ea.		50.75		50.75	door frames and drawer fronts,
Clean	Ea.	1.82	18.20		20.02	hardwood veneer on raised door
Paint	Ea.	2.08	51.50		53.58	panels, hinges and pulls.
Minimum Charge	Job		211		211	
42" w, 3 Doors / Drawer						
Demolish	Ea.		31.50		31.50	Includes material and labor to install
Install	Ea.	495	68.50		563.50	premium modular unit with solid
Demolish and Install	Ea.	495	100		595	hardwood faced frames, hardwood
Reinstall	Ea.		54.64		54.64	door frames and drawer fronts,
Clean	Ea.	2.33	22		24.33	hardwood veneer on raised door
Paint	Ea.	2.38	60		62.38	panels, hinges and pulls.
Minimum Charge	Job		211		211	
48" w, 3 Doors / Drawer						
Demolish	Ea.		34.50		34.50	Includes material and labor to install
Install	Ea.	550	74		624	premium modular unit with solid
Demolish and Install	Ea.	550	108.50		658.50	hardwood faced frames, hardwood
Reinstall	Ea.		59.18		59.18	door frames and drawer fronts,
Clean	Ea.	3.27	34.50		37.77	hardwood veneer on raised door
Paint	Ea.	3.56	90		93.56	panels, hinges and pulls.
Minimum Charge	Job		211		211	
Clean Interior						
Install	L.F.	.65	6.45		7.10	Includes labor and materials to clean
Minimum Charge	Job		155		155	base cabinetry.
Premium Hardwood						
24" Wide, 2 Door						
Demolish	Ea.		17.20		17.20	Includes material and labor to install
Install	Ea.	390	42.50		432.50	premium modular unit with solid
Demolish and Install	Ea.	390	59.70		449.70	hardwood faced frames, hardwood
Reinstall	Ea.		33.82		33.82	door frames and drawer fronts,
Clean	Ea.	1.25	12.40		13.65	hardwood veneer on raised door
Paint	Ea.	1.36	40		41.36	panels, hinges and pulls.
Minimum Charge	Job		211		211	
30" Wide, 2 Door						
Demolish	Ea.		21.50		21.50	Includes material and labor to install
Install	Ea.	595	53		648	premium modular unit with solid
Demolish and Install	Ea.	595	74.50		669.50	hardwood faced frames, hardwood
Reinstall	Ea.		42.28		42.28	door frames and drawer fronts,
Clean	Ea.	1.63	15.50		17.13	hardwood veneer on raised door
Paint	Ea.	1.72	45		46.72	panels, hinges and pulls.
Minimum Charge	Job		211		211	
36" Wide, 2 Door						
Demolish	Ea.		26.50		26.50	Includes material and labor to install
Install	Ea.	825	63.50		888.50	premium modular unit with solid
Demolish and Install	Ea.	825	90		915	hardwood faced frames, hardwood
Reinstall	Ea.		50.75		50.75	door frames and drawer fronts,
Clean	Ea.	1.82	18.20		20.02	hardwood veneer on raised door
Paint	Ea.	2.08	60		62.08	panels, hinges and pulls.
Minimum Charge	Job		211		211	

Cabinets and Countertops

Vanity Cabinets		Unit	Material	Labor	Equip.	Total	Specification
42" w, 3 Doors / Drawer							
	Demolish	Ea.		31.50		31.50	Includes material and labor to install
	Install	Ea.	535	68.50		603.50	premium modular unit with solid
	Demolish and Install	Ea.	535	100		635	hardwood faced frames, hardwood
	Reinstall	Ea.		54.64		54.64	door frames and drawer fronts,
	Clean	Ea.	2.33	22		24.33	hardwood veneer on raised door
	Paint	Ea.	2.38	72		74.38	panels, hinges and pulls.
	Minimum Charge	Job		211		211	
48" w, 3 Doors / Drawer							
	Demolish	Ea.		34.50		34.50	Includes material and labor to install
	Install	Ea.	595	74		669	premium modular unit with solid
	Demolish and Install	Ea.	595	108.50		703.50	hardwood faced frames, hardwood
	Reinstall	Ea.		59.18		59.18	door frames and drawer fronts,
	Clean	Ea.	3.27	34.50		37.77	hardwood veneer on raised door
	Paint	Ea.	3.56	90		93.56	panels, hinges and pulls.
	Minimum Charge	Job		211		211	
Clean Interior							
	Install	L.F.	.65	6.45		7.10	Includes labor and materials to clean
	Minimum Charge	Job		155		155	base cabinetry.
Strip and Refinish							
	Install	L.F.	4.61	45		49.61	Includes labor and material to strip,
	Minimum Charge	Job		180		180	prep and refinish exterior of cabinets.

Vanity Cabinet Tops		Unit	Material	Labor	Equip.	Total	Specification
Cultured Marble Top							
	Demolish	L.F.		5.75		5.75	Includes material and labor to install
	Install	L.F.	45	24.50		69.50	cultured marble vanity top with
	Demolish and Install	L.F.	45	30.25		75.25	integral sink, 22" deep.
	Reinstall	L.F.		19.52		19.52	
	Clean	L.F.	.12	1.17		1.29	
	Minimum Charge	Job		186		186	
Laminated Top							
	Demolish	L.F.		3.44		3.44	Includes material and labor to install
	Install	L.F.	45.50	15.10		60.60	one piece laminated top with 4"
	Demolish and Install	L.F.	45.50	18.54		64.04	backsplash.
	Reinstall	L.F.		12.08		12.08	
	Clean	L.F.	.08	.77		.85	
	Minimum Charge	Job		211		211	
Solid Surface							
	Demolish	L.F.		5.75		5.75	Includes material and labor to install
	Install	L.F.	82	31.50		113.50	solid surface vanity top with integral
	Demolish and Install	L.F.	82	37.25		119.25	sink installed, 22" deep.
	Reinstall	L.F.		25.05		25.05	
	Clean	L.F.	.12	1.17		1.29	
	Minimum Charge	Job		211		211	

Roller Handle

Paint Brush

Paint Preparation, Walls		Unit	Material	Labor	Equip.	Total	Specification
Cover / Protect Floors							
	Install	S.F.	.03	.14		.17	Includes material and labor to install
	Minimum Charge	Job		181		181	plastic masking sheet on large uninterrupted areas.
Cover / Protect Walls							
	Install	S.F.	.03	.14		.17	Includes material and labor to install
	Minimum Charge	Job		181		181	plastic masking sheet on walls.
Clean Walls							
Light							
	Clean	S.F.		.11		.11	Includes labor to wash gypsum
	Minimum Charge	Job		181		181	drywall or plaster wall surfaces.
Heavy							
	Clean	S.F.		.14		.14	Includes labor and material for heavy
	Minimum Charge	Job		181		181	cleaning (multiple applications) with detergent and solvent.
Prep Walls							
	Paint	S.F.	.04	.18		.22	Includes labor and material to prepare
	Minimum Charge	Job		181		181	for painting including scraping, patching and puttying.
Sand Walls							
	Paint	S.F.		.12		.12	Includes labor to sand gypsum drywall
	Minimum Charge	Job		181		181	wall surfaces.
Seal Walls							
Large Area							
	Paint	S.F.	.07	.27		.34	Includes labor and material to paint
	Minimum Charge	Job		181		181	with primer / sealer by roller.
Spot							
	Paint	S.F.	.07	.31		.38	Includes labor and material to seal
	Minimum Charge	Job		181		181	existing drywall with shellac-based material.

Paint Preparation, Ceilings		Unit	Material	Labor	Equip.	Total	Specification
Cover / Protect Floors							
	Install	S.F.	.03	.14		.17	Includes material and labor to install
	Minimum Charge	Job		181		181	plastic masking sheet on large uninterrupted areas.

Painting

Paint Preparation, Ceilings

Paint Preparation, Ceilings		Unit	Material	Labor	Equip.	Total	Specification
Cover / Protect Walls							
	Install	S.F.	.03	.14		.17	Includes material and labor to install plastic masking sheet on walls.
	Minimum Charge	Job		181		181	
Prep Ceiling							
	Paint	S.F.	.08	.69		.77	Includes labor and material to fill in hairline cracks up to 1/8" with filler, sand and prep for paint.
	Minimum Charge	Job		181		181	
Sand Ceiling							
	Paint	S.F.		.17		.17	Includes labor to sand gypsum drywall ceiling surfaces.
	Minimum Charge	Job		181		181	
Seal Ceiling							
Large Area							
	Paint	S.F.	.07	.28		.35	Includes labor and material to paint ceiling, one coat flat latex.
	Minimum Charge	Job		180		180	
Spot							
	Paint	S.F.	.07	.31		.38	Includes labor and material to seal existing drywall with shellac-based material.
	Minimum Charge	Job		180		180	

Paint / Texture, Walls

Paint / Texture, Walls		Unit	Material	Labor	Equip.	Total	Specification
Texture Walls							
Spray							
	Install	S.F.	.04	.45		.49	Includes labor and material to install by spray, texture finish.
	Minimum Charge	Job		180		180	
Trowel							
	Install	S.F.		.25		.25	Includes labor and material to apply hand troweled texture.
	Minimum Charge	Job		180		180	
Paint Walls 1 Coat							
	Paint	S.F.	.07	.28		.35	Includes labor and material to paint, one coat flat latex.
	Minimum Charge	Job		180		180	
Paint Walls 2 Coats							
	Paint	S.F.	.13	.45		.58	Includes labor and material to paint, two coats flat latex.
	Minimum Charge	Job		180		180	
Paint Walls 3 Coats							
	Paint	S.F.	.20	.56		.76	Includes labor and material to paint, three coats flat latex.
	Minimum Charge	Job		180		180	
Clean and Seal							
	Paint	S.F.	.50	.96		1.46	Includes labor and material to do light cleaning, sealing with 2 coats of latex paint.
	Minimum Charge	Job		180		180	

Paint / Texture, Ceilings

Paint / Texture, Ceilings		Unit	Material	Labor	Equip.	Total	Specification
Texture Ceiling							
Spray							
	Install	S.F.	.04	.45		.49	Includes labor and material to install by spray, texture finish.
	Minimum Charge	Job		180		180	
Trowel							
	Install	S.F.		.25		.25	Includes labor and material to apply hand troweled texture.
	Minimum Charge	Job		180		180	

Painting

Paint / Texture, Ceilings		Unit	Material	Labor	Equip.	Total	Specification
Paint Ceiling, 1 Coat							
	Paint	S.F.	.07	.28		.35	Includes labor and material to paint ceiling, one coat flat latex.
	Minimum Charge	Job		180		180	
Paint Ceiling, 2 Coats							
	Paint	S.F.	.29	.45		.74	Includes labor and material to paint two coats flat latex paint.
	Minimum Charge	Job		180		180	
Paint Ceiling, 3 Coats							
	Paint	S.F.	.43	.53		.96	Includes labor and material to paint ceiling, three coats flat latex.
	Minimum Charge	Job		180		180	
Clean and Seal							
	Paint	S.F.	.43	.76		1.19	Includes labor and material for light cleaning, sealing and paint 3 coats of flat latex paint.
	Minimum Charge	Job		180		180	

Flooring

Plank Flooring

Wood Strip Flooring

Tile Flooring

Wood Plank Flooring	Unit	Material	Labor	Equip.	Total	Specification
Maple						
2-1/4″						
Demolish	S.F.		1.38		1.38	Includes material and labor to install
Install	S.F.	5.50	2.11		7.61	unfinished T&G maple flooring,
Demolish and Install	S.F.	5.50	3.49		8.99	25/32″ thick, 2-1/4″ wide, random
Reinstall	S.F.		1.69		1.69	3′ to 16′ lengths including felt
Clean	S.F.		.39		.39	underlayment, nailed in place over
Paint	S.F.	1.44	2.38		3.82	prepared subfloor.
Minimum Charge	Job		211		211	
3-1/4″						
Demolish	S.F.		1.38		1.38	Includes material and labor to install
Install	S.F.	4.05	2.11		6.16	unfinished T&G maple flooring,
Demolish and Install	S.F.	4.05	3.49		7.54	33/32″ thick, 3-1/4″ wide, random
Reinstall	S.F.		1.69		1.69	3′ to 16′ lengths including felt
Clean	S.F.		.39		.39	underlayment, nailed in place over
Paint	S.F.	1.44	2.38		3.82	prepared subfloor.
Minimum Charge	Job		211		211	
Sand and Finish						
Install	S.F.	.97	.77		1.74	Includes labor and material to sand (3
Minimum Charge	Job		211		211	passes) and finish and two coats of
						urethane existing floor.
Oak						
2-1/4″						
Demolish	S.F.		1.38		1.38	Includes material and labor to install
Install	S.F.	3.60	2.11		5.71	unfinished T&G oak flooring, 25/32″
Demolish and Install	S.F.	3.60	3.49		7.09	thick, 2-1/4″ wide, random 3′ to 16′
Reinstall	S.F.		1.69		1.69	lengths including felt underlayment,
Clean	S.F.		.39		.39	nailed in place over prepared
Paint	S.F.	1.44	2.38		3.82	subfloor.
Minimum Charge	Job		211		211	
3-1/4″						
Demolish	S.F.		1.38		1.38	Includes material and labor to install
Install	S.F.	4.87	1.49		6.36	unfinished T&G oak flooring, 25/32″
Demolish and Install	S.F.	4.87	2.87		7.74	thick, 3-1/4″ wide, random 3′ to 16′
Reinstall	S.F.		1.19		1.19	lengths including felt underlayment,
Clean	S.F.		.39		.39	nailed in place over prepared
Paint	S.F.	1.44	2.38		3.82	subfloor.
Minimum Charge	Job		211		211	

Flooring

Wood Plank Flooring		Unit	Material	Labor	Equip.	Total	Specification
Sand and Finish							
	Install	S.F.	1.44	2.38		3.82	Includes labor and material to sand
	Minimum Charge	Job		211		211	and finish (3 passes) with two coats of
							urethane on a new floor.
Walnut							
3″ to 7″							
	Demolish	S.F.		1.38		1.38	Includes material and labor to install
	Install	S.F.	9.50	1.49		10.99	unfinished T&G walnut flooring,
	Demolish and Install	S.F.	9.50	2.87		12.37	25/32″ thick, random 3′ to 16′
	Reinstall	S.F.		1.19		1.19	lengths including felt underlayment,
	Clean	S.F.		.39		.39	nailed in place over prepared
	Paint	S.F.	1.44	2.38		3.82	subfloor.
	Minimum Charge	Job		211		211	
Sand and Finish							
	Install	S.F.	1.44	2.38		3.82	Includes labor and material to sand
	Minimum Charge	Job		211		211	and finish (3 passes) with two coats of
							urethane on a new floor.
Pine							
2-1/4″							
	Demolish	S.F.		1.38		1.38	Includes material and labor to install
	Install	S.F.	7.60	1.59		9.19	unfinished T&G pine flooring, 25/32″
	Demolish and Install	S.F.	7.60	2.97		10.57	thick, 2-1/4″ wide, random 3′ to 16′
	Reinstall	S.F.		1.27		1.27	lengths including felt underlayment,
	Clean	S.F.		.39		.39	nailed in place over prepared
	Paint	S.F.	.97	.77		1.74	subfloor.
	Minimum Charge	Job		211		211	
Sand and Finish							
	Install	S.F.	1.44	2.38		3.82	Includes labor and material to sand
	Minimum Charge	Job		211		211	and finish (3 passes) with two coats of
							urethane on a new floor.
Prefinished Oak							
2-1/4″							
	Demolish	S.F.		1.38		1.38	Includes material and labor to install
	Install	S.F.	4.37	2.11		6.48	prefinished T&G oak flooring, 2-1/4″
	Demolish and Install	S.F.	4.37	3.49		7.86	wide, random 3′ to 16′ lengths
	Reinstall	S.F.		1.69		1.69	including felt underlayment, nailed in
	Clean	S.F.		.39		.39	place over prepared subfloor.
	Minimum Charge	Job		211		211	
3-1/4″							
	Demolish	S.F.		1.38		1.38	Includes material and labor to install
	Install	S.F.	4.87	1.94		6.81	prefinished T&G oak flooring, 25/32″
	Demolish and Install	S.F.	4.87	3.32		8.19	thick, 3″ to 7″ wide, random 3′ to 16′
	Reinstall	S.F.		1.55		1.55	lengths including felt underlayment,
	Clean	S.F.		.39		.39	nailed in place over prepared
	Minimum Charge	Job		211		211	subfloor.
Sand and Finish New							
	Install	S.F.	.97	1.05		2.02	Includes labor and material to sand (3
	Minimum Charge	Job		211		211	passes) and finish, two coats of
							urethane, new floor.
Refinish Existing							
	Install	S.F.	.97	.77		1.74	Includes labor and material to sand (3
	Minimum Charge	Job		211		211	passes) and finish and two coats of
							urethane existing floor.

Flooring

Wood Plank Flooring

Wood Plank Flooring		Unit	Material	Labor	Equip.	Total	Specification
Urethane Coat							
	Install	S.F.	.17	.19		.36	Includes labor and material to install
	Minimum Charge	Job		211		211	two coats of urethane, on existing floor.
Clean and Wax							
	Clean	S.F.	.03	.24		.27	Includes labor and material to wash
	Minimum Charge	Job		211		211	and wax a hardwood floor.
Underlayment Per S.F.							
	Demolish	S.F.		.62		.62	Cost includes material and labor to
	Install	S.F.	.80	.56		1.36	install 1/4" lauan subfloor, standard
	Demolish and Install	S.F.	.80	1.18		1.98	interior grade, nailed every 6".
	Clean	S.F.		.30		.30	
	Minimum Charge	Job		211		211	

Wood Parquet Tile Floor

Wood Parquet Tile Floor		Unit	Material	Labor	Equip.	Total	Specification
Oak							
9" x 9"							
	Demolish	S.F.		1.04		1.04	Includes material and labor to install
	Install	S.F.	5.10	3.38		8.48	prefinished oak 9" x 9" parquet block
	Demolish and Install	S.F.	5.10	4.42		9.52	flooring, 5/16" thick, installed in
	Reinstall	S.F.		2.71		2.71	mastic.
	Clean	S.F.		.39		.39	
	Paint	S.F.	1.44	2.38		3.82	
	Minimum Charge	Job		211		211	
13" x 13"							
	Demolish	S.F.		1.04		1.04	Includes material and labor to install
	Install	S.F.	9.35	2.64		11.99	prefinished oak 13" x 13" parquet
	Demolish and Install	S.F.	9.35	3.68		13.03	block flooring, 5/16" thick, installed in
	Reinstall	S.F.		2.11		2.11	mastic.
	Clean	S.F.		.39		.39	
	Paint	S.F.	1.44	2.38		3.82	
	Minimum Charge	Job		211		211	
Cherry							
9" x 9"							
	Demolish	S.F.		1.04		1.04	Includes material and labor to install
	Install	S.F.	8.55	3.38		11.93	prefinished cherry 9" x 9" parquet
	Demolish and Install	S.F.	8.55	4.42		12.97	block flooring, 5/16" thick, installed in
	Reinstall	S.F.		3.38		3.38	mastic.
	Clean	S.F.		.39		.39	
	Paint	S.F.	1.44	2.38		3.82	
	Minimum Charge	Job		211		211	
13" x 13"							
	Demolish	S.F.		1.04		1.04	Includes material and labor to install
	Install	S.F.	9.75	2.64		12.39	prefinished cherry 13" x 13" parquet
	Demolish and Install	S.F.	9.75	3.68		13.43	block flooring, 5/16" thick, installed in
	Reinstall	S.F.		2.64		2.64	mastic.
	Clean	S.F.		.39		.39	
	Paint	S.F.	1.44	2.38		3.82	
	Minimum Charge	Job		211		211	

Flooring

Wood Parquet Tile Floor	Unit	Material	Labor	Equip.	Total	Specification
Walnut						
9″ x 9″						
Demolish	S.F.		1.04		1.04	Includes material and labor to install
Install	S.F.	5.45	3.38		8.83	prefinished walnut 9″ x 9″ parquet
Demolish and Install	S.F.	5.45	4.42		9.87	flooring in mastic.
Reinstall	S.F.		2.71		2.71	
Clean	S.F.		.39		.39	
Paint	S.F.	1.44	2.38		3.82	
Minimum Charge	Job		211		211	
13″ x 13″						
Demolish	S.F.		1.04		1.04	Includes material and labor to install
Install	S.F.	10.10	2.64		12.74	prefinished walnut 13″ x 13″ parquet
Demolish and Install	S.F.	10.10	3.68		13.78	block flooring, 5/16″ thick, installed in
Reinstall	S.F.		2.64		2.64	mastic.
Clean	S.F.		.39		.39	
Paint	S.F.	1.44	2.38		3.82	
Minimum Charge	Job		211		211	
Acrylic Impregnated						
Oak 12″ x 12″						
Demolish	S.F.		1.04		1.04	Includes material and labor to install
Install	S.F.	8.15	3.02		11.17	red oak 12″ x 12″ acrylic
Demolish and Install	S.F.	8.15	4.06		12.21	impregnated parquet block flooring,
Reinstall	S.F.		3.02		3.02	5/16″ thick, installed in mastic.
Clean	S.F.		.39		.39	
Minimum Charge	Job		211		211	
Cherry 12″ x 12″						
Demolish	S.F.		1.04		1.04	Includes material and labor to install
Install	S.F.	11.95	3.02		14.97	prefinished cherry 12″ x 12″ acrylic
Demolish and Install	S.F.	11.95	4.06		16.01	impregnated parquet block flooring,
Reinstall	S.F.		3.02		3.02	5/16″ thick, installed in mastic.
Clean	S.F.		.39		.39	
Minimum Charge	Job		211		211	
Ash 12″ x 12″						
Demolish	S.F.		1.04		1.04	Includes material and labor to install
Install	S.F.	10.75	3.02		13.77	ash 12″ x 12″ acrylic impregnated
Demolish and Install	S.F.	10.75	4.06		14.81	parquet block flooring, 5/16″ thick in
Reinstall	S.F.		3.02		3.02	mastic.
Clean	S.F.		.39		.39	
Minimum Charge	Job		211		211	
Teak						
Demolish	S.F.		1.04		1.04	Includes material and labor to install
Install	S.F.	8.10	2.64		10.74	prefinished teak parquet block
Demolish and Install	S.F.	8.10	3.68		11.78	flooring, 5/16″ thick, and installed in
Reinstall	S.F.		2.64		2.64	mastic.
Clean	S.F.		.39		.39	
Paint	S.F.	1.44	2.38		3.82	
Minimum Charge	Job		211		211	
Clean and Wax						
Clean	S.F.	.03	.24		.27	Includes labor and material to wash
Minimum Charge	Job		211		211	and wax a hardwood floor.

Flooring

Ceramic Tile Flooring	Unit	Material	Labor	Equip.	Total	Specification
Economy Grade						
Demolish	S.F.		1.02		1.02	Cost includes material and labor to
Install	S.F.	8.70	7.30		16	install 4-1/4" x 4-1/4" ceramic tile in
Demolish and Install	S.F.	8.70	8.32		17.02	mortar bed including grout.
Clean	S.F.		.59		.59	
Minimum Charge	Job		186		186	
Average Grade						
Demolish	S.F.		1.02		1.02	Cost includes material and labor to
Install	S.F.	11.05	7.45		18.50	install 4-1/4" x 4-1/4" ceramic tile in
Demolish and Install	S.F.	11.05	8.47		19.52	mortar bed including grout.
Clean	S.F.		.59		.59	
Minimum Charge	Job		186		186	
Premium Grade						
Demolish	S.F.		1.02		1.02	Cost includes material and labor to
Install	S.F.	13.25	7.45		20.70	install 4-1/4" x 4-1/4" ceramic tile in
Demolish and Install	S.F.	13.25	8.47		21.72	mortar bed including grout.
Clean	S.F.		.59		.59	
Minimum Charge	Job		186		186	
Re-grout						
Install	S.F.	.15	2.98		3.13	Includes material and labor to regrout
Minimum Charge	Job		186		186	tile floors.

Hard Tile Flooring	Unit	Material	Labor	Equip.	Total	Specification
Thick Set Paver						
Brick						
Demolish	S.F.		2.15		2.15	Cost includes material and labor to
Install	S.F.	1.52	17.80		19.32	install 6" x 12" adobe brick paver with
Demolish and Install	S.F.	1.52	19.95		21.47	1/2" mortar joints.
Clean	S.F.		.59		.59	
Minimum Charge	Job		186		186	
Mexican Red						
Demolish	S.F.		2.15		2.15	Cost includes material and labor to
Install	S.F.	1.54	7.75		9.29	install 12" x 12" red Mexican paver
Demolish and Install	S.F.	1.54	9.90		11.44	tile in mortar bed with grout.
Clean	S.F.		.59		.59	
Minimum Charge	Job		186		186	
Saltillo						
Demolish	S.F.		1.02		1.02	Cost includes material and labor to
Install	S.F.	1.54	7.75		9.29	install 12" x 12"saltillo tile in mortar
Demolish and Install	S.F.	1.54	8.77		10.31	bed with grout.
Clean	S.F.		.59		.59	
Minimum Charge	Job		186		186	
Marble						
Premium Grade						
Demolish	S.F.		1.02		1.02	Cost includes material and labor to
Install	S.F.	21	16.15		37.15	install 3/8" x 12" x 12" marble, in
Demolish and Install	S.F.	21	17.17		38.17	mortar bed with grout.
Clean	S.F.		.59		.59	
Minimum Charge	Job		186		186	

Flooring

Hard Tile Flooring		Unit	Material	Labor	Equip.	Total	Specification
Terrazzo							
Gray Cement							
	Demolish	S.F.		1.36		1.36	Cost includes material and labor to
	Install	S.F.	3.58	9.05	4.15	16.78	install 1-3/4" terrazzo , #1 and #2
	Demolish and Install	S.F.	3.58	10.41	4.15	18.14	chips in gray Portland cement.
	Clean	S.F.		.59		.59	
	Minimum Charge	Job		340	156	496	
White Cement							
	Demolish	S.F.		1.36		1.36	Cost includes material and labor to
	Install	S.F.	3.98	9.05	4.15	17.18	install 1-3/4" terrazzo, #1 and #2
	Demolish and Install	S.F.	3.98	10.41	4.15	18.54	chips in white Portland cement.
	Clean	S.F.		.59		.59	
	Minimum Charge	Job		340	156	496	
Non-skid Gray							
	Demolish	S.F.		1.36		1.36	Cost includes material and labor to
	Install	S.F.	6.60	11.35	5.20	23.15	install 1-3/4" terrazzo, #1 and #2
	Demolish and Install	S.F.	6.60	12.71	5.20	24.51	chips in gray Portland cement with
	Clean	S.F.		.59		.59	light non-skid abrasive.
	Minimum Charge	Job		340	156	496	
Non-skid White							
	Demolish	S.F.		1.36		1.36	Cost includes material and labor to
	Install	S.F.	6.95	11.35	5.20	23.50	install 1-3/4" terrazzo, #1 and #2
	Demolish and Install	S.F.	6.95	12.71	5.20	24.86	chips in white Portland cement with
	Clean	S.F.		.59		.59	light non-skid abrasive.
	Minimum Charge	Job		340	156	496	
Gray w / Brass Divider							
	Demolish	S.F.		1.38		1.38	Cost includes material and labor to
	Install	S.F.	10	12.35	5.65	28	install 1-3/4" terrazzo, #1 and #2
	Demolish and Install	S.F.	10	13.73	5.65	29.38	chips in gray Portland cement with
	Clean	S.F.		.59		.59	brass strips 2' O.C. each way.
	Minimum Charge	Job		340	156	496	
Clean							
	Clean	S.F.	.14	.39		.53	Includes labor and material to clean
	Minimum Charge	Job		340	156	496	terrazzo flooring.
Slate							
	Demolish	S.F.		1.23		1.23	Includes material and labor to install
	Install	S.F.	7.70	3.70		11.40	slate tile flooring in thin set with grout.
	Demolish and Install	S.F.	7.70	4.93		12.63	
	Clean	S.F.		.59		.59	
	Minimum Charge	Job		186		186	
Granite							
	Demolish	S.F.		1.01		1.01	Includes material and labor to install
	Install	S.F.	24	3.72		27.72	granite tile flooring in thin set with
	Demolish and Install	S.F.	24	4.73		28.73	grout.
	Reinstall	S.F.		3.72		3.72	
	Clean	S.F.		.59		.59	
	Minimum Charge	Job		186		186	
Re-grout							
	Install	S.F.	.15	2.98		3.13	Includes material and labor to regrout
	Minimum Charge	Job		186		186	tile floors.

240

Flooring

Sheet Vinyl Per S.F.

Sheet Vinyl Per S.F.	Unit	Material	Labor	Equip.	Total	Specification
Economy Grade						
Demolish	S.F.		.49		.49	Includes material and labor to install
Install	S.F.	2.47	1.16		3.63	resilient sheet vinyl flooring.
Demolish and Install	S.F.	2.47	1.65		4.12	
Clean	S.F.		.39		.39	
Minimum Charge	Job		186		186	
Average Grade						
Demolish	S.F.		.49		.49	Includes material and labor to install
Install	S.F.	3.91	1.16		5.07	resilient sheet vinyl flooring.
Demolish and Install	S.F.	3.91	1.65		5.56	
Clean	S.F.		.39		.39	
Minimum Charge	Job		186		186	
Premium Grade						
Demolish	S.F.		.49		.49	Includes material and labor to install
Install	S.F.	6.15	1.16		7.31	resilient sheet vinyl flooring.
Demolish and Install	S.F.	6.15	1.65		7.80	
Clean	S.F.		.39		.39	
Minimum Charge	Job		186		186	
Clean and Wax						
Clean	S.F.	.03	.19		.22	Includes labor and material to wash
Minimum Charge	Job		186		186	and wax a vinyl tile floor.

Sheet Vinyl

Sheet Vinyl	Unit	Material	Labor	Equip.	Total	Specification
Economy Grade						
Demolish	S.Y.		4.44		4.44	Includes material and labor to install
Install	S.Y.	22.50	10.45		32.95	no wax sheet vinyl flooring.
Demolish and Install	S.Y.	22.50	14.89		37.39	
Clean	S.Y.		2.69		2.69	
Minimum Charge	Job		186		186	
Average Grade						
Demolish	S.Y.		4.44		4.44	Includes material and labor to install
Install	S.Y.	35	10.45		45.45	no wax sheet vinyl flooring.
Demolish and Install	S.Y.	35	14.89		49.89	
Clean	S.Y.		2.69		2.69	
Minimum Charge	Job		186		186	
Premium Grade						
Demolish	S.Y.		4.44		4.44	Includes material and labor to install
Install	S.Y.	55	10.45		65.45	no wax sheet vinyl flooring.
Demolish and Install	S.Y.	55	14.89		69.89	
Clean	S.Y.		2.69		2.69	
Minimum Charge	Job		186		186	
Clean and Wax						
Clean	S.F.	.03	.19		.22	Includes labor and material to wash
Minimum Charge	Job		186		186	and wax a vinyl tile floor.

Vinyl Tile

Vinyl Tile	Unit	Material	Labor	Equip.	Total	Specification
Economy Grade						
Demolish	S.F.		.69		.69	Includes material and labor to install
Install	S.F.	3.74	.74		4.48	no wax 12" x 12" vinyl tile.
Demolish and Install	S.F.	3.74	1.43		5.17	
Clean	S.F.		.39		.39	
Minimum Charge	Job		186		186	

Flooring

Vinyl Tile

Vinyl Tile		Unit	Material	Labor	Equip.	Total	Specification
Average Grade							
	Demolish	S.F.		.69		.69	Includes material and labor to install
	Install	S.F.	4.02	.74		4.76	no wax 12" x 12" vinyl tile.
	Demolish and Install	S.F.	4.02	1.43		5.45	
	Clean	S.F.		.39		.39	
	Minimum Charge	Job		186		186	
Premium Grade							
	Demolish	S.F.		.69		.69	Includes material and labor to install
	Install	S.F.	16.05	.74		16.79	no wax 12" x 12" vinyl tile.
	Demolish and Install	S.F.	16.05	1.43		17.48	
	Clean	S.F.		.39		.39	
	Minimum Charge	Job		186		186	
Clean and Wax							
	Clean	S.F.	.03	.19		.22	Includes labor and material to wash
	Minimum Charge	Job		186		186	and wax a vinyl tile floor.

Carpeting Per S.F.

Carpeting Per S.F.		Unit	Material	Labor	Equip.	Total	Specification
Economy Grade							
	Demolish	S.F.		.08		.08	Includes material and labor to install
	Install	S.F.	1.44	.72		2.16	carpet including tack strips and hot
	Demolish and Install	S.F.	1.44	.80		2.24	melt tape on seams.
	Clean	S.F.		.54		.54	
	Minimum Charge	Job		186		186	
Average Grade							
	Demolish	S.F.		.08		.08	Includes material and labor to install
	Install	S.F.	1.73	.72		2.45	carpet including tack strips and hot
	Demolish and Install	S.F.	1.73	.80		2.53	melt tape on seams.
	Clean	S.F.		.54		.54	
	Minimum Charge	Job		186		186	
Premium Grade							
	Demolish	S.F.		.08		.08	Includes material and labor to install
	Install	S.F.	2.78	.72		3.50	carpet including tack strips and hot
	Demolish and Install	S.F.	2.78	.80		3.58	melt tape on seams.
	Clean	S.F.		.54		.54	
	Minimum Charge	Job		186		186	
Indoor/Outdoor							
	Demolish	S.F.		.08		.08	Includes material and labor to install
	Install	S.F.	1.16	.88		2.04	indoor-outdoor carpet.
	Demolish and Install	S.F.	1.16	.96		2.12	
	Clean	S.F.		.54		.54	
	Minimum Charge	Job		186		186	
100% Wool							
Average Grade							
	Demolish	S.F.		.08		.08	Includes material and labor to install
	Install	S.F.	11.50	.74		12.24	carpet including tack strips and hot
	Demolish and Install	S.F.	11.50	.82		12.32	melt tape on seams.
	Clean	S.F.		.54		.54	
	Minimum Charge	Job		186		186	
Premium Grade							
	Demolish	S.F.		.08		.08	Includes material and labor to install
	Install	S.F.	12.80	.74		13.54	carpet including tack strips and hot
	Demolish and Install	S.F.	12.80	.82		13.62	melt tape on seams.
	Clean	S.F.		.54		.54	
	Minimum Charge	Job		186		186	

Flooring

Carpeting Per S.F.

Carpeting Per S.F.		Unit	Material	Labor	Equip.	Total	Specification
Luxury Grade							
	Demolish	S.F.		.08		.08	Includes material and labor to install
	Install	S.F.	18.30	.74		19.04	carpet including tack strips and hot
	Demolish and Install	S.F.	18.30	.82		19.12	melt tape on seams.
	Clean	S.F.		.54		.54	
	Minimum Charge	Job		186		186	
Berber							
Average Grade							
	Demolish	S.F.		.08		.08	Includes material and labor to install
	Install	S.F.	4.17	.88		5.05	carpet including tack strips and hot
	Demolish and Install	S.F.	4.17	.96		5.13	melt tape on seams.
	Clean	S.F.		.54		.54	
	Minimum Charge	Job		186		186	
Premium Grade							
	Demolish	S.F.		.08		.08	Includes material and labor to install
	Install	S.F.	8.05	.88		8.93	carpet including tack strips and hot
	Demolish and Install	S.F.	8.05	.96		9.01	melt tape on seams.
	Clean	S.F.		.54		.54	
	Minimum Charge	Job		186		186	

Carpeting

Carpeting		Unit	Material	Labor	Equip.	Total	Specification
Indoor/Outdoor							
	Demolish	S.Y.		1.15		1.15	Includes material and labor to install
	Install	S.Y.	10.40	7.90		18.30	indoor-outdoor carpet.
	Demolish and Install	S.Y.	10.40	9.05		19.45	
	Reinstall	S.Y.		6.33		6.33	
	Clean	S.Y.		4.84		4.84	
	Minimum Charge	Job		186		186	
Economy Grade							
	Demolish	S.Y.		1.15		1.15	Includes material and labor to install
	Install	S.Y.	12.95	6.55		19.50	carpet including tack strips and hot
	Demolish and Install	S.Y.	12.95	7.70		20.65	melt tape on seams.
	Reinstall	S.Y.		5.22		5.22	
	Clean	S.Y.		4.84		4.84	
	Minimum Charge	Job		186		186	
Average Grade							
	Demolish	S.Y.		1.15		1.15	Includes material and labor to install
	Install	S.Y.	15.55	6.55		22.10	carpet including tack strips and hot
	Demolish and Install	S.Y.	15.55	7.70		23.25	melt tape on seams.
	Reinstall	S.Y.		5.22		5.22	
	Clean	S.Y.		4.84		4.84	
	Minimum Charge	Job		186		186	
Premium Grade							
	Demolish	S.Y.		1.15		1.15	Includes material and labor to install
	Install	S.Y.	25.50	6.55		32.05	carpet including tack strips and hot
	Demolish and Install	S.Y.	25.50	7.70		33.20	melt tape on seams.
	Reinstall	S.Y.		5.22		5.22	
	Clean	S.Y.		4.84		4.84	
	Minimum Charge	Job		186		186	

Flooring

Carpeting		Unit	Material	Labor	Equip.	Total	Specification
100% Wool							
Average Grade							
	Demolish	S.Y.		1.15		1.15	Includes material and labor to install
	Install	S.Y.	103	6.65		109.65	wool carpet including tack strips and
	Demolish and Install	S.Y.	103	7.80		110.80	hot melt tape on seams.
	Reinstall	S.Y.		5.31		5.31	
	Clean	S.Y.		4.84		4.84	
	Minimum Charge	Job		186		186	
Premium Grade							
	Demolish	S.Y.		1.15		1.15	Includes material and labor to install
	Install	S.Y.	116	6.65		122.65	carpet including tack strips and hot
	Demolish and Install	S.Y.	116	7.80		123.80	melt tape on seams.
	Reinstall	S.Y.		5.31		5.31	
	Clean	S.Y.		4.84		4.84	
	Minimum Charge	Job		186		186	
Luxury Grade							
	Demolish	S.Y.		1.15		1.15	Includes material and labor to install
	Install	S.Y.	165	6.65		171.65	carpet including tack strips and hot
	Demolish and Install	S.Y.	165	7.80		172.80	melt tape on seams.
	Reinstall	S.Y.		5.31		5.31	
	Clean	S.Y.		4.84		4.84	
	Minimum Charge	Job		186		186	
Berber							
Average Grade							
	Demolish	S.Y.		1.15		1.15	Includes material and labor to install
	Install	S.Y.	37.50	7.90		45.40	berber carpet including tack strips and
	Demolish and Install	S.Y.	37.50	9.05		46.55	hot melt tape on seams.
	Reinstall	S.Y.		6.33		6.33	
	Clean	S.Y.		4.84		4.84	
	Minimum Charge	Job		186		186	
Premium Grade							
	Demolish	S.Y.		1.15		1.15	Includes material and labor to install
	Install	S.Y.	72.50	7.90		80.40	carpet including tack strips and hot
	Demolish and Install	S.Y.	72.50	9.05		81.55	melt tape on seams.
	Reinstall	S.Y.		6.33		6.33	
	Clean	S.Y.		4.84		4.84	
	Minimum Charge	Job		186		186	

Carpet Pad Per S.F.		Unit	Material	Labor	Equip.	Total	Specification
Urethane							
	Demolish	S.F.		.02		.02	Includes material and labor to install
	Install	S.F.	.72	.28		1	urethane carpet pad.
	Demolish and Install	S.F.	.72	.30		1.02	
	Minimum Charge	Job		186		186	
Foam Rubber Slab							
	Demolish	S.F.		.02		.02	Includes material and labor to install
	Install	S.F.	.64	.28		.92	carpet pad, 3/8" thick foam rubber.
	Demolish and Install	S.F.	.64	.30		.94	
	Minimum Charge	Job		186		186	
Waffle							
	Demolish	S.F.		.02		.02	Includes material and labor to install
	Install	S.F.	.41	.03		.44	rubber waffle carpet pad.
	Demolish and Install	S.F.	.41	.05		.46	
	Minimum Charge	Job		186		186	

Flooring

Carpet Pad Per S.F.

Carpet Pad Per S.F.	Unit	Material	Labor	Equip.	Total	Specification
Jute						
Demolish	S.F.		.02		.02	Includes material and labor to install
Install	S.F.	.79	.24		1.03	jute hair carpet pad.
Demolish and Install	S.F.	.79	.26		1.05	
Minimum Charge	Job		186		186	
Rebound						
Demolish	S.F.		.02		.02	Includes material and labor to install
Install	S.F.	.36	.21		.57	rebound carpet pad.
Demolish and Install	S.F.	.36	.23		.59	
Minimum Charge	Job		186		186	

Carpet Pad

Carpet Pad	Unit	Material	Labor	Equip.	Total	Specification
Urethane						
Demolish	S.Y.		.20		.20	Includes material and labor to install
Install	S.Y.	6.45	2.48		8.93	urethane carpet pad.
Demolish and Install	S.Y.	6.45	2.68		9.13	
Reinstall	S.Y.		1.98		1.98	
Minimum Charge	Job		186		186	
Foam Rubber Slab						
Demolish	S.Y.		.20		.20	Includes material and labor to install
Install	S.Y.	5.70	2.48		8.18	rubber slab carpet pad.
Demolish and Install	S.Y.	5.70	2.68		8.38	
Reinstall	S.Y.		1.98		1.98	
Minimum Charge	Job		186		186	
Waffle						
Demolish	S.Y.		.20		.20	Includes material and labor to install
Install	S.Y.	3.69	.28		3.97	rubber waffle carpet pad.
Demolish and Install	S.Y.	3.69	.48		4.17	
Reinstall	S.Y.		.22		.22	
Minimum Charge	Job		186		186	
Jute						
Demolish	S.Y.		.20		.20	Includes material and labor to install
Install	S.Y.	7.15	2.19		9.34	jute hair carpet pad.
Demolish and Install	S.Y.	7.15	2.39		9.54	
Reinstall	S.Y.		1.75		1.75	
Minimum Charge	Job		186		186	
Rebound						
Demolish	S.Y.		.20		.20	Includes material and labor to install
Install	S.Y.	3.23	1.85		5.08	rebound carpet pad.
Demolish and Install	S.Y.	3.23	2.05		5.28	
Reinstall	S.Y.		1.48		1.48	
Minimum Charge	Job		186		186	

Stair Components

Stair Components	Unit	Material	Labor	Equip.	Total	Specification
Carpeting						
Demolish	Ea.		2.08		2.08	Includes material and labor to install
Install	Riser	56.50	9.80		66.30	carpet and pad on stairs.
Demolish and Install	Riser	56.50	11.88		68.38	
Reinstall	Riser		7.83		7.83	
Clean	Ea.		4.84		4.84	
Minimum Charge	Job		186		186	

Finish Mechanical

Tub/Shower

Stainless Steel Double Sink

Water Closet

Commode	Unit	Material	Labor	Equip.	Total	Specification
Floor Mounted w/Tank						
Demolish	Ea.		48.50		48.50	Includes material and labor to install a
Install	Ea.	253	165		418	toilet with valve, seat and cover.
Demolish and Install	Ea.	253	213.50		466.50	
Reinstall	Ea.		131.86		131.86	
Clean	Ea.	.33	15.50		15.83	
Minimum Charge	Job		243		243	
Wall w / Flush Valve						
Demolish	Ea.		48.50		48.50	Includes material and labor to install a
Install	Ea.	775	151		926	wall hung toilet with flush valve, seat
Demolish and Install	Ea.	775	199.50		974.50	and cover.
Reinstall	Ea.		120.50		120.50	
Clean	Ea.	.33	15.50		15.83	
Minimum Charge	Job		243		243	
Designer Floor Mount						
Demolish	Ea.		48.50		48.50	Includes material and labor to install a
Install	Ea.	1150	165		1315	toilet with valve, seat and cover.
Demolish and Install	Ea.	1150	213.50		1363.50	
Reinstall	Ea.		131.86		131.86	
Clean	Ea.	.33	15.50		15.83	
Minimum Charge	Job		243		243	
Urinal						
Demolish	Ea.		48.50		48.50	Includes material and labor to install a
Install	Ea.	335	291		626	wall hung urinal with flush valve.
Demolish and Install	Ea.	335	339.50		674.50	
Reinstall	Ea.		232.96		232.96	
Clean	Ea.	.33	15.50		15.83	
Minimum Charge	Job		243		243	
Bidet						
Vitreous China						
Demolish	Ea.		48.50		48.50	Includes material and labor to install a
Install	Ea.	595	175		770	vitreous china bidet complete with
Demolish and Install	Ea.	595	223.50		818.50	trim.
Reinstall	Ea.		139.78		139.78	
Clean	Ea.	.33	15.50		15.83	
Minimum Charge	Job		243		243	

Finish Mechanical

Commode

		Unit	Material	Labor	Equip.	Total	Specification
Chrome Fittings							
	Install	Ea.	209	60.50		269.50	Includes labor and materials to install chrome fittings for a bidet.
Brass Fittings							
	Install	Ea.	230	60.50		290.50	Includes labor and materials to install brass fittings for a bidet.
Seat							
	Demolish	Ea.		20		20	Includes material and labor to install a toilet seat and cover.
	Install	Ea.	32	20		52	
	Demolish and Install	Ea.	32	40		72	
	Reinstall	Ea.		16.17		16.17	
	Clean	Ea.	.17	6.45		6.62	
	Minimum Charge	Job		243		243	
Rough-in							
	Install	Ea.	440	310		750	Includes labor and materials to install water supply pipe with valves and drain, waste and vent pipe with all couplings, hangers and fasteners necessary.

Sink (assembly)

		Unit	Material	Labor	Equip.	Total	Specification
Single							
Porcelain							
	Demolish	Ea.		30.50		30.50	Includes material and labor to install a single bowl enamel finished cast iron kitchen sink with faucet and drain.
	Install	Ea.	615	156		771	
	Demolish and Install	Ea.	615	186.50		801.50	
	Reinstall	Ea.		124.80		124.80	
	Clean	Ea.	.21	11.05		11.26	
	Minimum Charge	Job		243		243	
Stainless Steel							
	Demolish	Ea.		30.50		30.50	Includes material and labor to install a single bowl stainless steel kitchen sink with faucet and drain.
	Install	Ea.	710	156		866	
	Demolish and Install	Ea.	710	186.50		896.50	
	Reinstall	Ea.		124.80		124.80	
	Clean	Ea.	.21	11.05		11.26	
	Minimum Charge	Job		243		243	
Double							
Porcelain							
	Demolish	Ea.		34.50		34.50	Includes material and labor to install a double bowl enamel finished cast iron kitchen sink with faucet and drain.
	Install	Ea.	1275	182		1457	
	Demolish and Install	Ea.	1275	216.50		1491.50	
	Reinstall	Ea.		145.60		145.60	
	Clean	Ea.	.33	15.50		15.83	
	Minimum Charge	Job		243		243	
Stainless Steel							
	Demolish	Ea.		34.50		34.50	Includes material and labor to install a double bowl stainless steel kitchen sink with faucet and drain.
	Install	Ea.	455	182		637	
	Demolish and Install	Ea.	455	216.50		671.50	
	Reinstall	Ea.		145.60		145.60	
	Clean	Ea.	.33	15.50		15.83	
	Minimum Charge	Job		243		243	
Bar							
	Demolish	Ea.		30.50		30.50	Includes material and labor to install a small single bowl stainless steel bar / vegetable / kitchen sink with faucet and drain.
	Install	Ea.	190	121		311	
	Demolish and Install	Ea.	190	151.50		341.50	
	Reinstall	Ea.		97.04		97.04	
	Clean	Ea.	.21	11.05		11.26	
	Minimum Charge	Job		243		243	

Finish Mechanical

Sink (assembly)

	Unit	Material	Labor	Equip.	Total	Specification
Floor						
Demolish	Ea.		97		97	Includes material and labor to install
Install	Ea.	1050	199		1249	an enameled cast iron floor mounted
Demolish and Install	Ea.	1050	296		1346	corner service sink with faucet and
Reinstall	Ea.		158.84		158.84	drain.
Clean	Ea.	.33	7.75		8.08	
Minimum Charge	Job		243		243	
Pedestal						
Demolish	Ea.		30.50		30.50	Includes material and labor to install a
Install	Ea.	455	137		592	vitreous china pedestal lavatory with
Demolish and Install	Ea.	455	167.50		622.50	faucet set and pop-up drain.
Reinstall	Ea.		109.20		109.20	
Clean	Ea.	.21	11.05		11.26	
Minimum Charge	Job		243		243	
Vanity Lavatory						
Demolish	Ea.		30.50		30.50	Includes material and labor to install a
Install	Ea.	294	162		456	vitreous china lavatory with faucet set
Demolish and Install	Ea.	294	192.50		486.50	and pop-up drain.
Reinstall	Ea.		129.42		129.42	
Clean	Ea.	.21	11.05		11.26	
Minimum Charge	Job		243		243	
Rough-in						
Install	Ea.	495	410		905	Includes labor and materials to install copper supply pipe with valves and drain, waste and vent pipe with all couplings, hangers and fasteners necessary.
Laundry						
Demolish	Ea.		31		31	Includes material and labor to install a
Install	Ea.	167	134		301	laundry sink (wall mounted or with
Demolish and Install	Ea.	167	165		332	legs) including faucet and pop up
Clean	Ea.	.21	11.05		11.26	drain.
Minimum Charge	Job		243		243	

Sink Only

	Unit	Material	Labor	Equip.	Total	Specification
Single						
Porcelain						
Demolish	Ea.		30.50		30.50	Includes material and labor to install a
Install	Ea.	530	173		703	single bowl enamel finished cast iron
Demolish and Install	Ea.	530	203.50		733.50	kitchen sink.
Reinstall	Ea.		138.63		138.63	
Clean	Ea.	.21	11.05		11.26	
Minimum Charge	Job		243		243	
Stainless Steel						
Demolish	Ea.		30.50		30.50	Includes material and labor to install a
Install	Ea.	625	121		746	single bowl stainless steel kitchen sink.
Demolish and Install	Ea.	625	151.50		776.50	
Reinstall	Ea.		97.04		97.04	
Clean	Ea.	.21	11.05		11.26	
Minimum Charge	Job		243		243	

Finish Mechanical

Sink Only		Unit	Material	Labor	Equip.	Total	Specification
Double							
Porcelain							
	Demolish	Ea.		34.50		34.50	Includes material and labor to install a
	Install	Ea.	240	202		442	double bowl enamel finished cast iron
	Demolish and Install	Ea.	240	236.50		476.50	kitchen sink.
	Reinstall	Ea.		161.73		161.73	
	Clean	Ea.	.33	15.50		15.83	
	Minimum Charge	Job		243		243	
Stainless Steel							
	Demolish	Ea.		34.50		34.50	Includes material and labor to install a
	Install	Ea.	935	157		1092	double bowl stainless steel kitchen
	Demolish and Install	Ea.	935	191.50		1126.50	sink.
	Reinstall	Ea.		125.21		125.21	
	Clean	Ea.	.33	15.50		15.83	
	Minimum Charge	Job		243		243	
Bar							
	Demolish	Ea.		30.50		30.50	Includes material and labor to install a
	Install	Ea.	75.50	121		196.50	small single bowl stainless steel bar /
	Demolish and Install	Ea.	75.50	151.50		227	vegetable / kitchen sink.
	Reinstall	Ea.		97.04		97.04	
	Clean	Ea.	.21	11.05		11.26	
	Minimum Charge	Job		243		243	
Floor							
	Demolish	Ea.		97		97	Includes material and labor to install
	Install	Ea.	925	221		1146	an enameled cast iron floor mounted
	Demolish and Install	Ea.	925	318		1243	corner service sink.
	Reinstall	Ea.		176.44		176.44	
	Clean	Ea.	.33	7.75		8.08	
	Minimum Charge	Job		243		243	
Pedestal							
	Demolish	Ea.		30.50		30.50	Includes material and labor to install a
	Install	Ea.	340	152		492	vitreous china pedestal lavatory.
	Demolish and Install	Ea.	340	182.50		522.50	
	Reinstall	Ea.		121.30		121.30	
	Clean	Ea.	.21	11.05		11.26	
	Minimum Charge	Job		243		243	
Vanity Lavatory							
	Demolish	Ea.		30.50		30.50	Includes material and labor to install a
	Install	Ea.	168	152		320	vitreous china lavatory.
	Demolish and Install	Ea.	168	182.50		350.50	
	Reinstall	Ea.		121.30		121.30	
	Clean	Ea.	.21	11.05		11.26	
	Minimum Charge	Job		243		243	
Rough-in							
	Install	Ea.	440	380		820	Includes labor and materials to install copper supply pipe with valves and drain, waste and vent pipe with all couplings, hangers and fasteners necessary.
Faucet Sink							
Good Grade							
	Demolish	Ea.		24.50		24.50	Includes material and labor to install a
	Install	Ea.	67.50	33.50		101	sink faucet and fittings.
	Demolish and Install	Ea.	67.50	58		125.50	
	Reinstall	Ea.		26.79		26.79	
	Clean	Ea.	.13	9.70		9.83	
	Minimum Charge	Job		243		243	

Finish Mechanical

Sink Only

Sink Only	Unit	Material	Labor	Equip.	Total	Specification
Better Grade						
Demolish	Ea.		24.50		24.50	Includes material and labor to install a
Install	Ea.	114	33.50		147.50	sink faucet and fittings.
Demolish and Install	Ea.	114	58		172	
Reinstall	Ea.		26.79		26.79	
Clean	Ea.	.13	9.70		9.83	
Minimum Charge	Job		243		243	
Premium Grade						
Demolish	Ea.		24.50		24.50	Includes material and labor to install a
Install	Ea.	222	42		264	sink faucet and fittings.
Demolish and Install	Ea.	222	66.50		288.50	
Reinstall	Ea.		33.40		33.40	
Clean	Ea.	.13	9.70		9.83	
Minimum Charge	Job		243		243	
Drain & Basket						
Demolish	Ea.		15.15		15.15	Includes material and labor to install a
Install	Ea.	16.35	30.50		46.85	sink drain and basket.
Demolish and Install	Ea.	16.35	45.65		62	
Reinstall	Ea.		24.26		24.26	
Clean	Ea.	.13	9.70		9.83	
Minimum Charge	Job		243		243	

Bathtub

Bathtub	Unit	Material	Labor	Equip.	Total	Specification
Enameled						
Steel						
Demolish	Ea.		87.50		87.50	Includes labor and material to install a
Install	Ea.	540	159		699	formed steel bath tub with spout,
Demolish and Install	Ea.	540	246.50		786.50	mixing valve, shower head and
Reinstall	Ea.		127.07		127.07	pop-up drain.
Clean	Ea.	.65	26		26.65	
Minimum Charge	Job		243		243	
Cast Iron						
Demolish	Ea.		87.50		87.50	Includes material and labor to install a
Install	Ea.	1200	199		1399	cast iron bath tub with spout, mixing
Demolish and Install	Ea.	1200	286.50		1486.50	valve, shower head and pop-up drain.
Reinstall	Ea.		158.84		158.84	
Clean	Ea.	.65	26		26.65	
Minimum Charge	Job		243		243	
Fiberglass						
Demolish	Ea.		87.50		87.50	Includes material and labor to install
Install	Ea.	1950	159		2109	an acrylic soaking tub with spout,
Demolish and Install	Ea.	1950	246.50		2196.50	mixing valve, shower head and
Reinstall	Ea.		127.07		127.07	pop-up drain.
Clean	Ea.	.65	26		26.65	
Minimum Charge	Job		243		243	
Institutional						
Demolish	Ea.		87.50		87.50	Includes material and labor to install a
Install	Ea.	1375	291		1666	hospital / institutional type bathtub
Demolish and Install	Ea.	1375	378.50		1753.50	with spout, mixing valve, shower head
Reinstall	Ea.		232.96		232.96	and pop-up drain.
Clean	Ea.	.65	26		26.65	
Minimum Charge	Job		243		243	

Finish Mechanical

Bathtub		Unit	Material	Labor	Equip.	Total	Specification
Whirlpool (Acrylic)							
	Demolish	Ea.		270		270	Includes material and labor to install a
	Install	Ea.	1875	875		2750	molded fiberglass whirlpool tub.
	Demolish and Install	Ea.	1875	1145		3020	
	Reinstall	Ea.		698.88		698.88	
	Clean	Ea.	.65	26		26.65	
	Minimum Charge	Job		243		243	
Pump / Motor							
	Demolish	Ea.		67.50		67.50	Includes material and labor to install a
	Install	Ea.	254	146		400	whirlpool tub pump / motor.
	Demolish and Install	Ea.	254	213.50		467.50	
	Reinstall	Ea.		116.56		116.56	
	Clean	Ea.	.17	19.35		19.52	
	Minimum Charge	Job		243		243	
Heater / Motor							
	Demolish	Ea.		67.50		67.50	Includes material and labor to install a
	Install	Ea.	365	216		581	whirlpool tub heater.
	Demolish and Install	Ea.	365	283.50		648.50	
	Reinstall	Ea.		172.52		172.52	
	Clean	Ea.	.17	19.35		19.52	
	Minimum Charge	Job		243		243	
Thermal Cover							
	Demolish	Ea.		10.80		10.80	Includes material and labor to install a
	Install	Ea.	235	30.50		265.50	whirlpool tub thermal cover.
	Demolish and Install	Ea.	235	41.30		276.30	
	Reinstall	Ea.		24.26		24.26	
	Clean	Ea.	1.63	19.35		20.98	
	Minimum Charge	Job		243		243	
Tub / Shower Combination							
	Demolish	Ea.		77		77	Includes material and labor to install a
	Install	Ea.	895	218		1113	fiberglass module tub with shower
	Demolish and Install	Ea.	895	295		1190	surround with spout, mixing valve,
	Reinstall	Ea.		174.72		174.72	shower head and pop-up drain.
	Clean	Ea.	.81	38.50		39.31	
	Minimum Charge	Job		243		243	
Accessories							
Sliding Door							
	Demolish	Ea.		21.50		21.50	Includes material and labor to install a
	Install	Ea.	595	98.50		693.50	48" aluminum framed tempered glass
	Demolish and Install	Ea.	595	120		715	shower door.
	Reinstall	Ea.		78.96		78.96	
	Clean	Ea.	1.36	12.90		14.26	
	Minimum Charge	Job		243		243	
Shower Head							
	Demolish	Ea.		6.05		6.05	Includes material and labor to install a
	Install	Ea.	57	20		77	water saving shower head.
	Demolish and Install	Ea.	57	26.05		83.05	
	Reinstall	Ea.		16.17		16.17	
	Clean	Ea.	.01	3.87		3.88	
	Minimum Charge	Job		243		243	
Faucet Set							
	Demolish	Ea.		30.50		30.50	Includes material and labor to install a
	Install	Ea.	95	60.50		155.50	combination spout / diverter for a
	Demolish and Install	Ea.	95	91		186	bathtub.
	Reinstall	Ea.		48.52		48.52	
	Clean	Ea.	.13	9.70		9.83	
	Minimum Charge	Job		243		243	

Finish Mechanical

Bathtub

	Unit	Material	Labor	Equip.	Total	Specification
Shower Rod						
Demolish	Ea.		3.58		3.58	Includes material and labor to install a
Install	Ea.	33.50	32.50		66	chrome plated shower rod.
Demolish and Install	Ea.	33.50	36.08		69.58	
Reinstall	Ea.		26.02		26.02	
Clean	Ea.		3.87		3.87	
Minimum Charge	Job		243		243	
Rough-in						
Install	Ea.	525	420		945	Includes labor and materials to install copper supply pipe with valves and drain, waste and vent pipe with all couplings, hangers and fasteners necessary.

Shower

	Unit	Material	Labor	Equip.	Total	Specification
Fiberglass						
32″ x 32″						
Demolish	Ea.		182		182	Includes material and labor to install a
Install	Ea.	605	159		764	fiberglass shower stall with door,
Demolish and Install	Ea.	605	341		946	mixing valve and shower head and
Reinstall	Ea.		127.07		127.07	drain fitting.
Clean	Ea.	.65	26		26.65	
Minimum Charge	Job		243		243	
36″ x 36″						
Demolish	Ea.		182		182	Includes material and labor to install a
Install	Ea.	620	159		779	fiberglass shower stall with door,
Demolish and Install	Ea.	620	341		961	mixing valve and shower head and
Reinstall	Ea.		127.07		127.07	drain fitting.
Clean	Ea.	.65	26		26.65	
Minimum Charge	Job		243		243	
35″ x 60″						
Demolish	Ea.		182		182	Includes material and labor to install a
Install	Ea.	835	218		1053	handicap fiberglass shower stall with
Demolish and Install	Ea.	835	400		1235	door & seat, mixing valve and shower
Reinstall	Ea.		174.72		174.72	head and drain fitting.
Clean	Ea.	.65	26		26.65	
Minimum Charge	Job		243		243	
Accessories						
Single Door						
Demolish	Ea.		21.50		21.50	Includes material and labor to install a
Install	Ea.	158	52		210	shower door.
Demolish and Install	Ea.	158	73.50		231.50	
Reinstall	Ea.		41.67		41.67	
Clean	Ea.	1.36	12.90		14.26	
Minimum Charge	Job		243		243	
Shower Head						
Demolish	Ea.		6.05		6.05	Includes material and labor to install a
Install	Ea.	57	20		77	water saving shower head.
Demolish and Install	Ea.	57	26.05		83.05	
Reinstall	Ea.		16.17		16.17	
Clean	Ea.	.01	3.87		3.88	
Minimum Charge	Job		243		243	

Finish Mechanical

Shower

Shower		Unit	Material	Labor	Equip.	Total	Specification
Faucet Set							
	Demolish	Ea.		30.50		30.50	Includes material and labor to install a
	Install	Ea.	95	60.50		155.50	combination spout / diverter for a
	Demolish and Install	Ea.	95	91		186	bathtub.
	Reinstall	Ea.		48.52		48.52	
	Clean	Ea.	.13	9.70		9.83	
	Minimum Charge	Job		243		243	
Shower Pan							
	Demolish	Ea.		27		27	Includes material and labor to install
	Install	Ea.	118	60.50		178.50	the base portion of a fiberglass
	Demolish and Install	Ea.	118	87.50		205.50	shower unit.
	Reinstall	Ea.		48.52		48.52	
	Clean	Ea.	.65	26		26.65	
	Minimum Charge	Job		243		243	
Rough-in							
	Install	Ea.	655	425		1080	Includes labor and materials to install copper supply pipe with valves and drain, waste and vent pipe with all couplings, hangers and fasteners necessary.

Ductwork

Ductwork		Unit	Material	Labor	Equip.	Total	Specification
Return Grill							
Under 10″ Wide							
	Demolish	Ea.		10.40		10.40	Includes material and labor to install a
	Install	Ea.	36.50	26		62.50	10″ x 10″ air return register.
	Demolish and Install	Ea.	36.50	36.40		72.90	
	Reinstall	Ea.		20.84		20.84	
	Clean	Ea.	.17	8.60		8.77	
	Minimum Charge	Job		234		234	
12″ to 20″ Wide							
	Demolish	Ea.		10.40		10.40	Includes material and labor to install a
	Install	Ea.	66.50	27.50		94	16″ x 16″ air return register.
	Demolish and Install	Ea.	66.50	37.90		104.40	
	Reinstall	Ea.		22.06		22.06	
	Clean	Ea.	.17	8.60		8.77	
	Minimum Charge	Job		234		234	
Over 20″ Wide							
	Demolish	Ea.		15.80		15.80	Includes material and labor to install a
	Install	Ea.	141	42.50		183.50	24″ x 24″ air return register.
	Demolish and Install	Ea.	141	58.30		199.30	
	Reinstall	Ea.		34.09		34.09	
	Clean	Ea.	.17	9.70		9.87	
	Minimum Charge	Job		234		234	
Supply Grill							
Baseboard Type							
	Demolish	Ea.		7.45		7.45	Includes material and labor to install a
	Install	Ea.	6.65	23.50		30.15	14″ x 6″ baseboard air supply
	Demolish and Install	Ea.	6.65	30.95		37.60	register.
	Reinstall	Ea.		18.75		18.75	
	Clean	Ea.	.17	8.60		8.77	
	Minimum Charge	Job		234		234	

Finish Mechanical

Ductwork		Unit	Material	Labor	Equip.	Total	Specification
10″ Wide							
	Demolish	Ea.		8.70		8.70	Includes material and labor to install a
	Install	Ea.	14.20	23.50		37.70	10″ x 6″ air supply register.
	Demolish and Install	Ea.	14.20	32.20		46.40	
	Reinstall	Ea.		18.75		18.75	
	Clean	Ea.	.17	8.60		8.77	
	Minimum Charge	Job		234		234	
12″ to 15″ Wide							
	Demolish	Ea.		8.70		8.70	Includes material and labor to install a
	Install	Ea.	17.20	27.50		44.70	14″ x 8″ air supply register.
	Demolish and Install	Ea.	17.20	36.20		53.40	
	Reinstall	Ea.		22.06		22.06	
	Clean	Ea.	.17	8.60		8.77	
	Minimum Charge	Job		234		234	
18″ to 24″ Wide							
	Demolish	Ea.		10.40		10.40	Includes material and labor to install a
	Install	Ea.	27.50	36		63.50	24″ x 8″ air supply register.
	Demolish and Install	Ea.	27.50	46.40		73.90	
	Reinstall	Ea.		28.85		28.85	
	Clean	Ea.	.17	9.70		9.87	
	Minimum Charge	Job		234		234	
30″ to 36″ Wide							
	Demolish	Ea.		13		13	Includes material and labor to install a
	Install	Ea.	36	33.50		69.50	30″ x 8″ air supply register.
	Demolish and Install	Ea.	36	46.50		82.50	
	Reinstall	Ea.		26.79		26.79	
	Clean	Ea.	.17	19.35		19.52	
	Minimum Charge	Job		234		234	
Diffuser							
12″ Louver							
	Demolish	Ea.		10.40		10.40	Includes material and labor to install a
	Install	Ea.	50.50	33.50		84	12″ aluminum louvered diffuser.
	Demolish and Install	Ea.	50.50	43.90		94.40	
	Reinstall	Ea.		26.79		26.79	
	Clean	Ea.	.17	8.60		8.77	
	Minimum Charge	Job		234		234	
14″ to 20″ Louver							
	Demolish	Ea.		10.40		10.40	Includes material and labor to install a
	Install	Ea.	56.50	47		103.50	14″ to 20″ aluminum louvered diffuser.
	Demolish and Install	Ea.	56.50	57.40		113.90	
	Reinstall	Ea.		37.50		37.50	
	Clean	Ea.	.17	9.70		9.87	
	Minimum Charge	Job		234		234	
25″ to 32″ Louver							
	Demolish	Ea.		15.80		15.80	Includes material and labor to install a
	Install	Ea.	73.50	67		140.50	25″ to 32″ aluminum louvered diffuser.
	Demolish and Install	Ea.	73.50	82.80		156.30	
	Reinstall	Ea.		53.58		53.58	
	Clean	Ea.	.17	12.90		13.07	
	Minimum Charge	Job		234		234	
12″ Round							
	Demolish	Ea.		14.45		14.45	Includes material and labor to install a
	Install	Ea.	46	39		85	12″ diameter aluminum diffuser with a
	Demolish and Install	Ea.	46	53.45		99.45	butterfly damper.
	Reinstall	Ea.		31.25		31.25	
	Clean	Ea.	.17	8.60		8.77	
	Minimum Charge	Job		234		234	

Finish Mechanical

Ductwork	Unit	Material	Labor	Equip.	Total	Specification
14″ to 20″ Round						
Demolish	Ea.		19.30		19.30	Includes material and labor to install a
Install	Ea.	158	52		210	20″ diameter aluminum diffuser with a
Demolish and Install	Ea.	158	71.30		229.30	butterfly damper.
Reinstall	Ea.		41.67		41.67	
Clean	Ea.	.17	9.70		9.87	
Minimum Charge	Job		234		234	
Flush-out / Sanitize						
Install	L.F.	.65	3.07		4.52	Includes labor and material costs to
Minimum Charge	Job		234		234	clean ductwork.

Finish Electrical

Lighting Fixture   Duplex Receptacle

Electrical Per S.F.	Unit	Material	Labor	Equip.	Total	Specification
Residential						
Light Fixtures						
Install	S.F.	.29	.23		.52	Includes labor and material to replace house light fixtures. Apply S.F. cost to floor area including garages but not basements.

Circuits	Unit	Material	Labor	Equip.	Total	Specification
120V						
Outlet with Wiring						
Demolish	Ea.		25.50		25.50	Includes material and labor to install general purpose outlet with cover, up to 20 feet of #12/2 wire.
Install	Ea.	12.05	31.50		43.55	
Demolish and Install	Ea.	12.05	57		69.05	
Reinstall	Ea.		25.25		25.25	
Clean	Ea.		1.94		1.94	
Minimum Charge	Job		230		230	
GFCI Outlet w / Wiring						
Demolish	Ea.		25.50		25.50	Cost includes material and labor to install 120V GFCI receptacle and wall plate, with up to 20 feet of #12/2 wire.
Install	Ea.	51.50	37.50		89	
Demolish and Install	Ea.	51.50	63		114.50	
Reinstall	Ea.		29.84		29.84	
Clean	Ea.		1.94		1.94	
Minimum Charge	Job		230		230	
Exterior Outlet w / Wiring						
Demolish	Ea.		25.50		25.50	Cost includes material and labor to install 110V waterproof external receptacle and cover plate with up to 20' of #12/2 wire.
Install	Ea.	36.50	38.50		75	
Demolish and Install	Ea.	36.50	64		100.50	
Reinstall	Ea.		30.61		30.61	
Clean	Ea.		1.94		1.94	
Minimum Charge	Job		230		230	
240V Outlet w / Wiring						
Demolish	Ea.		25.50		25.50	Cost includes material and labor to install 240V receptacle and cover plate with up to 20' of #12/2 wire.
Install	Ea.	71.50	71.50		143	
Demolish and Install	Ea.	71.50	97		168.50	
Reinstall	Ea.		57.31		57.31	
Clean	Ea.		1.94		1.94	
Minimum Charge	Job		230		230	

Finish Electrical

Circuits

	Unit	Material	Labor	Equip.	Total	Specification
TV / Cable Jack w / Wiring						
Demolish	Ea.		25.50		25.50	Includes material and labor to install
Install	Ea.	21.50	28.50		50	TV/cable outlet box and cover plate
Demolish and Install	Ea.	21.50	54		75.50	including 20′ of wiring.
Reinstall	Ea.		22.96		22.96	
Clean	Ea.		1.94		1.94	
Minimum Charge	Job		230		230	
Phone Jack with Wiring						
Demolish	Ea.		25.50		25.50	Includes material and labor to install
Install	Ea.	12.25	17.65		29.90	phone jack and cover plate including
Demolish and Install	Ea.	12.25	43.15		55.40	up to 20′ of telephone cable.
Reinstall	Ea.		14.13		14.13	
Clean	Ea.		1.21		1.21	
Minimum Charge	Job		230		230	

Light Switch

	Unit	Material	Labor	Equip.	Total	Specification
Single						
Switch						
Demolish	Ea.		5.10		5.10	Includes material and labor to install
Install	Ea.	7.90	22		29.90	single switch and wall plate. Wiring is
Demolish and Install	Ea.	7.90	27.10		35	not included.
Reinstall	Ea.		17.49		17.49	
Clean	Ea.		1.94		1.94	
Minimum Charge	Job		230		230	
Plate						
Demolish	Ea.		2.55		2.55	Includes material and labor to install a
Install	Ea.	.42	5.75		6.17	one gang brown wall switch plate.
Demolish and Install	Ea.	.42	8.30		8.72	
Reinstall	Ea.		4.59		4.59	
Clean	Ea.	.07	.39		.46	
Minimum Charge	Job		230		230	
Double						
Switch						
Demolish	Ea.		10.20		10.20	Includes material and labor to install
Install	Ea.	8.35	33		41.35	double switch and wall plate. Wiring
Demolish and Install	Ea.	8.35	43.20		51.55	is not included.
Reinstall	Ea.		26.24		26.24	
Clean	Ea.	.07	6.20		6.27	
Minimum Charge	Job		230		230	
Plate						
Demolish	Ea.		2.55		2.55	Includes material and labor to install a
Install	Ea.	.83	8.65		9.48	two gang brown wall switch plate.
Demolish and Install	Ea.	.83	11.20		12.03	
Reinstall	Ea.		6.93		6.93	
Clean	Ea.	.07	.39		.46	
Minimum Charge	Job		230		230	
Triple						
Switch						
Demolish	Ea.		15.45		15.45	Includes material and labor to install
Install	Ea.	26.50	38.50		65	triple switch and wall plate. Wiring is
Demolish and Install	Ea.	26.50	53.95		80.45	not included.
Reinstall	Ea.		30.61		30.61	
Clean	Ea.	.07	9.30		9.37	
Minimum Charge	Job		230		230	

Finish Electrical

Light Switch		Unit	Material	Labor	Equip.	Total	Specification
Plate							
	Demolish	Ea.		2.55		2.55	Includes material and labor to install a
	Install	Ea.	1.22	14.35		15.57	three gang brown wall switch plate.
	Demolish and Install	Ea.	1.22	16.90		18.12	
	Reinstall	Ea.		11.48		11.48	
	Clean	Ea.	.07	.39		.46	
	Minimum Charge	Job		230		230	
Dimmer							
Switch							
	Demolish	Ea.		5.10		5.10	Includes material and labor to install
	Install	Ea.	27.50	27		54.50	dimmer switch. Wiring is not included.
	Demolish and Install	Ea.	27.50	32.10		59.60	
	Reinstall	Ea.		21.61		21.61	
	Clean	Ea.		1.94		1.94	
	Minimum Charge	Job		230		230	
Plate							
	Demolish	Ea.		2.55		2.55	Includes material and labor to install a
	Install	Ea.	.42	5.75		6.17	one gang brown wall switch plate.
	Demolish and Install	Ea.	.42	8.30		8.72	
	Reinstall	Ea.		4.59		4.59	
	Clean	Ea.	.07	.39		.46	
	Minimum Charge	Job		230		230	
With Wiring							
Single							
	Demolish	Ea.		25.50		25.50	Includes material and labor to install
	Install	Ea.	13.40	27		40.40	single switch and plate including up to
	Demolish and Install	Ea.	13.40	52.50		65.90	20' of romex cable.
	Reinstall	Ea.		21.48		21.48	
	Clean	Ea.		1.94		1.94	
	Minimum Charge	Job		230		230	
Double							
	Demolish	Ea.		25.50		25.50	Includes material and labor to install
	Install	Ea.	30.50	46		76.50	double switch and plate including up
	Demolish and Install	Ea.	30.50	71.50		102	to 20' of romex cable.
	Reinstall	Ea.		36.74		36.74	
	Clean	Ea.	.07	6.20		6.27	
	Minimum Charge	Job		230		230	
Triple							
	Demolish	Ea.		38.50		38.50	Includes material and labor to install
	Install	Ea.	34	51.50		85.50	triple switch and plate including up to
	Demolish and Install	Ea.	34	90		124	20' of romex cable.
	Reinstall	Ea.		41.32		41.32	
	Clean	Ea.	.07	9.30		9.37	
	Minimum Charge	Job		230		230	
Dimmer							
	Demolish	Ea.		25.50		25.50	Includes material and labor to install
	Install	Ea.	33	31.50		64.50	dimmer switch and cover plate
	Demolish and Install	Ea.	33	57		90	including up to 20' of romex cable.
	Reinstall	Ea.		25.25		25.25	
	Clean	Ea.		1.94		1.94	
	Minimum Charge	Job		230		230	

Finish Electrical

Cover for Outlet / Switch

	Unit	Material	Labor	Equip.	Total	Specification
High Grade						
Demolish	Ea.		2.55		2.55	Includes material and labor to install
Install	Ea.	2.75	5.75		8.50	wall plate, stainless steel, 1 gang.
Demolish and Install	Ea.	2.75	8.30		11.05	
Reinstall	Ea.		4.59		4.59	
Clean	Ea.	.07	.39		.46	
Minimum Charge	Job		230		230	
Deluxe Grade						
Demolish	Ea.		2.55		2.55	Includes material and labor to install
Install	Ea.	4.88	5.75		10.63	wall plate, brushed brass, 1 gang.
Demolish and Install	Ea.	4.88	8.30		13.18	
Reinstall	Ea.		4.59		4.59	
Clean	Ea.	.07	.39		.46	
Minimum Charge	Job		230		230	

Residential Light Fixture

	Unit	Material	Labor	Equip.	Total	Specification
Ceiling						
Good Quality						
Demolish	Ea.		10.65		10.65	Includes material and labor to install
Install	Ea.	40.50	11.50		52	standard quality incandescent ceiling
Demolish and Install	Ea.	40.50	22.15		62.65	light fixture. Wiring is not included.
Reinstall	Ea.		9.18		9.18	
Clean	Ea.	.08	26		26.08	
Minimum Charge	Job		230		230	
Better Quality						
Demolish	Ea.		10.65		10.65	Includes material and labor to install
Install	Ea.	102	24		126	custom quality incandescent ceiling
Demolish and Install	Ea.	102	34.65		136.65	light fixture. Wiring is not included.
Reinstall	Ea.		19.33		19.33	
Clean	Ea.	.08	26		26.08	
Minimum Charge	Job		230		230	
Premium Quality						
Demolish	Ea.		10.65		10.65	Includes material and labor to install
Install	Ea.	299	24		323	deluxe quality incandescent ceiling
Demolish and Install	Ea.	299	34.65		333.65	light fixture. Wiring is not included.
Reinstall	Ea.		19.33		19.33	
Clean	Ea.	.08	26		26.08	
Minimum Charge	Job		230		230	
Recessed Ceiling						
Demolish	Ea.		21.50		21.50	Includes material and labor to install
Install	Ea.	51	15.30		66.30	standard quality recessed eyeball
Demolish and Install	Ea.	51	36.80		87.80	spotlight with housing. Wiring is not
Reinstall	Ea.		12.25		12.25	included.
Clean	Ea.	.03	3.22		3.25	
Minimum Charge	Job		230		230	
Recessed Eyeball						
Demolish	Ea.		21.50		21.50	Includes material and labor to install
Install	Ea.	76.50	16.40		92.90	custom quality recessed eyeball
Demolish and Install	Ea.	76.50	37.90		114.40	spotlight with housing. Wiring is not
Reinstall	Ea.		13.12		13.12	included.
Clean	Ea.	.03	3.22		3.25	
Minimum Charge	Job		230		230	

Finish Electrical

Residential Light Fixture

	Unit	Material	Labor	Equip.	Total	Specification
Recessed Shower Type						Includes material and labor to install
Demolish	Ea.		21.50		21.50	deluxe quality recessed eyeball
Install	Ea.	63.50	15.30		78.80	spotlight with housing. Wiring is not
Demolish and Install	Ea.	63.50	36.80		100.30	included.
Reinstall	Ea.		12.25		12.25	
Clean	Ea.	.03	3.22		3.25	
Minimum Charge	Job		230		230	
Run Wiring						Includes labor and material to install a
Install	Ea.	19.35	18.35		37.70	electric fixture utility box with
Minimum Charge	Job		230		230	non-metallic cable.
Fluorescent						
Good Quality						Includes material and labor to install
Demolish	Ea.		23		23	standard quality 12" wide by 48" long
Install	Ea.	41.50	65.50		107	decorative surface mounted fluorescent
Demolish and Install	Ea.	41.50	88.50		130	fixture w/acrylic diffuser.
Reinstall	Ea.		52.48		52.48	
Clean	Ea.	.03	3.22		3.25	
Minimum Charge	Job		230		230	
Better Quality						Includes material and labor to install
Demolish	Ea.		23		23	an interior fluorescent lighting fixture.
Install	Ea.	84.50	74		158.50	
Demolish and Install	Ea.	84.50	97		181.50	
Reinstall	Ea.		59.25		59.25	
Clean	Ea.	.03	3.22		3.25	
Minimum Charge	Job		230		230	
Premium Quality						Includes material and labor to install
Demolish	Ea.		23		23	deluxe quality 24" wide by 48" long
Install	Ea.	88.50	86.50		175	decorative surface mounted fluorescent
Demolish and Install	Ea.	88.50	109.50		198	fixture w/acrylic diffuser.
Reinstall	Ea.		69.31		69.31	
Clean	Ea.	.03	3.22		3.25	
Minimum Charge	Job		230		230	
Run Wiring						Includes labor and material to install a
Install	Ea.	19.35	18.35		37.70	electric fixture utility box with
Minimum Charge	Job		230		230	non-metallic cable.
Exterior						
Good Quality						Includes material and labor to install
Demolish	Ea.		20.50		20.50	standard quality incandescent outdoor
Install	Ea.	46	28.50		74.50	light fixture. Wiring is not included.
Demolish and Install	Ea.	46	49		95	
Reinstall	Ea.		22.96		22.96	
Clean	Ea.	.17	12.90		13.07	
Minimum Charge	Job		230		230	
Better Quality						Includes material and labor to install
Demolish	Ea.		20.50		20.50	custom quality incandescent outdoor
Install	Ea.	116	28.50		144.50	light fixture. Wiring is not included.
Demolish and Install	Ea.	116	49		165	
Reinstall	Ea.		22.96		22.96	
Clean	Ea.	.17	12.90		13.07	
Minimum Charge	Job		230		230	
Premium Quality						Includes material and labor to install
Demolish	Ea.		20.50		20.50	deluxe quality incandescent outdoor
Install	Ea.	174	115		289	light fixture. Wiring is not included.
Demolish and Install	Ea.	174	135.50		309.50	
Reinstall	Ea.		91.84		91.84	
Clean	Ea.	.17	12.90		13.07	
Minimum Charge	Job		230		230	

Finish Electrical

Residential Light Fixture	Unit	Material	Labor	Equip.	Total	Specification
Outdoor w / Pole						
Demolish	Ea.		20.50		20.50	Includes material and labor to install
Install	Ea.	129	115		244	post lantern type incandescent outdoor
Demolish and Install	Ea.	129	135.50		264.50	light fixture. Wiring is not included.
Reinstall	Ea.		91.84		91.84	
Clean	Ea.	.17	12.90		13.07	
Minimum Charge	Job		230		230	
Run Wiring						
Install	Ea.	19.35	18.35		37.70	Includes labor and material to install a
Minimum Charge	Job		230		230	electric fixture utility box with
						non-metallic cable.
Track Lighting						
Demolish	L.F.		7.10		7.10	Includes material and labor to install
Install	L.F.	18.35	19.15		37.50	track lighting.
Demolish and Install	L.F.	18.35	26.25		44.60	
Reinstall	L.F.		15.31		15.31	
Clean	L.F.	.17	6.20		6.37	
Minimum Charge	Job		230		230	

Commercial Light Fixture	Unit	Material	Labor	Equip.	Total	Specification
Fluorescent						
2 Bulb, 2'						
Demolish	Ea.		20.50		20.50	Includes material and labor to install a
Install	Ea.	38	57.50		95.50	2' long surface mounted fluorescent
Demolish and Install	Ea.	38	78		116	light fixture with two 40 watt bulbs.
Reinstall	Ea.		45.92		45.92	
Clean	Ea.	.81	22		22.81	
Minimum Charge	Job		230		230	
2 Bulb, 4'						
Demolish	Ea.		20.50		20.50	Includes material and labor to install a
Install	Ea.	69	65.50		134.50	4' long surface mounted fluorescent
Demolish and Install	Ea.	69	86		155	light fixture with two 40 watt bulbs.
Reinstall	Ea.		52.48		52.48	
Clean	Ea.	.81	22		22.81	
Minimum Charge	Job		230		230	
3 Bulb, 4'						
Demolish	Ea.		20.50		20.50	Includes material and labor to install a
Install	Ea.	103	65.50		168.50	4' long pendant mounted fluorescent
Demolish and Install	Ea.	103	86		189	light fixture with three 34 watt bulbs.
Reinstall	Ea.		52.48		52.48	
Clean	Ea.	.81	22		22.81	
Minimum Charge	Job		230		230	
4 Bulb, 4'						
Demolish	Ea.		20.50		20.50	Includes material and labor to install a
Install	Ea.	94.50	70.50		165	4' long pendant mounted fluorescent
Demolish and Install	Ea.	94.50	91		185.50	light fixture with four 34 watt bulbs.
Reinstall	Ea.		56.52		56.52	
Clean	Ea.	.81	22		22.81	
Minimum Charge	Job		230		230	
2 Bulb, 8'						
Demolish	Ea.		25.50		25.50	Includes material and labor to install a
Install	Ea.	106	86.50		192.50	8' long surface mounted fluorescent
Demolish and Install	Ea.	106	112		218	light fixture with two 110 watt bulbs.
Reinstall	Ea.		69.31		69.31	
Clean	Ea.	.81	22		22.81	
Minimum Charge	Job		230		230	

Finish Electrical

Commercial Light Fixture		Unit	Material	Labor	Equip.	Total	Specification
Run Wiring							
	Install	Ea.	19.35	18.35		37.70	Includes labor and material to install a
	Minimum Charge	Job		230		230	electric fixture utility box with
							non-metallic cable.
Ceiling, Recessed							
Eyeball Type							
	Demolish	Ea.		21.50		21.50	Includes material and labor to install
	Install	Ea.	76.50	16.40		92.90	custom quality recessed eyeball
	Demolish and Install	Ea.	76.50	37.90		114.40	spotlight with housing. Wiring is not
	Reinstall	Ea.		13.12		13.12	included.
	Clean	Ea.	.03	3.22		3.25	
	Minimum Charge	Job		230		230	
Shower Type							
	Demolish	Ea.		21.50		21.50	Includes material and labor to install
	Install	Ea.	63.50	15.30		78.80	deluxe quality recessed eyeball
	Demolish and Install	Ea.	63.50	36.80		100.30	spotlight with housing. Wiring is not
	Reinstall	Ea.		12.25		12.25	included.
	Clean	Ea.	.03	3.22		3.25	
	Minimum Charge	Job		230		230	
Run Wiring							
	Install	Ea.	19.35	18.35		37.70	Includes labor and material to install a
	Minimum Charge	Job		230		230	electric fixture utility box with
							non-metallic cable.
Ceiling Grid Type							
2 Tube							
	Demolish	Ea.		30		30	Cost includes material and labor to
	Install	Ea.	54.50	86.50		141	install 2 lamp fluorescent light fixture
	Demolish and Install	Ea.	54.50	116.50		171	in 2' x 4' ceiling grid system including
	Reinstall	Ea.		69.31		69.31	lens and tubes.
	Clean	Ea.	.81	22		22.81	
	Minimum Charge	Job		230		230	
4 Tube							
	Demolish	Ea.		34		34	Cost includes material and labor to
	Install	Ea.	62.50	97.50		160	install 4 lamp fluorescent light fixture
	Demolish and Install	Ea.	62.50	131.50		194	in 2' x 4' ceiling grid system including
	Reinstall	Ea.		78.16		78.16	lens and tubes.
	Clean	Ea.	.81	22		22.81	
	Minimum Charge	Job		230		230	
Acrylic Diffuser							
	Demolish	Ea.		5.30		5.30	Includes material and labor to install
	Install	Ea.	65	13.90		78.90	fluorescent light fixture w/acrylic
	Demolish and Install	Ea.	65	19.20		84.20	diffuser.
	Reinstall	Ea.		11.13		11.13	
	Clean	Ea.	.52	3.87		4.39	
	Minimum Charge	Job		230		230	
Louver Diffuser							
	Demolish	Ea.		6.40		6.40	Includes material and labor to install
	Install	Ea.	72.50	13.90		86.40	fluorescent light fixture w/louver
	Demolish and Install	Ea.	72.50	20.30		92.80	diffuser.
	Reinstall	Ea.		11.13		11.13	
	Clean	Ea.	.52	3.87		4.39	
	Minimum Charge	Job		230		230	
Run Wiring							
	Install	Ea.	19.35	18.35		37.70	Includes labor and material to install a
	Minimum Charge	Job		230		230	electric fixture utility box with
							non-metallic cable.

Finish Electrical

Commercial Light Fixture		Unit	Material	Labor	Equip.	Total	Specification
Track Lighting							
	Demolish	L.F.		7.10		7.10	Includes material and labor to install
	Install	L.F.	18.35	19.15		37.50	track lighting.
	Demolish and Install	L.F.	18.35	26.25		44.60	
	Reinstall	L.F.		15.31		15.31	
	Clean	L.F.	.17	6.20		6.37	
	Minimum Charge	Job		230		230	
Emergency							
	Demolish	Ea.		36.50		36.50	Includes material and labor to install
	Install	Ea.	150	115		265	battery-powered emergency lighting
	Demolish and Install	Ea.	150	151.50		301.50	unit with two (2) lights, including
	Reinstall	Ea.		91.84		91.84	wiring.
	Clean	Ea.	.41	22		22.41	
	Minimum Charge	Job		230		230	
Run Wiring							
	Install	Ea.	19.35	18.35		37.70	Includes labor and material to install a
	Minimum Charge	Job		230		230	electric fixture utility box with
							non-metallic cable.

Fan		Unit	Material	Labor	Equip.	Total	Specification
Ceiling Paddle							
Good Quality							
	Demolish	Ea.		28.50		28.50	Includes material and labor to install
	Install	Ea.	102	46		148	standard ceiling fan.
	Demolish and Install	Ea.	102	74.50		176.50	
	Reinstall	Ea.		36.74		36.74	
	Clean	Ea.	.33	19.35		19.68	
	Minimum Charge	Job		230		230	
Better Quality							
	Demolish	Ea.		28.50		28.50	Includes material and labor to install
	Install	Ea.	154	46		200	better ceiling fan.
	Demolish and Install	Ea.	154	74.50		228.50	
	Reinstall	Ea.		36.74		36.74	
	Clean	Ea.	.33	19.35		19.68	
	Minimum Charge	Job		230		230	
Premium Quality							
	Demolish	Ea.		28.50		28.50	Includes material and labor to install
	Install	Ea.	365	57.50		422.50	deluxe ceiling fan.
	Demolish and Install	Ea.	365	86		451	
	Reinstall	Ea.		45.92		45.92	
	Clean	Ea.	.33	19.35		19.68	
	Minimum Charge	Job		230		230	
Light Kit							
	Install	Ea.	45	57.50		102.50	Includes labor and material to install a
	Minimum Charge	Job		230		230	ceiling fan light kit.
Bathroom Exhaust							
Ceiling Mounted							
	Demolish	Ea.		23.50		23.50	Cost includes material and labor to
	Install	Ea.	58.50	46		104.50	install 60 CFM ceiling mounted
	Demolish and Install	Ea.	58.50	69.50		128	bathroom exhaust fan, up to 10' of
	Reinstall	Ea.		36.63		36.63	wiring with one wall switch.
	Clean	Ea.	.17	7.75		7.92	
	Minimum Charge	Job		230		230	

Finish Electrical

Fan		Unit	Material	Labor	Equip.	Total	Specification
Wall Mounted							
	Demolish	Ea.		23.50		23.50	Cost includes material and labor to
	Install	Ea.	74	46		120	install 110 CFM wall mounted
	Demolish and Install	Ea.	74	69.50		143.50	bathroom exhaust fan, up to 10' of
	Reinstall	Ea.		36.63		36.63	wiring with one wall switch.
	Clean	Ea.	.17	7.75		7.92	
	Minimum Charge	Job		230		230	
Lighted							
	Demolish	Ea.		23.50		23.50	Cost includes material and labor to
	Install	Ea.	54	46		100	install 50 CFM ceiling mounted lighted
	Demolish and Install	Ea.	54	69.50		123.50	exhaust fan, up to 10' of wiring with
	Reinstall	Ea.		36.63		36.63	one wall switch..
	Clean	Ea.	.17	7.75		7.92	
	Minimum Charge	Job		230		230	
Blower Heater w / Light							
	Demolish	Ea.		23.50		23.50	Includes material and labor to install
	Install	Ea.	111	76.50		187.50	ceiling blower type ventilator with
	Demolish and Install	Ea.	111	100		211	switch, light and snap-on grill. Includes
	Reinstall	Ea.		61.04		61.04	1300 watt/120 volt heater.
	Clean	Ea.	.17	7.75		7.92	
	Minimum Charge	Job		230		230	
Fan Heater w / Light							
	Demolish	Ea.		23.50		23.50	Includes material and labor to install
	Install	Ea.	153	76.50		229.50	circular ceiling fan heater unit, 1500
	Demolish and Install	Ea.	153	100		253	watt, surface mounted.
	Reinstall	Ea.		61.04		61.04	
	Clean	Ea.	.17	7.75		7.92	
	Minimum Charge	Job		230		230	
Whole House Exhaust							
36" Ceiling Mounted							
	Install	Ea.	1000	114		1114	Includes material and labor to install a
	Clean	Ea.	.04	5.65		5.69	whole house exhaust fan.
	Minimum Charge	Job		230		230	
Kitchen Exhaust							
Standard Model							
	Install	Ea.	82	30.50		112.50	Includes labor and materials to install
	Clean	Ea.	.04	5.65		5.69	a bathroom or kitchen vent fan.
	Minimum Charge	Job		230		230	
Low Noise Model							
	Install	Ea.	102	30.50		132.50	Includes labor and materials to install
	Clean	Ea.	.04	5.65		5.69	a bathroom or kitchen vent fan, low
	Minimum Charge	Job		230		230	noise model.

Low Voltage Systems		Unit	Material	Labor	Equip.	Total	Specification
Smoke Detector							
Pre-wired							
	Demolish	Ea.		10.65		10.65	Includes material and labor to install
	Install	Ea.	22	86		108	AC type smoke detector including
	Demolish and Install	Ea.	22	96.65		118.65	wiring.
	Reinstall	Ea.		68.92		68.92	
	Clean	Ea.	.07	9.70		9.77	
	Minimum Charge	Job		230		230	

Finish Electrical

Low Voltage Systems		Unit	Material	Labor	Equip.	Total	Specification
Run Wiring							
	Install	Ea.	19.35	18.35		37.70	Includes labor and material to install a
	Minimum Charge	Job		230		230	electric fixture utility box with
							non-metallic cable.
Door Bell / Chimes							
Good Quality							
	Demolish	Ea.		32		32	Includes material and labor to install
	Install	Ea.	59.50	115		174.50	surface mounted 2 note door bell.
	Demolish and Install	Ea.	59.50	147		206.50	Wiring is not included.
	Reinstall	Ea.		91.84		91.84	
	Clean	Ea.	.17	9.70		9.87	
	Minimum Charge	Job		230		230	
Better Quality							
	Demolish	Ea.		32		32	Includes material and labor to install
	Install	Ea.	84	115		199	standard quality surface mounted 2
	Demolish and Install	Ea.	84	147		231	note door bell. Wiring is not included.
	Reinstall	Ea.		91.84		91.84	
	Clean	Ea.	.17	9.70		9.87	
	Minimum Charge	Job		230		230	
Premium Quality							
	Demolish	Ea.		32		32	Includes material and labor to install
	Install	Ea.	105	115		220	deluxe quality surface mounted 2 note
	Demolish and Install	Ea.	105	147		252	door bell. Wiring is not included.
	Reinstall	Ea.		91.84		91.84	
	Clean	Ea.	.17	9.70		9.87	
	Minimum Charge	Job		230		230	
Run Wiring							
	Install	Ea.	19.35	18.35		37.70	Includes labor and material to install a
	Minimum Charge	Job		230		230	electric fixture utility box with
							non-metallic cable.
Intercom System							
Intercom (master)							
	Demolish	Ea.		85		85	Includes material and labor to install
	Install	Ea.	635	230		865	master intercom station with AM/FM
	Demolish and Install	Ea.	635	315		950	music and room monitoring. Wiring is
	Reinstall	Ea.		183.68		183.68	not included.
	Clean	Ea.	.17	26		26.17	
	Minimum Charge	Job		230		230	
Intercom (remote)							
	Demolish	Ea.		51		51	Includes material and labor to install
	Install	Ea.	57	57.50		114.50	remote station for master intercom
	Demolish and Install	Ea.	57	108.50		165.50	system. Wiring is not included.
	Reinstall	Ea.		45.92		45.92	
	Clean	Ea.	.17	9.70		9.87	
	Minimum Charge	Job		230		230	
Security System							
Alarm Panel							
	Demolish	Ea.		102		102	Includes material and labor to install a
	Install	Ea.	163	230		393	6 zone burglar alarm panel.
	Demolish and Install	Ea.	163	332		495	
	Reinstall	Ea.		183.68		183.68	
	Clean	Ea.	.17	9.70		9.87	
	Minimum Charge	Job		230		230	

Finish Electrical

Low Voltage Systems

	Unit	Material	Labor	Equip.	Total	Specification
Motion Detector						
Demolish	Ea.		73		73	Includes material and labor to install a
Install	Ea.	248	200		448	ultrasonic motion detector for a
Demolish and Install	Ea.	248	273		521	security system.
Reinstall	Ea.		159.72		159.72	
Clean	Ea.	.17	9.70		9.87	
Minimum Charge	Job		230		230	
Door Switch						
Demolish	Ea.		32		32	Includes material and labor to install a
Install	Ea.	150	86.50		236.50	door / window contact for a security
Demolish and Install	Ea.	150	118.50		268.50	system.
Reinstall	Ea.		69.31		69.31	
Clean	Ea.	.17	3.10		3.27	
Minimum Charge	Job		230		230	
Window Switch						
Demolish	Ea.		32		32	Includes material and labor to install a
Install	Ea.	150	86.50		236.50	door / window contact for a security
Demolish and Install	Ea.	150	118.50		268.50	system.
Reinstall	Ea.		69.31		69.31	
Clean	Ea.	.17	3.10		3.27	
Minimum Charge	Job		230		230	
Satellite System						
Dish						
Demolish	Ea.		142		142	Includes material, labor and
Install	Ea.	1275	385		1660	equipment to install 10' mesh dish.
Demolish and Install	Ea.	1275	527		1802	
Reinstall	Ea.		306.13		306.13	
Clean	Ea.	3.27	38.50		41.77	
Minimum Charge	Job		230		230	
Motor						
Demolish	Ea.		71		71	Includes material and labor to install a
Install	Ea.	400	191		591	television satellite dish motor unit.
Demolish and Install	Ea.	400	262		662	
Reinstall	Ea.		153.07		153.07	
Clean	Ea.	.17	9.70		9.87	
Minimum Charge	Job		230		230	
Television Antenna						
40' Pole						
Demolish	Ea.		79.50		79.50	Includes material and labor to install
Install	Ea.	30.50	144		174.50	stand-alone 1"-2" metal pole, 10' long.
Demolish and Install	Ea.	30.50	223.50		254	Cost does not include antenna.
Reinstall	Ea.		114.80		114.80	
Clean	Ea.	.65	38.50		39.15	
Paint	Ea.	8.05	45		53.05	
Minimum Charge	Job		230		230	
Rotor Unit						
Demolish	Ea.		64		64	Includes material and labor to install a
Install	Ea.	83	57.50		140.50	television antenna rotor.
Demolish and Install	Ea.	83	121.50		204.50	
Reinstall	Ea.		45.92		45.92	
Clean	Ea.	.65	38.50		39.15	
Minimum Charge	Job		230		230	
Single Booster						
Demolish	Ea.		64		64	Includes material and labor to install
Install	Ea.	40.50	57.50		98	standard grade VHF television signal
Demolish and Install	Ea.	40.50	121.50		162	booster unit and all hardware required
Reinstall	Ea.		45.92		45.92	for connection to TV cable.
Clean	Ea.	.03	4.84		4.87	
Minimum Charge	Job		230		230	

Gazebo   Gutter   Shutter

Window Treatment		Unit	Material	Labor	Equip.	Total	Specification
Drapery							
Good Grade							
	Install	L.F.	45.50	6.20		51.70	Includes material and labor to install
	Reinstall	L.F.		4.95		4.95	good quality drapery.
	Clean	L.F.	2.85	.15		3	
	Minimum Charge	Job		155		155	
Premium Grade							
	Install	L.F.	80.50	6.20		86.70	Includes material and labor to install
	Reinstall	L.F.		4.95		4.95	premium quality drapery.
	Clean	L.F.	2.85	.17		3.02	
	Minimum Charge	Job		155		155	
Lining							
	Install	L.F.	19.90	6.20		26.10	Includes material and labor to install
	Reinstall	L.F.		4.95		4.95	drapery lining.
	Clean	L.F.	1.43	.09		1.52	
	Minimum Charge	Job		155		155	
Valance Board							
	Demolish	L.F.		1.72		1.72	Includes material and labor to install a
	Install	L.F.	4.98	12.40		17.38	drapery valance board.
	Demolish and Install	L.F.	4.98	14.12		19.10	
	Reinstall	L.F.		9.91		9.91	
	Clean	L.F.	1.43	.31		1.74	
	Minimum Charge	Job		155		155	
Rod							
	Demolish	L.F.		4.30		4.30	Includes material and labor to install
	Install	L.F.	7.10	7.15		14.25	standard traverse drapery hardware.
	Demolish and Install	L.F.	7.10	11.45		18.55	
	Reinstall	L.F.		5.73		5.73	
	Clean	L.F.	.07	.28		.35	
	Paint	L.F.	.07	.69		.76	
	Minimum Charge	Job		155		155	
Blinds							
Mini-type							
	Demolish	S.F.		.92		.92	Includes material and labor to install
	Install	S.F.	6.05	.72		6.77	horizontal 1″ slat blinds.
	Demolish and Install	S.F.	6.05	1.64		7.69	
	Reinstall	S.F.		.57		.57	
	Clean	S.F.	.54	.03		.57	
	Minimum Charge	Job		155		155	

Improvements / Appliances / Treatments

Window Treatment		Unit	Material	Labor	Equip.	Total	Specification
Vertical Type							
	Demolish	S.F.		.92		.92	Includes material and labor to install
	Install	S.F.	9	.92		9.92	vertical 3" - 5" cloth or PVC blinds.
	Demolish and Install	S.F.	9	1.84		10.84	
	Reinstall	S.F.		.74		.74	
	Clean	S.F.	.29	.03		.32	
	Minimum Charge	Job		155		155	
Vertical Type per L.F.							
	Demolish	L.F.		1.43		1.43	Includes material and labor to install
	Install	L.F.	54.50	4.13		58.63	vertical 3" - 5" cloth or PVC blinds x
	Demolish and Install	L.F.	54.50	5.56		60.06	72" long.
	Reinstall	L.F.		3.30		3.30	
	Clean	L.F.	2.02	.22		2.24	
	Minimum Charge	Job		155		155	
Venetian Type							
	Demolish	S.F.		.92		.92	Includes material and labor to install
	Install	S.F.	9.40	.72		10.12	horizontal 2" slat blinds.
	Demolish and Install	S.F.	9.40	1.64		11.04	
	Reinstall	S.F.		.57		.57	
	Clean	S.F.	.29	.03		.32	
	Minimum Charge	Job		155		155	
Wood Shutter							
27" x 36", 4 Panel							
	Demolish	Ea.		29.50		29.50	Includes material and labor to install
	Install	Pr.	99	25		124	moveable louver, up to 33" x 36", 4
	Demolish and Install	Pr.	99	54.50		153.50	panel shutter.
	Reinstall	Pr.		19.90		19.90	
	Clean	Ea.	.07	9.45		9.52	
	Paint	Ea.	1.29	26		27.29	
	Minimum Charge	Job		211		211	
33" x 36", 4 Panel							
	Demolish	Ea.		29.50		29.50	Includes material and labor to install
	Install	Pr.	99	25		124	moveable louver, up to 33" x 36", 4
	Demolish and Install	Pr.	99	54.50		153.50	panel shutter.
	Reinstall	Pr.		19.90		19.90	
	Clean	Ea.	.07	9.45		9.52	
	Paint	Ea.	1.34	28		29.34	
	Minimum Charge	Job		211		211	
39" x 36", 4 Panel							
	Demolish	Ea.		29.50		29.50	Includes material and labor to install
	Install	Pr.	108	25		133	moveable louver, up to 47" x 36", 4
	Demolish and Install	Pr.	108	54.50		162.50	panel shutter.
	Reinstall	Pr.		19.90		19.90	
	Clean	Ea.	.07	9.45		9.52	
	Paint	Ea.	1.34	28		29.34	
	Minimum Charge	Job		211		211	
47" x 36", 4 Panel							
	Demolish	Ea.		29.50		29.50	Includes material and labor to install
	Install	Pr.	108	25		133	moveable louver, up to 47" x 36", 4
	Demolish and Install	Pr.	108	54.50		162.50	panel shutter.
	Reinstall	Pr.		19.90		19.90	
	Clean	Ea.	.07	9.45		9.52	
	Paint	Ea.	1.40	30		31.40	
	Minimum Charge	Job		211		211	

Improvements / Appliances / Treatments

Stairs

	Unit	Material	Labor	Equip.	Total	Specification
Concrete Steps						
Precast						
Demolish	Ea.		60	45	105	Includes material and labor to install
Install	Flight	480	145	42	667	precast concrete front entrance stairs,
Demolish and Install	Flight	480	205	87	772	4' wide, with 48" platform and 2
Clean	Ea.		4.30		4.30	risers.
Paint	Ea.	4.70	45		49.70	
Minimum Charge	Job		375		375	
Metal						
Concrete Filled						
Demolish	Ea.		17.20		17.20	Includes material and labor to install
Install	Riser	600	70	5.30	675.30	steel, cement filled metal pan stairs
Demolish and Install	Riser	600	87.20	5.30	692.50	with picket rail.
Clean	Ea.	.10	2.67		2.77	
Paint	Ea.	.69	2.67		3.36	
Minimum Charge	Job		258		258	
Landing						
Demolish	S.F.		4.30		4.30	Includes material and labor to install
Install	S.F.	127	8.20	.62	135.82	pre-erected conventional steel pan
Demolish and Install	S.F.	127	12.50	.62	140.12	landing.
Clean	S.F.	.10	.55		.65	
Paint	S.F.	.24	.77		1.01	
Minimum Charge	Job		258		258	
Railing						
Demolish	L.F.		2.37		2.37	Includes material and labor to install
Install	L.F.	16.60	11.90	.90	29.40	primed steel pipe.
Demolish and Install	L.F.	16.60	14.27	.90	31.77	
Clean	L.F.	.65	1.94		2.59	
Paint	L.F.	.77	6		6.77	
Minimum Charge	Job		258		258	

Elevator

	Unit	Material	Labor	Equip.	Total	Specification
2 Stop Residential						
Install	Ea.	12300	6225		18525	Includes labor, material and
Clean	Ea.	.79	19.35		20.14	equipment to install 2 stop elevator for
Minimum Charge	Job		625		625	residential use.
2 Stop Commercial						
Install	Ea.	45000	12500		57500	Includes labor, material and
Clean	Ea.	.79	19.35		20.14	equipment to install 2 stop elevator for
Minimum Charge	Job		625		625	apartment or commercial building with
						2,000 lb/13 passenger capacity.
3 Stop Commercial						
Install	Ea.	56500	19200		75700	Includes labor, material and
Clean	Ea.	.79	19.35		20.14	equipment to install 3 stop elevator for
Minimum Charge	Job		625		625	apartment or commercial building with
						2,000 lb/13 passenger capacity.
4 Stop Commercial						
Install	Ea.	69000	25900		94900	Includes labor, material and
Clean	Ea.	.79	19.35		20.14	equipment to install 4 stop elevator for
Minimum Charge	Job		625		625	apartment or commercial building with
						2,000 lb/13 passenger capacity.

Improvements / Appliances / Treatments

Elevator		Unit	Material	Labor	Equip.	Total	Specification
5 Stop Commercial							Includes labor, material and
	Install	Ea.	80500	32800		113300	equipment to install 5 stop elevator for
	Clean	Ea.	.79	19.35		20.14	apartment or commercial building with
	Minimum Charge	Job		625		625	2,000 lb/13 passenger capacity.

Rangetop		Unit	Material	Labor	Equip.	Total	Specification
4-burner							
	Demolish	Ea.		31.50		31.50	Cost includes material and labor to
	Install	Ea.	305	76.50		381.50	install 4-burner electric surface unit.
	Demolish and Install	Ea.	305	108		413	
	Reinstall	Ea.		61.23		61.23	
	Clean	Ea.	.04	28.50		28.54	
	Minimum Charge	Job		77.50		77.50	
6-burner							
	Demolish	Ea.		31.50		31.50	Includes material and labor to install a
	Install	Ea.	2300	92		2392	6-burner cooktop unit.
	Demolish and Install	Ea.	2300	123.50		2423.50	
	Reinstall	Ea.		73.47		73.47	
	Clean	Ea.	.04	28.50		28.54	
	Minimum Charge	Job		77.50		77.50	
Premium Brand w / Grill							
	Demolish	Ea.		31.50		31.50	Includes material and labor to install
	Install	Ea.	1100	92		1192	electric downdraft cooktop with grill.
	Demolish and Install	Ea.	1100	123.50		1223.50	
	Reinstall	Ea.		73.47		73.47	
	Clean	Ea.	.04	28.50		28.54	
	Minimum Charge	Job		77.50		77.50	

Rangehood		Unit	Material	Labor	Equip.	Total	Specification
Vented							
30″							
	Demolish	Ea.		28.50		28.50	Includes material and labor to install a
	Install	Ea.	69.50	92		161.50	vented range hood.
	Demolish and Install	Ea.	69.50	120.50		190	
	Reinstall	Ea.		73.47		73.47	
	Clean	Ea.	.04	8.10		8.14	
	Minimum Charge	Job		77.50		77.50	
36″							
	Demolish	Ea.		28.50		28.50	Includes material and labor to install a
	Install	Ea.	166	92		258	vented range hood.
	Demolish and Install	Ea.	166	120.50		286.50	
	Reinstall	Ea.		73.47		73.47	
	Clean	Ea.	.04	8.10		8.14	
	Minimum Charge	Job		77.50		77.50	
42″							
	Demolish	Ea.		28.50		28.50	Includes material and labor to install a
	Install	Ea.	179	108		287	vented range hood.
	Demolish and Install	Ea.	179	136.50		315.50	
	Reinstall	Ea.		86.02		86.02	
	Clean	Ea.	.04	8.10		8.14	
	Minimum Charge	Job		77.50		77.50	

Improvements / Appliances / Treatments

Rangehood		Unit	Material	Labor	Equip.	Total	Specification
Ventless							
30"							
	Demolish	Ea.		28.50		28.50	Includes material and labor to install a
	Install	Ea.	84	57.50		141.50	ventless range hood.
	Demolish and Install	Ea.	84	86		170	
	Reinstall	Ea.		45.92		45.92	
	Clean	Ea.	.04	8.10		8.14	
	Minimum Charge	Job		77.50		77.50	
36"							
	Demolish	Ea.		28.50		28.50	Includes material and labor to install a
	Install	Ea.	149	92		241	ventless range hood.
	Demolish and Install	Ea.	149	120.50		269.50	
	Reinstall	Ea.		73.47		73.47	
	Clean	Ea.	.04	8.10		8.14	
	Minimum Charge	Job		77.50		77.50	
42"							
	Demolish	Ea.		28.50		28.50	Includes material and labor to install a
	Install	Ea.	213	76.50		289.50	ventless range hood.
	Demolish and Install	Ea.	213	105		318	
	Reinstall	Ea.		61.23		61.23	
	Clean	Ea.	.04	8.10		8.14	
	Minimum Charge	Job		77.50		77.50	

Oven		Unit	Material	Labor	Equip.	Total	Specification
Free Standing							
	Demolish	Ea.		31.50		31.50	Cost includes material and labor to
	Install	Ea.	530	62		592	install 30" free standing electric range
	Demolish and Install	Ea.	530	93.50		623.50	with self-cleaning oven.
	Reinstall	Ea.		49.54		49.54	
	Clean	Ea.	.04	28.50		28.54	
	Minimum Charge	Job		77.50		77.50	
Single Wall Oven							
	Demolish	Ea.		31.50		31.50	Includes material and labor to install a
	Install	Ea.	675	115		790	built-in oven.
	Demolish and Install	Ea.	675	146.50		821.50	
	Reinstall	Ea.		91.84		91.84	
	Clean	Ea.	.04	28.50		28.54	
	Minimum Charge	Job		77.50		77.50	
Double Wall Oven							
	Demolish	Ea.		31.50		31.50	Includes material and labor to install a
	Install	Ea.	1550	211		1761	conventional double oven.
	Demolish and Install	Ea.	1550	242.50		1792.50	
	Reinstall	Ea.		169.12		169.12	
	Clean	Ea.	.04	28.50		28.54	
	Minimum Charge	Job		77.50		77.50	
Drop-in Type w / Range							
	Demolish	Ea.		31.50		31.50	Includes material and labor to install a
	Install	Ea.	815	76.50		891.50	built-in cooking range with oven.
	Demolish and Install	Ea.	815	108		923	
	Reinstall	Ea.		61.23		61.23	
	Clean	Ea.	.04	28.50		28.54	
	Minimum Charge	Job		77.50		77.50	

Improvements / Appliances / Treatments

Oven

Oven	Unit	Material	Labor	Equip.	Total	Specification
Hi/Lo w / Range, Microwave						
Demolish	Ea.		31.50		31.50	Includes material and labor to install a
Install	Ea.	1875	425		2300	double oven, one conventional, one
Demolish and Install	Ea.	1875	456.50		2331.50	microwave.
Reinstall	Ea.		338.24		338.24	
Clean	Ea.	.04	28.50		28.54	
Minimum Charge	Job		77.50		77.50	
Hi/Lo w/ Range / Double Oven						
Demolish	Ea.		31.50		31.50	Includes material and labor to install a
Install	Ea.	1550	211		1761	conventional double oven.
Demolish and Install	Ea.	1550	242.50		1792.50	
Reinstall	Ea.		169.12		169.12	
Clean	Ea.	.04	28.50		28.54	
Minimum Charge	Job		77.50		77.50	
Countertop Microwave						
Demolish	Ea.		10.75		10.75	Includes material and labor to install a
Install	Ea.	157	38.50		195.50	1.0 cubic foot microwave oven.
Demolish and Install	Ea.	157	49.25		206.25	
Reinstall	Ea.		30.96		30.96	
Clean	Ea.	.04	14.20		14.24	
Minimum Charge	Job		77.50		77.50	
Cabinet / Wall Mounted Microwave						
Demolish	Ea.		21.50		21.50	Includes material and labor to install a
Install	Ea.	770	115		885	space saver microwave oven.
Demolish and Install	Ea.	770	136.50		906.50	
Reinstall	Ea.		91.84		91.84	
Clean	Ea.	.04	14.20		14.24	
Minimum Charge	Job		77.50		77.50	

Dryer

Dryer	Unit	Material	Labor	Equip.	Total	Specification
Basic Grade						
Demolish	Ea.		14.35		14.35	Includes material and labor to install
Install	Ea.	390	245		635	electric or gas dryer. Vent kit not
Demolish and Install	Ea.	390	259.35		649.35	included.
Reinstall	Ea.		195.97		195.97	
Clean	Ea.	.04	28.50		28.54	
Minimum Charge	Job		77.50		77.50	
Good Grade						
Demolish	Ea.		14.35		14.35	Includes material and labor to install
Install	Ea.	490	245		735	electric or gas dryer. Vent kit not
Demolish and Install	Ea.	490	259.35		749.35	included.
Reinstall	Ea.		195.97		195.97	
Clean	Ea.	.04	28.50		28.54	
Minimum Charge	Job		77.50		77.50	
Premium Grade						
Demolish	Ea.		14.35		14.35	Includes material and labor to install
Install	Ea.	1600	365		1965	electric or gas dryer. Vent kit not
Demolish and Install	Ea.	1600	379.35		1979.35	included.
Reinstall	Ea.		293.95		293.95	
Clean	Ea.	.04	28.50		28.54	
Minimum Charge	Job		77.50		77.50	

Improvements / Appliances / Treatments

Dryer

Dryer	Unit	Material	Labor	Equip.	Total	Specification
Washer / Dryer Combination Unit						
Demolish	Ea.		14.35		14.35	Includes material and labor to install
Install	Ea.	1125	62		1187	electric washer/dryer combination
Demolish and Install	Ea.	1125	76.35		1201.35	unit, 27" wide, stackable or unitized.
Reinstall	Ea.		49.54		49.54	
Clean	Ea.	.04	28.50		28.54	
Minimum Charge	Job		77.50		77.50	

Washer

Washer	Unit	Material	Labor	Equip.	Total	Specification
Basic Grade						
Demolish	Ea.		14.35		14.35	Includes material and labor to install
Install	Ea.	530	162		692	washing machine. New hoses not
Demolish and Install	Ea.	530	176.35		706.35	included.
Reinstall	Ea.		129.39		129.39	
Clean	Ea.	.04	28.50		28.54	
Minimum Charge	Job		77.50		77.50	
Good Grade						
Demolish	Ea.		14.35		14.35	Includes material and labor to install
Install	Ea.	865	243		1108	washing machine. New hoses not
Demolish and Install	Ea.	865	257.35		1122.35	included.
Reinstall	Ea.		194.08		194.08	
Clean	Ea.	.04	28.50		28.54	
Minimum Charge	Job		77.50		77.50	
Premium Grade						
Demolish	Ea.		14.35		14.35	Includes material and labor to install
Install	Ea.	1600	485		2085	washing machine with digital
Demolish and Install	Ea.	1600	499.35		2099.35	computer readouts and large capacity.
Reinstall	Ea.		388.16		388.16	New hoses not included.
Clean	Ea.	.04	28.50		28.54	
Minimum Charge	Job		77.50		77.50	
Washer / Dryer Combination Unit						
Demolish	Ea.		14.35		14.35	Includes material and labor to install
Install	Ea.	1125	62		1187	electric washer/dryer combination
Demolish and Install	Ea.	1125	76.35		1201.35	unit, 27" wide, stackable or unitized.
Reinstall	Ea.		49.54		49.54	
Clean	Ea.	.04	28.50		28.54	
Minimum Charge	Job		77.50		77.50	

Refrigerator

Refrigerator	Unit	Material	Labor	Equip.	Total	Specification
12 Cubic Foot						
Demolish	Ea.		43		43	Includes material and labor to install a
Install	Ea.	430	62		492	refrigerator.
Demolish and Install	Ea.	430	105		535	
Reinstall	Ea.		49.54		49.54	
Clean	Ea.	.04	28.50		28.54	
Minimum Charge	Job		77.50		77.50	
16 Cubic Foot						
Demolish	Ea.		43		43	Includes material and labor to install a
Install	Ea.	550	69		619	refrigerator.
Demolish and Install	Ea.	550	112		662	
Reinstall	Ea.		55.04		55.04	
Clean	Ea.	.04	28.50		28.54	
Minimum Charge	Job		77.50		77.50	

Improvements / Appliances / Treatments

Refrigerator		Unit	Material	Labor	Equip.	Total	Specification
18 Cubic Foot							
	Demolish	Ea.		43		43	Includes material and labor to install a
	Install	Ea.	785	77.50		862.50	refrigerator.
	Demolish and Install	Ea.	785	120.50		905.50	
	Reinstall	Ea.		61.92		61.92	
	Clean	Ea.	.04	28.50		28.54	
	Minimum Charge	Job		77.50		77.50	
21 Cubic Foot							
	Demolish	Ea.		43		43	Includes material and labor to install a
	Install	Ea.	1100	88.50		1188.50	refrigerator.
	Demolish and Install	Ea.	1100	131.50		1231.50	
	Reinstall	Ea.		70.77		70.77	
	Clean	Ea.	.04	28.50		28.54	
	Minimum Charge	Job		77.50		77.50	
24 Cubic Foot							
	Demolish	Ea.		43		43	Includes material and labor to install a
	Install	Ea.	1450	88.50		1538.50	refrigerator.
	Demolish and Install	Ea.	1450	131.50		1581.50	
	Reinstall	Ea.		70.77		70.77	
	Clean	Ea.	.04	28.50		28.54	
	Minimum Charge	Job		77.50		77.50	
27 Cubic Foot							
	Demolish	Ea.		43		43	Includes material and labor to install a
	Install	Ea.	2850	62		2912	refrigerator.
	Demolish and Install	Ea.	2850	105		2955	
	Reinstall	Ea.		49.54		49.54	
	Clean	Ea.	.04	28.50		28.54	
	Minimum Charge	Job		77.50		77.50	
Ice Maker							
	Install	Ea.	61.50	60.50		122	Includes labor and material to install
	Minimum Charge	Job		77.50		77.50	an automatic ice maker in a
							refrigerator.

Freezer		Unit	Material	Labor	Equip.	Total	Specification
15 Cubic Foot							
	Demolish	Ea.		43		43	Cost includes material and labor to
	Install	Ea.	645	62		707	install 15 cubic foot freezer.
	Demolish and Install	Ea.	645	105		750	
	Reinstall	Ea.		49.54		49.54	
	Clean	Ea.	.04	28.50		28.54	
	Minimum Charge	Job		77.50		77.50	
18 Cubic Foot							
	Demolish	Ea.		43		43	Cost includes material and labor to
	Install	Ea.	855	62		917	install 18 cubic foot freezer.
	Demolish and Install	Ea.	855	105		960	
	Reinstall	Ea.		49.54		49.54	
	Clean	Ea.	.04	28.50		28.54	
	Minimum Charge	Job		77.50		77.50	
21 Cubic Foot							
	Demolish	Ea.		43		43	Includes material and labor to install a
	Install	Ea.	815	62		877	15 to 23 cubic foot deep freezer.
	Demolish and Install	Ea.	815	105		920	
	Reinstall	Ea.		49.54		49.54	
	Clean	Ea.	.04	28.50		28.54	
	Minimum Charge	Job		77.50		77.50	

Improvements / Appliances / Treatments

Freezer

	Unit	Material	Labor	Equip.	Total	Specification
24 Cubic Foot						
Demolish	Ea.		43		43	Includes material and labor to install a
Install	Ea.	900	124		1024	24 cubic foot deep freezer.
Demolish and Install	Ea.	900	167		1067	
Reinstall	Ea.		99.07		99.07	
Clean	Ea.	.04	28.50		28.54	
Minimum Charge	Job		77.50		77.50	
Chest-type						
Demolish	Ea.		43		43	Cost includes material and labor to
Install	Ea.	645	62		707	install 15 cubic foot freezer.
Demolish and Install	Ea.	645	105		750	
Clean	Ea.	.04	28.50		28.54	
Minimum Charge	Job		77.50		77.50	

Dishwasher

	Unit	Material	Labor	Equip.	Total	Specification
4 Cycle						
Demolish	Ea.		34.50		34.50	Includes material and labor to install
Install	Ea.	375	150		525	good grade automatic dishwasher.
Demolish and Install	Ea.	375	184.50		559.50	
Reinstall	Ea.		120		120	
Clean	Ea.	.04	28.50		28.54	
Minimum Charge	Job		77.50		77.50	

Garbage Disposal

	Unit	Material	Labor	Equip.	Total	Specification
In Sink Unit						
Demolish	Ea.		33.50		33.50	Cost includes material and labor to
Install	Ea.	160	60		220	install 1/2 HP custom disposal unit.
Demolish and Install	Ea.	160	93.50		253.50	
Reinstall	Ea.		48		48	
Clean	Ea.	.02	8.10		8.12	
Minimum Charge	Job		77.50		77.50	

Trash Compactor

	Unit	Material	Labor	Equip.	Total	Specification
4 to 1 Compaction						
Demolish	Ea.		32		32	Includes material and labor to install a
Install	Ea.	935	84.50		1019.50	trash compactor.
Demolish and Install	Ea.	935	116.50		1051.50	
Reinstall	Ea.		67.65		67.65	
Clean	Ea.	.04	19.35		19.39	
Minimum Charge	Job		77.50		77.50	

Kitchenette Unit

	Unit	Material	Labor	Equip.	Total	Specification
Range, Refrigerator and Sink						
Demolish	Ea.		128		128	Includes material and labor to install
Install	Ea.	4850	500		5350	range/refrigerator/sink unit, 72"
Demolish and Install	Ea.	4850	628		5478	wide, base unit only.
Reinstall	Ea.		400		400	
Clean	Ea.	.81	14.20		15.01	
Minimum Charge	Job		77.50		77.50	

Bath Accessories		Unit	Material	Labor	Equip.	Total	Specification
Towel Bar							
	Demolish	Ea.		7.15		7.15	Includes material and labor to install
	Install	Ea.	59.50	17.60		77.10	towel bar up to 24" long.
	Demolish and Install	Ea.	59.50	24.75		84.25	
	Reinstall	Ea.		14.09		14.09	
	Clean	Ea.		3.87		3.87	
	Minimum Charge	Job		211		211	
Toothbrush Holder							
	Demolish	Ea.		7.15		7.15	Includes material and labor to install a
	Install	Ea.	13.70	21		34.70	surface mounted tumbler and
	Demolish and Install	Ea.	13.70	28.15		41.85	toothbrush holder.
	Reinstall	Ea.		16.91		16.91	
	Clean	Ea.		3.87		3.87	
	Minimum Charge	Job		211		211	
Grab Rail							
	Demolish	Ea.		10.75		10.75	Includes material and labor to install a
	Install	Ea.	34	18.40		52.40	24" long grab bar. Blocking is not
	Demolish and Install	Ea.	34	29.15		63.15	included.
	Reinstall	Ea.		14.71		14.71	
	Clean	Ea.	.33	6.45		6.78	
	Minimum Charge	Job		211		211	
Soap Dish							
	Demolish	Ea.		7.15		7.15	Includes material and labor to install a
	Install	Ea.	6.05	21		27.05	surface mounted soap dish.
	Demolish and Install	Ea.	6.05	28.15		34.20	
	Reinstall	Ea.		16.91		16.91	
	Clean	Ea.		3.87		3.87	
	Minimum Charge	Job		211		211	
Toilet Paper Roller							
	Demolish	Ea.		7.15		7.15	Includes material and labor to install a
	Install	Ea.	19.60	14.10		33.70	surface mounted toilet tissue dispenser.
	Demolish and Install	Ea.	19.60	21.25		40.85	
	Reinstall	Ea.		11.27		11.27	
	Clean	Ea.	.33	9.70		10.03	
	Minimum Charge	Job		211		211	
Dispenser							
Soap							
	Demolish	Ea.		9.55		9.55	Includes material and labor to install a
	Install	Ea.	51	21		72	surface mounted soap dispenser.
	Demolish and Install	Ea.	51	30.55		81.55	
	Reinstall	Ea.		16.91		16.91	
	Clean	Ea.	.33	9.70		10.03	
	Minimum Charge	Job		211		211	
Towel							
	Demolish	Ea.		10.75		10.75	Includes material and labor to install a
	Install	Ea.	49	26.50		75.50	surface mounted towel dispenser.
	Demolish and Install	Ea.	49	37.25		86.25	
	Reinstall	Ea.		21.14		21.14	
	Clean	Ea.	.33	12.90		13.23	
	Minimum Charge	Job		211		211	
Seat Cover							
	Demolish	Ea.		7.15		7.15	Includes material and labor to install a
	Install	Ea.	37.50	28		65.50	surface mounted toilet seat cover
	Demolish and Install	Ea.	37.50	35.15		72.65	dispenser.
	Reinstall	Ea.		22.55		22.55	
	Clean	Ea.	.33	9.70		10.03	
	Minimum Charge	Job		211		211	

Improvements / Appliances / Treatments

Bath Accessories

		Unit	Material	Labor	Equip.	Total	Specification
Sanitary Napkin							
	Demolish	Ea.		18.10		18.10	Includes material and labor to install a
	Install	Ea.	35	44.50		79.50	surface mounted sanitary napkin
	Demolish and Install	Ea.	35	62.60		97.60	dispenser.
	Reinstall	Ea.		35.60		35.60	
	Clean	Ea.	.17	12.90		13.07	
	Minimum Charge	Job		211		211	
Partition							
Urinal							
	Demolish	Ea.		21.50		21.50	Includes material and labor to install a
	Install	Ea.	253	106		359	urinal screen.
	Demolish and Install	Ea.	253	127.50		380.50	
	Reinstall	Ea.		84.56		84.56	
	Clean	Ea.	.50	8.60		9.10	
	Minimum Charge	Job		211		211	
Toilet							
	Demolish	Ea.		63.50		63.50	Includes material and labor to install
	Install	Ea.	650	169		819	baked enamel standard size partition,
	Demolish and Install	Ea.	650	232.50		882.50	floor or ceiling mounted with one wall
	Reinstall	Ea.		135.30		135.30	and one door.
	Clean	Ea.	3.27	19.35		22.62	
	Minimum Charge	Job		211		211	
Handicap							
	Demolish	Ea.		63.50		63.50	Includes material and labor to install
	Install	Ea.	995	169		1164	baked enamel handicap partition,
	Demolish and Install	Ea.	995	232.50		1227.50	floor or ceiling mounted with one wall
	Reinstall	Ea.		135.30		135.30	and one door.
	Clean	Ea.	3.27	19.35		22.62	
	Minimum Charge	Job		211		211	

Fire Prevention / Protection

		Unit	Material	Labor	Equip.	Total	Specification
Fire Extinguisher							
5# Carbon Dioxide							
	Install	Ea.	199			199	Includes labor and material to install one wall mounted 5 lb. carbon dioxide factory charged unit, complete with hose, horn and wall mounting bracket.
10# Carbon Dioxide							
	Install	Ea.	287			287	Includes labor and material to install one wall mounted 10 lb. carbon dioxide factory charged unit, complete with hose, horn and wall mounting bracket.
15# Carbon Dioxide							
	Install	Ea.	370			370	Includes labor and material to install one wall mounted 15 lb. carbon dioxide factory charged unit, complete with hose, horn and wall mounting bracket.

Improvements / Appliances / Treatments

Fire Prevention / Protection	Unit	Material	Labor	Equip.	Total	Specification
5# Dry Chemical						Includes labor and material to install one wall mounted 5 lb. dry chemical factory charged unit, complete with hose, horn and wall mounting bracket.
Install	Ea.	54			54	
20# Dry Chemical						Includes labor and material to install one wall mounted 20 lb. dry chemical factory charged unit, complete with hose, horn and wall mounting bracket.
Install	Ea.	144			144	

Gutters	Unit	Material	Labor	Equip.	Total	Specification
Galvanized						Cost includes material and labor to install 5" galvanized steel gutter with enamel finish, fittings, hangers, and corners.
Demolish	L.F.		1.43		1.43	
Install	L.F.	2.22	3.91		6.13	
Demolish and Install	L.F.	2.22	5.34		7.56	
Reinstall	L.F.		3.13		3.13	
Clean	L.F.		.41		.41	
Paint	L.F.	.22	1.11		1.33	
Minimum Charge	Job		234		234	
Aluminum						Cost includes material and labor to install 5" aluminum gutter with enamel finish, fittings, hangers, and corners.
Demolish	L.F.		1.43		1.43	
Install	L.F.	2.92	3.91		6.83	
Demolish and Install	L.F.	2.92	5.34		8.26	
Reinstall	L.F.		3.13		3.13	
Clean	L.F.		.41		.41	
Paint	L.F.	.22	1.11		1.33	
Minimum Charge	Job		234		234	
Copper						Cost includes material and labor to install 5" half-round copper gutter with fittings, hangers, and corners.
Demolish	L.F.		1.43		1.43	
Install	L.F.	9.20	3.91		13.11	
Demolish and Install	L.F.	9.20	5.34		14.54	
Reinstall	L.F.		3.13		3.13	
Clean	L.F.		.41		.41	
Minimum Charge	Job		234		234	
Plastic						Cost includes material and labor to install 4" white vinyl gutter fittings, hangers, and corners.
Demolish	L.F.		1.43		1.43	
Install	L.F.	1.28	3.84		5.12	
Demolish and Install	L.F.	1.28	5.27		6.55	
Reinstall	L.F.		3.07		3.07	
Clean	L.F.		.41		.41	
Paint	L.F.	.22	1.11		1.33	
Minimum Charge	Job		234		234	

Downspouts	Unit	Material	Labor	Equip.	Total	Specification
Galvanized						Cost includes material and labor to install 4" diameter downspout with enamel finish.
Demolish	L.F.		.98		.98	
Install	L.F.	2.87	3.23		6.10	
Demolish and Install	L.F.	2.87	4.21		7.08	
Reinstall	L.F.		2.59		2.59	
Clean	L.F.		.41		.41	
Paint	L.F.	.22	1.11		1.33	
Minimum Charge	Job		234		234	

Improvements / Appliances / Treatments

Downspouts		Unit	Material	Labor	Equip.	Total	Specification
Aluminum							
	Demolish	L.F.		.98		.98	Includes material and labor to install
	Install	L.F.	3.17	3.35		6.52	aluminum enameled downspout.
	Demolish and Install	L.F.	3.17	4.33		7.50	
	Reinstall	L.F.		2.68		2.68	
	Clean	L.F.		.41		.41	
	Paint	L.F.	.22	1.11		1.33	
	Minimum Charge	Job		234		234	
Copper							
	Demolish	L.F.		.98		.98	Includes material and labor to install
	Install	L.F.	9.40	2.47		11.87	round copper downspout.
	Demolish and Install	L.F.	9.40	3.45		12.85	
	Reinstall	L.F.		1.97		1.97	
	Clean	L.F.		.41		.41	
	Paint	L.F.	.22	1.11		1.33	
	Minimum Charge	Job		234		234	
Plastic							
	Demolish	L.F.		.98		.98	Cost includes material and labor to
	Install	L.F.	2.29	2.23		4.52	install 2" x 3" vinyl rectangular
	Demolish and Install	L.F.	2.29	3.21		5.50	downspouts.
	Reinstall	L.F.		1.79		1.79	
	Clean	L.F.		.41		.41	
	Paint	L.F.	.22	1.11		1.33	
	Minimum Charge	Job		234		234	

Shutters		Unit	Material	Labor	Equip.	Total	Specification
Exterior							
25" High							
	Demolish	Ea.		10.75		10.75	Includes material and labor to install
	Install	Pr.	141	42.50		183.50	pair of exterior shutters 25" x 16",
	Demolish and Install	Pr.	141	53.25		194.25	unpainted, installed as fixed shutters
	Reinstall	Pr.		42.28		42.28	on concrete or wood frame.
	Clean	Ea.	.07	9.45		9.52	
	Paint	Ea.	1.71	19.40		21.11	
	Minimum Charge	Job		211		211	
39" High							
	Demolish	Ea.		10.75		10.75	Includes material and labor to install
	Install	Pr.	197	42.50		239.50	pair of shutters 39" x 16", unpainted,
	Demolish and Install	Pr.	197	53.25		250.25	installed as fixed shutters on concrete
	Reinstall	Pr.		42.28		42.28	or wood frame.
	Clean	Ea.	.07	9.45		9.52	
	Paint	Ea.	2.65	26		28.65	
	Minimum Charge	Job		211		211	
51" High							
	Demolish	Ea.		10.75		10.75	Includes material and labor to install
	Install	Pr.	248	42.50		290.50	pair of shutters 51" x 16", unpainted,
	Demolish and Install	Pr.	248	53.25		301.25	installed as fixed shutters on concrete
	Reinstall	Pr.		42.28		42.28	or wood frame.
	Clean	Ea.	.07	9.45		9.52	
	Paint	Ea.	3.45	27		30.45	
	Minimum Charge	Job		211		211	

Shutters		Unit	Material	Labor	Equip.	Total	Specification
59" High							
	Demolish	Ea.		10.75		10.75	Includes material and labor to install
	Install	Pr.	330	42.50		372.50	pair of shutters 59" x 16", unpainted,
	Demolish and Install	Pr.	330	53.25		383.25	installed as fixed shutters on concrete
	Reinstall	Pr.		42.28		42.28	or wood frame.
	Clean	Ea.	.07	9.45		9.52	
	Paint	Ea.	3.98	30		33.98	
	Minimum Charge	Job		211		211	
67" High							
	Demolish	Ea.		10.75		10.75	Includes material and labor to install
	Install	Pr.	350	47		397	pair of shutters 67" x 16", unpainted,
	Demolish and Install	Pr.	350	57.75		407.75	installed as fixed shutters on concrete
	Reinstall	Pr.		46.98		46.98	or wood frame.
	Clean	Ea.	.07	9.45		9.52	
	Paint	Ea.	4.55	33.50		38.05	
	Minimum Charge	Job		211		211	
Interior Wood							
23" x 24", 2 Panel							
	Demolish	Ea.		10.75		10.75	Cost includes material and labor to
	Install	Ea.	86	25		111	install 23" x 24", 4 panel interior
	Demolish and Install	Ea.	86	35.75		121.75	moveable pine shutter, 1-1/4" louver,
	Reinstall	Ea.		19.90		19.90	knobs and hooks and hinges.
	Clean	Ea.	.07	9.45		9.52	
	Paint	Ea.	1.29	26		27.29	
	Minimum Charge	Job		211		211	
27" x 24", 4 Panel							
	Demolish	Ea.		10.75		10.75	Cost includes material and labor to
	Install	Ea.	86	25		111	install 27" x 24", 4 panel interior
	Demolish and Install	Ea.	86	35.75		121.75	moveable pine shutter, 1-1/4" louver,
	Reinstall	Ea.		19.90		19.90	knobs and hooks and hinges.
	Clean	Ea.	.07	9.45		9.52	
	Paint	Ea.	1.29	26		27.29	
	Minimum Charge	Job		211		211	
31" x 24", 4 Panel							
	Demolish	Ea.		10.75		10.75	Cost includes material and labor to
	Install	Ea.	94.50	25		119.50	install 31" x 24", 4 panel interior
	Demolish and Install	Ea.	94.50	35.75		130.25	moveable pine shutter, 1-1/4" louver,
	Reinstall	Ea.		19.90		19.90	knobs and hooks and hinges.
	Clean	Ea.	.07	9.45		9.52	
	Paint	Ea.	1.29	26		27.29	
	Minimum Charge	Job		211		211	
35" x 24", 4 Panel							
	Demolish	Ea.		10.75		10.75	Cost includes material and labor to
	Install	Ea.	101	25		126	install 35" x 24", 4 panel interior
	Demolish and Install	Ea.	101	35.75		136.75	moveable pine shutter, 1-1/4" louver,
	Reinstall	Ea.		19.90		19.90	knobs and hooks and hinges.
	Clean	Ea.	.07	9.45		9.52	
	Paint	Ea.	1.29	26		27.29	
	Minimum Charge	Job		211		211	
39" x 24", 4 Panel							
	Demolish	Ea.		10.75		10.75	Cost includes material and labor to
	Install	Ea.	112	25		137	install 39" x 24", 4 panel interior
	Demolish and Install	Ea.	112	35.75		147.75	moveable pine shutter, 1-1/4" louver,
	Reinstall	Ea.		19.90		19.90	knobs and hooks and hinges.
	Clean	Ea.	.07	9.45		9.52	
	Paint	Ea.	1.29	26		27.29	
	Minimum Charge	Job		211		211	

Improvements / Appliances / Treatments

Shutters

	Unit	Material	Labor	Equip.	Total	Specification
47" x 24", 4 Panel						
Demolish	Ea.		10.75		10.75	Cost includes material and labor to
Install	Ea.	128	25		153	install 47" x 24", 4 panel interior
Demolish and Install	Ea.	128	35.75		163.75	moveable pine shutter, 1-1/4" louver,
Reinstall	Ea.		19.90		19.90	knobs and hooks and hinges.
Clean	Ea.	.07	9.45		9.52	
Paint	Ea.	1.34	28		29.34	
Minimum Charge	Job		211		211	

Awning

	Unit	Material	Labor	Equip.	Total	Specification
Standard Aluminum						
36" Wide x 30" Deep						
Demolish	Ea.		10.75		10.75	Includes material and labor to install
Install	Ea.	208	35		243	one aluminum awning with
Demolish and Install	Ea.	208	45.75		253.75	weather-resistant finish.
Reinstall	Ea.		28.19		28.19	
Clean	Ea.		12.90		12.90	
Minimum Charge	Job		211		211	
48" Wide x 30" Deep						
Demolish	Ea.		10.75		10.75	Includes material and labor to install
Install	Ea.	233	42.50		275.50	one aluminum awning with
Demolish and Install	Ea.	233	53.25		286.25	weather-resistant finish.
Reinstall	Ea.		33.82		33.82	
Clean	Ea.		12.90		12.90	
Minimum Charge	Job		211		211	
60" Wide x 30" Deep						
Demolish	Ea.		21.50		21.50	Includes material and labor to install
Install	Ea.	273	47		320	one aluminum awning with
Demolish and Install	Ea.	273	68.50		341.50	weather-resistant finish.
Reinstall	Ea.		37.58		37.58	
Clean	Ea.		12.90		12.90	
Minimum Charge	Job		211		211	
72" Wide x 30" Deep						
Demolish	Ea.		21.50		21.50	Includes material and labor to install
Install	Ea.	310	53		363	one aluminum awning with
Demolish and Install	Ea.	310	74.50		384.50	weather-resistant finish.
Reinstall	Ea.		42.28		42.28	
Clean	Ea.		12.90		12.90	
Minimum Charge	Job		211		211	
Roll-up Type						
Demolish	S.F.		.77		.77	Includes material and labor to install
Install	S.F.	11.35	4.23		15.58	an aluminum roll-up awning.
Demolish and Install	S.F.	11.35	5		16.35	
Reinstall	S.F.		3.38		3.38	
Clean	S.F.	.03	.65		.68	
Minimum Charge	Job		211		211	
Canvas						
30" Wide						
Demolish	Ea.		6.45		6.45	Includes material and labor to install a
Install	Ea.	205	35		240	canvas window awning.
Demolish and Install	Ea.	205	41.45		246.45	
Reinstall	Ea.		28.19		28.19	
Clean	Ea.		3.87		3.87	
Minimum Charge	Job		211		211	

Improvements / Appliances / Treatments

Awning

	Unit	Material	Labor	Equip.	Total	Specification
36" Wide						
Demolish	Ea.		6.45		6.45	Includes material and labor to install a
Install	Ea.	224	42.50		266.50	canvas window awning.
Demolish and Install	Ea.	224	48.95		272.95	
Reinstall	Ea.		33.82		33.82	
Clean	Ea.		4.24		4.24	
Minimum Charge	Job		211		211	
42" Wide						
Demolish	Ea.		6.45		6.45	Includes material and labor to install a
Install	Ea.	233	53		286	canvas window awning.
Demolish and Install	Ea.	233	59.45		292.45	
Reinstall	Ea.		42.28		42.28	
Clean	Ea.		5.05		5.05	
Minimum Charge	Job		211		211	
48" Wide						
Demolish	Ea.		6.45		6.45	Includes material and labor to install a
Install	Ea.	246	60.50		306.50	canvas window awning.
Demolish and Install	Ea.	246	66.95		312.95	
Reinstall	Ea.		48.32		48.32	
Clean	Ea.		5.85		5.85	
Minimum Charge	Job		211		211	
Cloth Canopy Cover						
8' x 10'						
Demolish	Ea.		355		355	Includes material and labor to install
Install	Ea.	430	211		641	one 8' x 10' cloth canopy patio cover
Demolish and Install	Ea.	430	566		996	with front bar and tension support
Reinstall	Ea.		169.12		169.12	rafters and 9" valance.
Clean	Ea.	2.61	51.50		54.11	
Minimum Charge	Job		425		425	
8' x 15'						
Demolish	Ea.		355		355	Includes material and labor to install
Install	Ea.	590	211		801	one 8' x 15' cloth canopy patio cover
Demolish and Install	Ea.	590	566		1156	with front bar and tension support
Reinstall	Ea.		169.12		169.12	rafters and 9" valance.
Clean	Ea.	2.61	77.50		80.11	
Minimum Charge	Job		425		425	

Vent

	Unit	Material	Labor	Equip.	Total	Specification
Small Roof Turbine						
Demolish	Ea.		10.75		10.75	Includes material and labor to install
Install	Ea.	71	42		113	spinner ventilator, wind driven,
Demolish and Install	Ea.	71	52.75		123.75	galvanized, 4" neck diam, 180 CFM.
Reinstall	Ea.		33.75		33.75	
Clean	Ea.		6.90		6.90	
Minimum Charge	Job		243		243	
Large Roof Turbine						
Demolish	Ea.		8.60		8.60	Includes material and labor to install
Install	Ea.	82	60.50		142.50	spinner ventilator, wind driven,
Demolish and Install	Ea.	82	69.10		151.10	galvanized, 8" neck diam, 360 CFM.
Reinstall	Ea.		48.21		48.21	
Clean	Ea.		7.75		7.75	
Minimum Charge	Job		243		243	

Improvements / Appliances / Treatments

Vent		Unit	Material	Labor	Equip.	Total	Specification
Gable Louvered							
	Demolish	Ea.		21.50		21.50	Includes material and labor to install a
	Install	Ea.	49.50	14.10		63.60	vinyl gable end vent.
	Demolish and Install	Ea.	49.50	35.60		85.10	
	Reinstall	Ea.		11.27		11.27	
	Clean	Ea.		4.13		4.13	
	Paint	Ea.	1.08	5.55		6.63	
	Minimum Charge	Job		211		211	
Ridge Vent							
	Demolish	L.F.		1.11		1.11	Includes material and labor to install a
	Install	L.F.	3.28	3.02		6.30	mill finish aluminum ridge vent strip.
	Demolish and Install	L.F.	3.28	4.13		7.41	
	Minimum Charge	Job		211		211	
Foundation							
	Demolish	Ea.		7.70		7.70	Includes material and labor to install
	Install	Ea.	17.80	13.85		31.65	galvanized foundation block vent.
	Demolish and Install	Ea.	17.80	21.55		39.35	
	Reinstall	Ea.		11.06		11.06	
	Clean	Ea.	.17	7.75		7.92	
	Minimum Charge	Job		211		211	
Frieze							
	Demolish	L.F.		.76		.76	Includes material and labor to install
	Install	L.F.	1.28	1.69		2.97	pine frieze board.
	Demolish and Install	L.F.	1.28	2.45		3.73	
	Reinstall	L.F.		1.35		1.35	
	Clean	L.F.	.03	.26		.29	
	Minimum Charge	Job		211		211	
Water Heater Cap							
	Demolish	Ea.		16.85		16.85	Includes material and labor to install a
	Install	Ea.	34	58.50		92.50	galvanized steel water heater cap.
	Demolish and Install	Ea.	34	75.35		109.35	
	Reinstall	Ea.		46.88		46.88	
	Clean	Ea.	1.14	19.35		20.49	

Cupolas		Unit	Material	Labor	Equip.	Total	Specification
22" x 22"							
	Demolish	Ea.		204		204	Includes material and labor to install a
	Install	Ea.	335	106		441	cedar cupola with aluminum roof
	Demolish and Install	Ea.	335	310		645	covering.
	Reinstall	Ea.		84.56		84.56	
	Minimum Charge	Job		211		211	
29" x 29"							
	Demolish	Ea.		204		204	Includes material and labor to install a
	Install	Ea.	450	106		556	cedar cupola with aluminum roof
	Demolish and Install	Ea.	450	310		760	covering.
	Reinstall	Ea.		84.56		84.56	
	Minimum Charge	Job		211		211	
35" x 35"							
	Demolish	Ea.		204		204	Includes material and labor to install a
	Install	Ea.	750	141		891	cedar cupola with aluminum roof
	Demolish and Install	Ea.	750	345		1095	covering.
	Reinstall	Ea.		112.75		112.75	
	Minimum Charge	Job		211		211	

Improvements / Appliances / Treatments

Cupolas		Unit	Material	Labor	Equip.	Total	Specification
47" x 47"							
	Demolish	Ea.		204		204	Includes material and labor to install a
	Install	Ea.	1325	211		1536	cedar cupola with aluminum roof
	Demolish and Install	Ea.	1325	415		1740	covering.
	Reinstall	Ea.		169.12		169.12	
	Minimum Charge	Job		211		211	
Weather Vane							
	Install	Ea.	166	53		219	Includes min. labor and equip. to
	Minimum Charge	Job		211		211	install residential type weathervane.

Aluminum Carport		Unit	Material	Labor	Equip.	Total	Specification
20' x 20' Complete							
	Demolish	Ea.		215		215	Includes material, labor and
	Install	Ea.	8625	700	180	9505	equipment to install an aluminum
	Demolish and Install	Ea.	8625	915	180	9720	carport. Foundations are not included.
	Reinstall	Ea.		560.83	144.19	705.02	
	Clean	Ea.	.17	77.50		77.67	
	Minimum Charge	Job		425		425	
24' x 24' Complete							
	Demolish	Ea.		215		215	Includes material, labor and
	Install	Ea.	11800	935	240	12975	equipment to install an aluminum
	Demolish and Install	Ea.	11800	1150	240	13190	carport. Foundations are not included.
	Reinstall	Ea.		747.78	192.26	940.04	
	Clean	Ea.	.17	155		155.17	
	Minimum Charge	Job		425		425	
Metal Support Posts							
	Demolish	Ea.		2.87		2.87	Includes material and labor to install
	Install	Ea.	19.20	53		72.20	metal support posts for a carport.
	Demolish and Install	Ea.	19.20	55.87		75.07	
	Reinstall	Ea.		42.28		42.28	
	Clean	Ea.	.17	3.22		3.39	
	Paint	Ea.	.21	7.20		7.41	
	Minimum Charge	Job		425		425	

Storage Shed		Unit	Material	Labor	Equip.	Total	Specification
Aluminum							
Pre-fab (8' x 10')							
	Demolish	Ea.		62.50		62.50	Includes material and labor to install
	Install	Ea.	1950	610		2560	an aluminum storage shed.
	Demolish and Install	Ea.	1950	672.50		2622.50	
	Clean	Ea.	1.09	38.50		39.59	
	Minimum Charge	Ea.		515		515	
Pre-fab (10' x 12')							
	Demolish	Ea.		119		119	Includes material and labor to install
	Install	Ea.	2425	610		3035	an aluminum storage shed.
	Demolish and Install	Ea.	2425	729		3154	
	Clean	Ea.	1.09	43		44.09	
	Minimum Charge	Ea.		515		515	

Improvements / Appliances / Treatments

Storage Shed

Storage Shed		Unit	Material	Labor	Equip.	Total	Specification
Wood							
Pre-fab (8′ x 10′)							
	Demolish	Ea.		62.50		62.50	Includes material and labor to install a
	Install	Ea.	1925	610		2535	pre-fabricated wood storage shed.
	Demolish and Install	Ea.	1925	672.50		2597.50	
	Clean	Ea.	1.09	38.50		39.59	
	Minimum Charge	Ea.		515		515	
Pre-fab (10′ x 12′)							
	Demolish	Ea.		119		119	Includes material and labor to install a
	Install	Ea.	2425	610		3035	pre-fabricated wood storage shed.
	Demolish and Install	Ea.	2425	729		3154	
	Clean	Ea.	1.09	43		44.09	
	Minimum Charge	Ea.		515		515	

Gazebo

Gazebo		Unit	Material	Labor	Equip.	Total	Specification
Good Grade							
	Demolish	S.F.		3.44		3.44	Includes material and labor to install a
	Install	S.F.	50.50	31.50		82	gazebo.
	Demolish and Install	S.F.	50.50	34.94		85.44	
	Reinstall	S.F.		25.37		25.37	
	Clean	S.F.	.13	2.06		2.19	
	Minimum Charge	Job		211		211	
Custom Grade							
	Demolish	S.F.		3.44		3.44	Includes material and labor to install a
	Install	S.F.	73	31.50		104.50	gazebo.
	Demolish and Install	S.F.	73	34.94		107.94	
	Reinstall	S.F.		25.37		25.37	
	Clean	S.F.	.13	2.06		2.19	
	Minimum Charge	Job		211		211	
Bench Seating							
	Demolish	L.F.		5.75		5.75	Includes material and labor to install
	Install	L.F.	1.86	21		22.86	pressure treated bench seating.
	Demolish and Install	L.F.	1.86	26.75		28.61	
	Reinstall	L.F.		16.91		16.91	
	Clean	L.F.	.13	1.55		1.68	
	Paint	L.F.	.63	2.89		3.52	
	Minimum Charge	Job		211		211	
Re-screen							
	Install	S.F.	1.77	1.69		3.46	Includes labor and material to
	Minimum Charge	Job		211		211	re-screen wood frame.

Pool

Pool		Unit	Material	Labor	Equip.	Total	Specification
Screen Enclosure							
Re-screen							
	Install	SF Wall		5.15		5.15	Includes material and labor to install a
	Minimum Charge	Job		211		211	screen enclosure for a pool.
Screen Door							
	Demolish	Ea.		8.40		8.40	Includes material and labor to install
	Install	Ea.	340	70.50		410.50	wood screen door with aluminum cloth
	Demolish and Install	Ea.	340	78.90		418.90	screen. Frame and hardware not
	Reinstall	Ea.		56.37		56.37	included.
	Clean	Ea.	.12	11.35		11.47	
	Paint	Ea.	4.99	14.45		19.44	
	Minimum Charge	Job		211		211	

Pool	Unit	Material	Labor	Equip.	Total	Specification
Heater / Motor						
Demolish	Ea.		43		43	Includes material and labor to install
Install	Ea.	2500	540		3040	swimming pool heater, not incl. base
Demolish and Install	Ea.	2500	583		3083	or pad, elec., 12 kW, 4,800 gal
Reinstall	Ea.		430.46		430.46	pool, incl. pump.
Clean	Ea.	3.27	38.50		41.77	
Minimum Charge	Job		246		246	
Filter System						
Demolish	Ea.		69		69	Includes material, labor and
Install	Total	1800	390		2190	equipment to install sand filter system
Demolish and Install	Total	1800	459		2259	tank including hook-up to existing
Reinstall	Total		310.53		310.53	equipment. Pump and motor not
Minimum Charge	Job		243		243	included.
Pump / Motor						
Demolish	Ea.		43		43	Includes material and labor to install a
Install	Ea.	790	97		887	22 GPM pump.
Demolish and Install	Ea.	790	140		930	
Reinstall	Ea.		77.63		77.63	
Clean	Ea.	3.27	38.50		41.77	
Minimum Charge	Job		246		246	
Pump Water						
Install	Day		435	96.50	531.50	Includes labor and equipment to pump
Minimum Charge	Job		195		195	water from a swimming pool.
Replaster						
Install	SF Surf	1.46	10.15		11.61	Includes labor and material to
Minimum Charge	Job		195		195	replaster gunite pool with a surface
						area of 500 to 600 S.F.
Solar Panels						
Demolish	S.F.		1.80		1.80	Includes material and labor to install
Install	S.F.	35	3.19		38.19	solar panel.
Demolish and Install	S.F.	35	4.99		39.99	
Reinstall	S.F.		2.55		2.55	
Minimum Charge	Ea.		515		515	

Construction Clean-up	Unit	Material	Labor	Equip.	Total	Specification
Final						
Clean	S.F.		.27		.27	Includes labor for final site/job
Minimum Charge	Job		155		155	clean-up including detail cleaning.

Location Factors

Costs shown in *RSMeans Residential Cost Data* are based on national averages for materials and installation. To adjust these costs to a specific location, simply multiply the base cost by the factor for that city. The data is arranged alphabetically by state and postal zip code numbers. For a city not listed, use the factor for a nearby city with similar economic characteristics.

STATE	CITY	Residential
ALABAMA		
350-352	Birmingham	.86
354	Tuscaloosa	.77
355	Jasper	.72
356	Decatur	.77
357-358	Huntsville	.83
359	Gadsden	.73
360-361	Montgomery	.75
362	Anniston	.76
363	Dothan	.75
364	Evergreen	.73
365-366	Mobile	.81
367	Selma	.72
368	Phenix City	.73
369	Butler	.73
ALASKA		
995-996	Anchorage	1.22
997	Fairbanks	1.24
998	Juneau	1.21
999	Ketchikan	1.27
ARIZONA		
850,853	Phoenix	.86
851,852	Mesa/Tempe	.84
855	Globe	.76
856-857	Tucson	.83
859	Show Low	.78
860	Flagstaff	.85
863	Prescott	.76
864	Kingman	.83
865	Chambers	.76
ARKANSAS		
716	Pine Bluff	.78
717	Camden	.69
718	Texarkana	.73
719	Hot Springs	.70
720-722	Little Rock	.82
723	West Memphis	.78
724	Jonesboro	.76
725	Batesville	.74
726	Harrison	.74
727	Fayetteville	.70
728	Russellville	.76
729	Fort Smith	.81
CALIFORNIA		
900-902	Los Angeles	1.08
903-905	Inglewood	1.05
906-908	Long Beach	1.04
910-912	Pasadena	1.04
913-916	Van Nuys	1.06
917-918	Alhambra	1.07
919-921	San Diego	1.03
922	Palm Springs	1.03
923-924	San Bernardino	1.04
925	Riverside	1.06
926-927	Santa Ana	1.05
928	Anaheim	1.06
930	Oxnard	1.07
931	Santa Barbara	1.07
932-933	Bakersfield	1.05
934	San Luis Obispo	1.06
935	Mojave	1.03
936-938	Fresno	1.11
939	Salinas	1.14
940-941	San Francisco	1.25
942,956-958	Sacramento	1.13
943	Palo Alto	1.16
944	San Mateo	1.22
945	Vallejo	1.17
946	Oakland	1.22
947	Berkeley	1.24
948	Richmond	1.23
949	San Rafael	1.21
950	Santa Cruz	1.16
951	San Jose	1.22
952	Stockton	1.11
953	Modesto	1.10

STATE	CITY	Residential
CALIFORNIA (CONT'D)		
954	Santa Rosa	1.19
955	Eureka	1.11
959	Marysville	1.10
960	Redding	1.11
961	Susanville	1.10
COLORADO		
800-802	Denver	.90
803	Boulder	.92
804	Golden	.88
805	Fort Collins	.88
806	Greeley	.86
807	Fort Morgan	.90
808-809	Colorado Springs	.87
810	Pueblo	.89
811	Alamosa	.85
812	Salida	.88
813	Durango	.88
814	Montrose	.86
815	Grand Junction	.92
816	Glenwood Springs	.89
CONNECTICUT		
060	New Britain	1.10
061	Hartford	1.10
062	Willimantic	1.10
063	New London	1.10
064	Meriden	1.10
065	New Haven	1.10
066	Bridgeport	1.11
067	Waterbury	1.10
068	Norwalk	1.12
069	Stamford	1.12
D.C.		
200-205	Washington	.94
DELAWARE		
197	Newark	1.03
198	Wilmington	1.03
199	Dover	1.02
FLORIDA		
320,322	Jacksonville	.82
321	Daytona Beach	.87
323	Tallahassee	.79
324	Panama City	.78
325	Pensacola	.82
326,344	Gainesville	.81
327-328,347	Orlando	.88
329	Melbourne	.88
330-332,340	Miami	.87
333	Fort Lauderdale	.87
334,349	West Palm Beach	.87
335-336,346	Tampa	.91
337	St. Petersburg	.81
338	Lakeland	.88
339,341	Fort Myers	.87
342	Sarasota	.90
GEORGIA		
300-303,399	Atlanta	.87
304	Statesboro	.70
305	Gainesville	.77
306	Athens	.75
307	Dalton	.75
308-309	Augusta	.81
310-312	Macon	.79
313-314	Savannah	.81
315	Waycross	.79
316	Valdosta	.75
317,398	Albany	.77
318-319	Columbus	.80
HAWAII		
967	Hilo	1.18
968	Honolulu	1.20

Location Factors

STATE	CITY	Residential
STATES & POSS.		
969	Guam	.92
IDAHO		
832	Pocatello	.84
833	Twin Falls	.70
834	Idaho Falls	.72
835	Lewiston	.92
836-837	Boise	.85
838	Coeur d'Alene	.90
ILLINOIS		
600-603	North Suburban	1.17
604	Joliet	1.20
605	South Suburban	1.17
606-608	Chicago	1.22
609	Kankakee	1.11
610-611	Rockford	1.11
612	Rock Island	.96
613	La Salle	1.09
614	Galesburg	1.01
615-616	Peoria	1.05
617	Bloomington	1.03
618-619	Champaign	1.05
620-622	East St. Louis	1.01
623	Quincy	1.01
624	Effingham	1.02
625	Decatur	1.04
626-627	Springfield	1.04
628	Centralia	1.02
629	Carbondale	.98
INDIANA		
460	Anderson	.91
461-462	Indianapolis	.94
463-464	Gary	1.04
465-466	South Bend	.91
467-468	Fort Wayne	.90
469	Kokomo	.91
470	Lawrenceburg	.87
471	New Albany	.87
472	Columbus	.92
473	Muncie	.92
474	Bloomington	.94
475	Washington	.91
476-477	Evansville	.91
478	Terre Haute	.91
479	Lafayette	.93
IOWA		
500-503,509	Des Moines	.91
504	Mason City	.78
505	Fort Dodge	.76
506-507	Waterloo	.83
508	Creston	.85
510-511	Sioux City	.85
512	Sibley	.74
513	Spencer	.75
514	Carroll	.80
515	Council Bluffs	.86
516	Shenandoah	.81
520	Dubuque	.85
521	Decorah	.78
522-524	Cedar Rapids	.92
525	Ottumwa	.83
526	Burlington	.88
527-528	Davenport	.94
KANSAS		
660-662	Kansas City	.97
664-666	Topeka	.77
667	Fort Scott	.88
668	Emporia	.82
669	Belleville	.80
670-672	Wichita	.77
673	Independence	.87
674	Salina	.79
675	Hutchinson	.79
676	Hays	.81
677	Colby	.82
678	Dodge City	.81
679	Liberal	.80
KENTUCKY		
400-402	Louisville	.91
403-405	Lexington	.90

STATE	CITY	Residential
KENTUCKY (CONT'D)		
406	Frankfort	.89
407-409	Corbin	.82
410	Covington	.95
411-412	Ashland	.94
413-414	Campton	.89
415-416	Pikeville	.92
417-418	Hazard	.88
420	Paducah	.88
421-422	Bowling Green	.89
423	Owensboro	.90
424	Henderson	.91
425-426	Somerset	.84
427	Elizabethtown	.87
LOUISIANA		
700-701	New Orleans	.87
703	Thibodaux	.82
704	Hammond	.76
705	Lafayette	.82
706	Lake Charles	.83
707-708	Baton Rouge	.83
710-711	Shreveport	.76
712	Monroe	.72
713-714	Alexandria	.72
MAINE		
039	Kittery	.91
040-041	Portland	.96
042	Lewiston	.96
043	Augusta	.93
044	Bangor	.95
045	Bath	.92
046	Machias	.96
047	Houlton	.96
048	Rockland	.96
049	Waterville	.91
MARYLAND		
206	Waldorf	.89
207-208	College Park	.87
209	Silver Spring	.88
210-212	Baltimore	.91
214	Annapolis	.90
215	Cumberland	.89
216	Easton	.84
217	Hagerstown	.88
218	Salisbury	.83
219	Elkton	.90
MASSACHUSETTS		
010-011	Springfield	1.05
012	Pittsfield	1.04
013	Greenfield	1.04
014	Fitchburg	1.13
015-016	Worcester	1.14
017	Framingham	1.16
018	Lowell	1.16
019	Lawrence	1.18
020-022, 024	Boston	1.22
023	Brockton	1.16
025	Buzzards Bay	1.14
026	Hyannis	1.14
027	New Bedford	1.16
MICHIGAN		
480,483	Royal Oak	1.02
481	Ann Arbor	1.03
482	Detroit	1.06
484-485	Flint	.97
486	Saginaw	.93
487	Bay City	.94
488-489	Lansing	.95
490	Battle Creek	.93
491	Kalamazoo	.92
492	Jackson	.93
493,495	Grand Rapids	.89
494	Muskegon	.90
496	Traverse City	.83
497	Gaylord	.81
498-499	Iron mountain	.90
MINNESOTA		
550-551	Saint Paul	1.11
553-555	Minneapolis	1.14
556-558	Duluth	1.07

Location Factors

STATE	CITY	Residential
MINNESOTA (CONT'D)		
559	Rochester	1.05
560	Mankato	1.00
561	Windom	.93
562	Willmar	.98
563	St. Cloud	1.07
564	Brainerd	.97
565	Detroit Lakes	.96
566	Bemidji	.95
567	Thief River Falls	.95
MISSISSIPPI		
386	Clarksdale	.75
387	Greenville	.82
388	Tupelo	.77
389	Greenwood	.78
390-392	Jackson	.82
393	Meridian	.80
394	Laurel	.77
395	Biloxi	.80
396	Mccomb	.75
397	Columbus	.76
MISSOURI		
630-631	St. Louis	1.03
633	Bowling Green	.96
634	Hannibal	.93
635	Kirksville	.89
636	Flat River	.96
637	Cape Girardeau	.89
638	Sikeston	.88
639	Poplar Bluff	.87
640-641	Kansas City	1.05
644-645	St. Joseph	.97
646	Chillicothe	.96
647	Harrisonville	.99
648	Joplin	.90
650-651	Jefferson City	.93
652	Columbia	.92
653	Sedalia	.92
654-655	Rolla	.97
656-658	Springfield	.89
MONTANA		
590-591	Billings	.88
592	Wolf Point	.86
593	Miles City	.87
594	Great Falls	.89
595	Havre	.84
596	Helena	.87
597	Butte	.88
598	Missoula	.88
599	Kalispell	.86
NEBRASKA		
680-681	Omaha	.91
683-685	Lincoln	.84
686	Columbus	.84
687	Norfolk	.89
688	Grand Island	.88
689	Hastings	.90
690	Mccook	.82
691	North Platte	.88
692	Valentine	.82
693	Alliance	.82
NEVADA		
889-891	Las Vegas	1.03
893	Ely	1.00
894-895	Reno	.93
897	Carson City	.93
898	Elko	.92
NEW HAMPSHIRE		
030	Nashua	.97
031	Manchester	.97
032-033	Concord	.96
034	Keene	.87
035	Littleton	.91
036	Charleston	.85
037	Claremont	.86
038	Portsmouth	.96

STATE	CITY	Residential
NEW JERSEY		
070-071	Newark	1.13
072	Elizabeth	1.15
073	Jersey City	1.12
074-075	Paterson	1.13
076	Hackensack	1.12
077	Long Branch	1.11
078	Dover	1.13
079	Summit	1.13
080,083	Vineland	1.10
081	Camden	1.11
082,084	Atlantic City	1.14
085-086	Trenton	1.11
087	Point Pleasant	1.10
088-089	New Brunswick	1.13
NEW MEXICO		
870-872	Albuquerque	.82
873	Gallup	.82
874	Farmington	.82
875	Santa Fe	.83
877	Las Vegas	.82
878	Socorro	.82
879	Truth/Consequences	.81
880	Las Cruces	.81
881	Clovis	.82
882	Roswell	.82
883	Carrizozo	.82
884	Tucumcari	.83
NEW YORK		
100-102	New York	1.35
103	Staten Island	1.30
104	Bronx	1.31
105	Mount Vernon	1.12
106	White Plains	1.15
107	Yonkers	1.16
108	New Rochelle	1.16
109	Suffern	1.11
110	Queens	1.31
111	Long Island City	1.34
112	Brooklyn	1.35
113	Flushing	1.33
114	Jamaica	1.33
115,117,118	Hicksville	1.20
116	Far Rockaway	1.32
119	Riverhead	1.21
120-122	Albany	.96
123	Schenectady	.98
124	Kingston	1.05
125-126	Poughkeepsie	1.17
127	Monticello	1.05
128	Glens Falls	.91
129	Plattsburgh	.95
130-132	Syracuse	.97
133-135	Utica	.94
136	Watertown	.92
137-139	Binghamton	.97
140-142	Buffalo	1.05
143	Niagara Falls	1.03
144-146	Rochester	.97
147	Jamestown	.90
148-149	Elmira	.91
NORTH CAROLINA		
270,272-274	Greensboro	.85
271	Winston-Salem	.87
275-276	Raleigh	.86
277	Durham	.85
278	Rocky Mount	.82
279	Elizabeth City	.80
280	Gastonia	.85
281-282	Charlotte	.85
283	Fayetteville	.90
284	Wilmington	.86
285	Kinston	.81
286	Hickory	.80
287-288	Asheville	.82
289	Murphy	.83
NORTH DAKOTA		
580-581	Fargo	.77
582	Grand Forks	.72
583	Devils Lake	.76
584	Jamestown	.70
585	Bismarck	.76

Location Factors

STATE	CITY	Residential
NORTH DAKOTA (CONT'D)		
586	Dickinson	.74
587	Minot	.82
588	Williston	.74
OHIO		
430-432	Columbus	.95
433	Marion	.90
434-436	Toledo	.99
437-438	Zanesville	.90
439	Steubenville	.93
440	Lorain	.95
441	Cleveland	1.00
442-443	Akron	.97
444-445	Youngstown	.94
446-447	Canton	.92
448-449	Mansfield	.91
450	Hamilton	.93
451-452	Cincinnati	.93
453-454	Dayton	.93
455	Springfield	.94
456	Chillicothe	.97
457	Athens	.91
458	Lima	.92
OKLAHOMA		
730-731	Oklahoma City	.77
734	Ardmore	.76
735	Lawton	.79
736	Clinton	.75
737	Enid	.75
738	Woodward	.75
739	Guymon	.66
740-741	Tulsa	.75
743	Miami	.85
744	Muskogee	.69
745	Mcalester	.72
746	Ponca City	.75
747	Durant	.75
748	Shawnee	.74
749	Poteau	.76
OREGON		
970-972	Portland	.99
973	Salem	.98
974	Eugene	.99
975	Medford	.97
976	Klamath Falls	.98
977	Bend	1.00
978	Pendleton	.98
979	Vale	.96
PENNSYLVANIA		
150-152	Pittsburgh	1.01
153	Washington	.96
154	Uniontown	.94
155	Bedford	.91
156	Greensburg	.96
157	Indiana	.93
158	Dubois	.92
159	Johnstown	.92
160	Butler	.93
161	New Castle	.93
162	Kittanning	.94
163	Oil City	.90
164-165	Erie	.94
166	Altoona	.88
167	Bradford	.90
168	State College	.91
169	Wellsboro	.92
170-171	Harrisburg	.95
172	Chambersburg	.89
173-174	York	.93
175-176	Lancaster	.92
177	Williamsport	.84
178	Sunbury	.93
179	Pottsville	.93
180	Lehigh Valley	1.02
181	Allentown	1.04
182	Hazleton	.93
183	Stroudsburg	.92
184-185	Scranton	.97
186-187	Wilkes-Barre	.94
188	Montrose	.91
189	Doylestown	1.06

STATE	CITY	Residential
PENNSYLVANIA (CONT'D)		
190-191	Philadelphia	1.16
193	Westchester	1.11
194	Norristown	1.10
195-196	Reading	.98
PUERTO RICO		
009	San Juan	.72
RHODE ISLAND		
028	Newport	1.09
029	Providence	1.09
SOUTH CAROLINA		
290-292	Columbia	.81
293	Spartanburg	.80
294	Charleston	.83
295	Florence	.76
296	Greenville	.79
297	Rock Hill	.78
298	Aiken	.99
299	Beaufort	.78
SOUTH DAKOTA		
570-571	Sioux Falls	.76
572	Watertown	.73
573	Mitchell	.74
574	Aberdeen	.75
575	Pierre	.74
576	Mobridge	.73
577	Rapid City	.76
TENNESSEE		
370-372	Nashville	.84
373-374	Chattanooga	.82
375,380-381	Memphis	.82
376	Johnson City	.71
377-379	Knoxville	.75
382	Mckenzie	.74
383	Jackson	.77
384	Columbia	.77
385	Cookeville	.71
TEXAS		
750	Mckinney	.72
751	Waxahackie	.78
752-753	Dallas	.84
754	Greenville	.69
755	Texarkana	.72
756	Longview	.67
757	Tyler	.73
758	Palestine	.75
759	Lufkin	.78
760-761	Fort Worth	.83
762	Denton	.74
763	Wichita Falls	.77
764	Eastland	.74
765	Temple	.74
766-767	Waco	.77
768	Brownwood	.72
769	San Angelo	.72
770-772	Houston	.84
773	Huntsville	.77
774	Wharton	.72
775	Galveston	.82
776-777	Beaumont	.81
778	Bryan	.73
779	Victoria	.71
780	Laredo	.74
781-782	San Antonio	.81
783-784	Corpus Christi	.78
785	Mc Allen	.75
786-787	Austin	.78
788	Del Rio	.72
789	Giddings	.78
790-791	Amarillo	.79
792	Childress	.79
793-794	Lubbock	.79
795-796	Abilene	.75
797	Midland	.76
798-799,885	El Paso	.74
UTAH		
840-841	Salt Lake City	.80
842,844	Ogden	.79
843	Logan	.79

Location Factors

STATE	CITY	Residential
UTAH (CONT'D)		
845	Price	.77
846-847	Provo	.79
VERMONT		
050	White River Jct.	.84
051	Bellows Falls	.90
052	Bennington	.90
053	Brattleboro	.92
054	Burlington	.89
056	Montpelier	.90
057	Rutland	.89
058	St. Johnsbury	.85
059	Guildhall	.85
VIRGINIA		
220-221	Fairfax	1.03
222	Arlington	1.05
223	Alexandria	1.07
224-225	Fredericksburg	1.04
226	Winchester	1.03
227	Culpeper	1.03
228	Harrisonburg	.88
229	Charlottesville	.92
230-232	Richmond	.97
233-235	Norfolk	.99
236	Newport News	.98
237	Portsmouth	.92
238	Petersburg	.96
239	Farmville	.86
240-241	Roanoke	.97
242	Bristol	.85
243	Pulaski	.83
244	Staunton	.88
245	Lynchburg	.95
246	Grundy	.82
WASHINGTON		
980-981,987	Seattle	1.02
982	Everett	1.03
983-984	Tacoma	1.00
985	Olympia	.99
986	Vancouver	.97
988	Wenatchee	.93
989	Yakima	.97
990-992	Spokane	.97
993	Richland	.97
994	Clarkston	.92
WEST VIRGINIA		
247-248	Bluefield	.90
249	Lewisburg	.90
250-253	Charleston	.96
254	Martinsburg	.87
255-257	Huntington	.97
258-259	Beckley	.90
260	Wheeling	.92
261	Parkersburg	.92
262	Buckhannon	.92
263-264	Clarksburg	.91
265	Morgantown	.92
266	Gassaway	.92
267	Romney	.90
268	Petersburg	.91
WISCONSIN		
530,532	Milwaukee	1.08
531	Kenosha	1.02
534	Racine	1.00
535	Beloit	.97
537	Madison	.97
538	Lancaster	.95
539	Portage	.94
540	New Richmond	.98
541-543	Green Bay	1.01
544	Wausau	.95
545	Rhinelander	.94
546	La Crosse	.95
547	Eau Claire	.98
548	Superior	.96
549	Oshkosh	.90
WYOMING		
820	Cheyenne	.80
821	Yellowstone Nat. Pk.	.80
822	Wheatland	.77

STATE	CITY	Residential
WYOMING (CONT'D)		
823	Rawlins	.81
824	Worland	.80
825	Riverton	.80
826	Casper	.74
827	Newcastle	.80
828	Sheridan	.80
829-831	Rock Springs	.84
CANADIAN FACTORS (reflect Canadian currency)		
ALBERTA		
	Calgary	1.17
	Edmonton	1.15
	Fort McMurray	1.15
	Lethbridge	1.13
	Lloydminster	1.08
	Medicine Hat	1.08
	Red Deer	1.08
BRITISH COLUMBIA		
	Kamloops	1.07
	Prince George	1.06
	Vancouver	1.11
	Victoria	1.06
MANITOBA		
	Brandon	1.03
	Portage la Prairie	1.03
	Winnipeg	1.03
NEW BRUNSWICK		
	Bathurst	.95
	Dalhousie	.96
	Fredericton	1.02
	Moncton	.96
	Newcastle	.96
	St. John	1.03
NEWFOUNDLAND		
	Corner Brook	.96
	St Johns	1.05
NORTHWEST TERRITORIES		
	Yellowknife	1.09
NOVA SCOTIA		
	Bridgewater	.98
	Dartmouth	.98
	Halifax	1.02
	New Glasgow	.98
	Sydney	.97
	Truro	.98
	Yarmouth	.98
ONTARIO		
	Barrie	1.14
	Brantford	1.15
	Cornwall	1.14
	Hamilton	1.15
	Kingston	1.14
	Kitchener	1.09
	London	1.13
	North Bay	1.12
	Oshawa	1.13
	Ottawa	1.14
	Owen Sound	1.13
	Peterborough	1.13
	Sarnia	1.15
	Sault Ste Marie	1.08
	St. Catharines	1.10
	Sudbury	1.07
	Thunder Bay	1.12
	Timmins	1.12
	Toronto	1.17
	Windsor	1.11
PRINCE EDWARD ISLAND		
	Charlottetown	.92
	Summerside	.92
QUEBEC		
	Cap-de-la-Madeleine	1.14
	Charlesbourg	1.14
	Chicoutimi	1.16
	Gatineau	1.13
	Granby	1.13

Location Factors

STATE	CITY	Residential
QUEBEC (CONT'D)		
	Hull	1.13
	Joliette	1.14
	Laval	1.13
	Montreal	1.18
	Quebec	1.17
	Rimouski	1.16
	Rouyn-Noranda	1.13
	Saint Hyacinthe	1.13
	Sherbrooke	1.13
	Sorel	1.14
	St Jerome	1.13
	Trois Rivieres	1.14
SASKATCHEWAN		
	Moose Jaw	.95
	Prince Albert	.94
	Regina	1.07
	Saskatoon	1.06
YUKON		
	Whitehorse	.96

Abbreviations

A	Area
ASTM	American Society for Testing and Materials
B.F.	Board feet
Carp.	Carpenter
C.F.	Cubic feet
CWJ	Composite wood joist
C.Y.	Cubic yard
Ea.	Each
Equip.	Equipment
Exp.	Exposure
Ext.	Exterior
F	Fahrenheit
Ft.	Foot, feet
Gal.	Gallon
Hr.	Hour
in.	Inch, inches
Inst.	Installation
Int.	Interior
Lb.	Pound
L.F.	Linear feet
LVL	Laminated veneer lumber
Mat.	Material
Max.	Maximum
MBF	Thousand board feet
MBM	Thousand feet board measure
MSF	Thousand square feet
Min.	Minimum
O.C.	On center
O&P	Overhead and profit
OWJ	Open web wood joist
Oz.	Ounce
Pr.	Pair
Quan.	Quantity
S.F.	Square foot
Sq.	Square, 100 square feet
S.Y.	Square yard
V.L.F.	Vertical linear feet
'	Foot, feet
"	Inch, inches
°	Degrees

Notes

RSMeans Building Construction Cost Data 2013

Offers you unchallenged unit price reliability in an easy-to-use format. Whether used for verifying complete, finished estimates or for periodic checks, it supplies more cost facts better and faster than any comparable source. Over 23,000 unit prices have been updated for 2013. The City Cost Indexes and Location Factors cover over 930 areas, for indexing to any project location in North America. Order and get *RSMeans Quarterly Update Service* FREE. You'll have year-long access to the RSMeans Estimating **HOTLINE** FREE with your subscription. Expert assistance when using RSMeans data is just a phone call away.

$186.95 per copy | Available Sept. 2012 | Catalog no. 60013

RSMeans Green Building Cost Data 2013

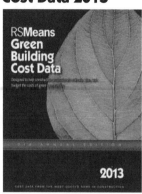

Estimate, plan, and budget the costs of green building for both new commercial construction and renovation work with this first edition of RSMeans Green Building Cost Data. More than 9,000 unit costs for a wide array of green building products plus assemblies costs. Easily identified cross references to LEED and Green Globes building rating systems criteria.

$155.95 per copy | Available Nov. 2012 | Catalog no. 60553

RSMeans Mechanical Cost Data 2013

Total unit and systems price guidance for mechanical construction. . . materials, parts, fittings, and complete labor cost information. Includes prices for piping, heating, air conditioning, ventilation, and all related construction.

Plus new 2013 unit costs for:

• Thousands of installed HVAC/controls assemblies components

• "On-site" Location Factors for over 930 cities and towns in the U.S. and Canada

• Crews, labor, and equipment

$183.95 per copy | Available Oct. 2012 | Catalog no. 60023

RSMeans Facilities Construction Cost Data 2013

For the maintenance and construction of commercial, industrial, municipal, and institutional properties. Costs are shown for new and remodeling construction and are broken down into materials, labor, equipment, and overhead and profit. Special emphasis is given to sections on mechanical, electrical, furnishings, site work, building maintenance, finish work, and demolition.

More than 47,000 unit costs, plus assemblies costs and a comprehensive Reference Section are included.

$476.95 per copy | Available Nov. 2012 | Catalog no. 60203

RSMeans Square Foot Costs 2013

Accurate and Easy To Use

• **Updated price information** based on nationwide figures from suppliers, estimators, labor experts, and contractors

• "How-to-Use" sections, with **clear examples** of commercial, residential, industrial, and institutional structures

• Realistic graphics, offering true-to-life illustrations of building projects

• Extensive information on using square foot cost data, including sample estimates and alternate pricing methods

$197.95 per copy | Available Oct. 2012 | Catalog no. 60053

RSMeans Commercial Renovation Cost Data 2013

Commercial/Multifamily Residential

Use this valuable tool to estimate commercial and multifamily residential renovation and remodeling.

Includes: New costs for hundreds of unique methods, materials, and conditions that only come up in repair and remodeling, PLUS:

• Unit costs for more than 17,000 construction components

• Installed costs for more than 2,000 assemblies

• More than 930 "on-site" localization factors for the U.S. and Canada

$152.95 per copy | Available Oct. 2012 | Catalog no. 60043

RSMeans Electrical Cost Data 2013

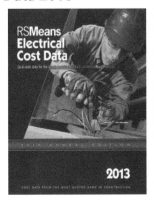

Pricing information for every part of electrical cost planning. More than 13,000 unit and systems costs with design tables; clear specifications and drawings; engineering guides; illustrated estimating procedures; complete labor-hour and materials costs for better scheduling and procurement; and the latest electrical products and construction methods.

- A variety of special electrical systems including cathodic protection
- Costs for maintenance, demolition, HVAC/mechanical, specialties, equipment, and more

$188.95 per copy | Available Oct. 2012 | Catalog no. 60033

RSMeans Electrical Change Order Cost Data 2013

RSMeans Electrical Change Order Cost Data provides you with electrical unit prices exclusively for pricing change orders—based on the recent, direct experience of contractors and suppliers. Analyze and check your own change order estimates against the experience others have had doing the same work. It also covers productivity analysis and change order cost justifications. With useful information for calculating the effects of change orders and dealing with their administration.

$183.95 per copy | Available Dec. 2012 | Catalog no. 60233

RSMeans Assemblies Cost Data 2013

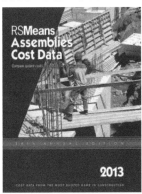

RSMeans Assemblies Cost Data takes the guesswork out of preliminary or conceptual estimates. Now you don't have to try to calculate the assembled cost by working up individual component costs. We've done all the work for you.

Presents detailed illustrations, descriptions, specifications, and costs for every conceivable building assembly—over 350 types in all—arranged in the easy-to-use UNIFORMAT II system. Each illustrated "assembled" cost includes a complete grouping of materials and associated installation costs, including the installing contractor's overhead and profit.

$305.95 per copy | Available Sept. 2012 | Catalog no. 60063

RSMeans Open Shop Building Construction Cost Data 2013

The latest costs for accurate budgeting and estimating of new commercial and residential construction. . . renovation work. . . change orders. . . cost engineering.

RSMeans Open Shop "BCCD" will assist you to:

- Develop benchmark prices for change orders
- Plug gaps in preliminary estimates and budgets
- Estimate complex projects
- Substantiate invoices on contracts
- Price ADA-related renovations

$160.95 per copy | Available Dec. 2012 | Catalog no. 60153

RSMeans Residential Cost Data 2013

Contains square foot costs for 28 basic home models with the look of today, plus hundreds of custom additions and modifications you can quote right off the page. Includes costs for more than 700 residential systems. Complete with blank estimating forms, sample estimates, and step-by-step instructions.

Now contains line items for cultured stone and brick, PVC trim lumber, and TPO roofing.

$134.95 per copy | Available Oct. 2012 | Catalog no. 60173

RSMeans Site Work & Landscape Cost Data 2013

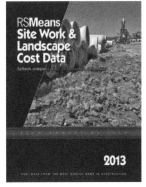

Includes unit and assemblies costs for earthwork, sewerage, piped utilities, site improvements, drainage, paving, trees and shrubs, street openings/repairs, underground tanks, and more. Contains 78 types of assemblies costs for accurate conceptual estimates.

Includes:

- Estimating for infrastructure improvements
- Environmentally-oriented construction
- ADA-mandated handicapped access
- Hazardous waste line items

$181.95 per copy | Available Dec. 2012 | Catalog no. 60283

RSMeans Facilities Maintenance & Repair Cost Data 2013

RSMeans Facilities Maintenance & Repair Cost Data gives you a complete system to manage and plan your facility repair and maintenance costs and budget efficiently. Guidelines for auditing a facility and developing an annual maintenance plan. Budgeting is included, along with reference tables on cost and management, and information on frequency and productivity of maintenance operations.

The only nationally recognized source of maintenance and repair costs. Developed in cooperation with the Civil Engineering Research Laboratory (CERL) of the Army Corps of Engineers.

$417.95 per copy | Available Nov. 2012 | Catalog no. 60303

RSMeans Concrete & Masonry Cost Data 2013

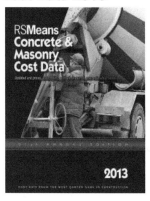

Provides you with cost facts for virtually all concrete/masonry estimating needs, from complicated formwork to various sizes and face finishes of brick and block—all in great detail. The comprehensive Unit Price Section contains more than 8,000 selected entries. Also contains an Assemblies [Cost] Section, and a detailed Reference Section that supplements the cost data.

$171.95 per copy | Available Dec. 2012 | Catalog no. 60113

RSMeans Construction Cost Indexes 2013

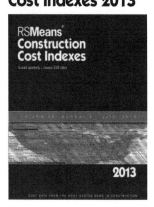

What materials and labor costs will change unexpectedly this year? By how much?

- Breakdowns for 318 major cities
- National averages for 30 key cities
- Expanded five major city indexes
- Historical construction cost indexes

$362.00 per year (subscription) | Catalog no. 50143
$90.50 individual quarters | Catalog no. 60143 A,B,C,D

RSMeans Light Commercial Cost Data 2013

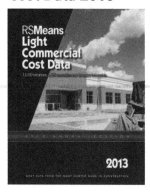

Specifically addresses the light commercial market, which is a specialized niche in the construction industry. Aids you, the owner/designer/contractor, in preparing all types of estimates—from budgets to detailed bids. Includes new advances in methods and materials.

Assemblies Section allows you to evaluate alternatives in the early stages of design/planning.

Over 13,000 unit costs ensure that you have the prices you need. . . when you need them.

$139.95 per copy | Available Nov. 2012 | Catalog no. 60183

RSMeans Labor Rates for the Construction Industry 2013

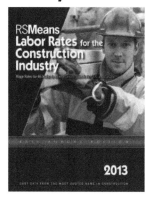

Complete information for estimating labor costs, making comparisons, and negotiating wage rates by trade for more than 300 U.S. and Canadian cities. With 46 construction trades listed by local union number in each city, and historical wage rates included for comparison. Each city chart lists the county and is alphabetically arranged with handy visual flip tabs for quick reference.

$415.95 per copy | Available Dec. 2012 | Catalog no. 60123

RSMeans Interior Cost Data 2013

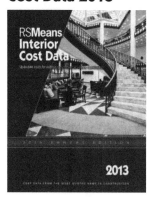

Provides you with prices and guidance needed to make accurate interior work estimates. Contains costs on materials, equipment, hardware, custom installations, furnishings, and labor costs. . . for new and remodel commercial and industrial interior construction, including updated information on office furnishings, and reference information.

$192.95 per copy | Available Nov. 2012 | Catalog no. 60093

Annual Cost Guides

Cost Guides also available in CD and at rsmeansonline.com

For more information visit the RSMeans website at **www.rsmeans.com**

Unit prices according to the latest MasterFormat!

RSMeans Heavy Construction Cost Data 2013

A comprehensive guide to heavy construction costs. Includes costs for highly specialized projects such as tunnels, dams, highways, airports, and waterways. Information on labor rates, equipment, and materials costs is included. Features unit price costs, systems costs, and numerous reference tables for costs and design.

$186.95 per copy | Available Dec. 2012 | Catalog no. 60163

RSMeans Plumbing Cost Data 2013

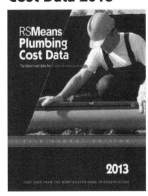

Comprehensive unit prices and assemblies for plumbing, irrigation systems, commercial and residential fire protection, point-of-use water heaters, and the latest approved materials. This publication and its companion, *RSMeans Mechanical Cost Data*, provide full-range cost estimating coverage for all the mechanical trades.

Now contains updated costs for potable water and radiant heat systems and high efficiency fixtures.

$188.95 per copy | Available Oct. 2012 | Catalog no. 60213

2013 RSMeans Seminar Schedule

Note: call for exact dates and details.

Location	Dates	Location	Dates
Las Vegas, NV	March	Las Vegas, NV	October
Washington, DC	April	Orlando, FL	November
San Francisco, CA	June	San Diego, CA	December
Washington, DC	September		
Dallas, TX	September		

1-800-334-3509, Press 1

2013 Regional Seminar Schedule

Note: call for exact dates and details.

Location	Dates	Location	Dates
Seattle, WA	January and August	Columbus, GA	June
Dallas/Ft. Worth, TX	January	El Segundo, CA	August
Austin, TX	February	Jacksonville FL	September
Oklahoma City, OK	February	St. Louis, MO	October
Anchorage, AK	March and September	Houston, TX	October
Phoenix, AZ	March	Vancouver, BC	November
San Bernardino, CA	April	Baltimore, MD	November
Kansas City, MO	April	San Antonio, TX	December
Toronto	May	Raleigh, NC	December
Denver, CO	May		
Bethesda, MD	June		

1-800-334-3509, Press 1

Professional Development

For more information visit the RSMeans website at **www.rsmeans.com**

Registration Information

Register early... Save up to $100! Register 30 days before the start date of a seminar and save $100 off your total fee. *Note: This discount can be applied only once per order. It cannot be applied to team discount registrations or any other special offer.*

How to register Register by phone today! The RSMeans toll-free number for making reservations is **1-800-334-3509, Press 1.**

Individual seminar registration fee - $935. *Means CostWorks®* **training registration fee - $375.** To register by mail, complete the registration form and return, with your full fee, to: RSMeans Seminars, 700 Longwater Drive, Norwell, MA 02061.

Government pricing All federal government employees save off the regular seminar price. Other promotional discounts cannot be combined with the government discount.

Team discount program for two to four seminar registrations. Call for pricing: 1-800-334-3509, Press 1

Multiple course discounts When signing up for two or more courses, call for pricing.

Refund policy Cancellations will be accepted up to ten business days prior to the seminar start. There are no refunds for cancellations received later than ten working days prior to the first day of the seminar. A $150 processing fee will be applied for all cancellations. Written notice of cancellation is required. Substitutions can be made at any time before the session starts. **No-shows are subject to the full seminar fee.**

AACE approved courses Many seminars described and offered here have been approved for 14 hours (1.4 recertification credits) of credit by the AACE

International Certification Board toward meeting the continuing education requirements for recertification as a Certified Cost Engineer/Certified Cost Consultant.

AIA Continuing Education We are registered with the AIA Continuing Education System (AIA/CES) and are committed to developing quality learning activities in accordance with the CES criteria. Many seminars meet the AIA/CES criteria for Quality Level 2. AIA members may receive (14) learning units (LUs) for each two-day RSMeans course.

Daily course schedule The first day of each seminar session begins at 8:30 a.m. and ends at 4:30 p.m. The second day begins at 8:00 a.m. and ends at 4:00 p.m. Participants are urged to bring a hand-held calculator, since many actual problems will be worked out in each session.

Continental breakfast Your registration includes the cost of a continental breakfast, and a morning and afternoon refreshment break. These informal segments allow you to discuss topics of mutual interest with other seminar attendees. (You are free to make your own lunch and dinner arrangements.)

Hotel/transportation arrangements RSMeans arranges to hold a block of rooms at most host hotels. To take advantage of special group rates when making your reservation, be sure to mention that you are attending the RSMeans seminar. You are, of course, free to stay at the lodging place of your choice. (**Hotel reservations and transportation arrangements should be made directly by seminar attendees.**)

Important Class sizes are limited, so please register as soon as possible.

Note: Pricing subject to change.

Registration Form

ADDS-1000

Call 1-800-334-3509, Press 1 to register or FAX this form 1-800-632-6732. Visit our website: www.rsmeans.com

Please register the following people for the RSMeans construction seminars as shown here. We understand that we must make our own hotel reservations if overnight stays are necessary.

☐ Full payment of $_____enclosed.

☐ Bill me.

Please print name of registrant(s).

(To appear on certificate of completion)

_____

_____

_____

_____

P.O. #: _____
GOVERNMENT AGENCIES MUST SUPPLY PURCHASE ORDER NUMBER OR
TRAINING FORM.

Firm name_____

Address_____

City/State/Zip_____

Telephone no._____ Fax no._____

E-mail address_____

Charge registration(s) to: ☐ MasterCard ☐ VISA ☐ American Express

Account no._____ Exp. date_____

Cardholder's signature_____

Seminar name_____

Seminar City _____

Please mail check to: RSMeans Seminars, 700 Longwater Drive, Norwell, MA 02061 USA

Professional Development

For more information visit the RSMeans website at **www.rsmeans.com**

RSMeans Online™ Training

Construction estimating is vital to the decision-making process at each state of every project. Means CostWorks.com works the way you do. It's systematic, flexible and intuitive. In this one day class you will see how you can estimate any phase of any project faster and better.

Some of what you'll learn:
- Customizing Means CostWorks.com
- Making the most of RSMeans "Circle Reference" numbers
- How to integrate your cost data
- Generate reports, exporting estimates to MS Excel, sharing, collaborating and more

Also available: RSMeans Online™ training webinar

Facilities Construction Estimating

In this *two-day* course, professionals working in facilities management can get help with their daily challenges to establish budgets for all phases of a project.

Some of what you'll learn:
- Determining the full scope of a project
- Identifying the scope of risks and opportunities
- Creative solutions to estimating issues
- Organizing estimates for presentation and discussion
- Special techniques for repair/remodel and maintenance projects
- Negotiating project change orders

Who should attend: facility managers, engineers, contractors, facility tradespeople, planners, and project managers.

Scheduling with Microsoft Project

Two days of hands-on training gives you a look at the basics to putting it all together with MS Project to create a complete project schedule. Learn to work better and faster, manage changes and updates, enhance tracking, generate actionable reports, and boost control of your time and budget.

Some of what you'll learn:
- Defining task/activity relationships: link tasks, establish predecessor/successor relationships and create logs
- Baseline scheduling: the essential what/when
- Using MS Project to determine the critical path
- Integrate RSMeans data with MS Project
- The dollar-loaded schedule: enhance the reliability of this "must-know" technique for scheduling public projects.

Who should attend: contractors' project-management teams, architects and engineers, project owners and their representatives, and those interested in improving their project planning prod scheduling and management skills.

Maintenance & Repair Estimating for Facilities

This *two-day* course teaches attendees how to plan, budget, and estimate the cost of ongoing and preventive maintenance and repair for existing buildings and grounds.

Some of what you'll learn:
- The most financially favorable maintenance, repair, and replacement scheduling and estimating
- Auditing and value engineering facilities
- Preventive planning and facilities upgrading
- Determining both in-house and contract-out service costs
- Annual, asset-protecting M&R plan

Who should attend: facility managers, maintenance supervisors, buildings and grounds superintendents, plant managers, planners, estimators, and others involved in facilities planning and budgeting.

Practical Project Management for Construction Professionals

In this *two-day* course, acquire the essential knowledge and develop the skills to effectively and efficiently execute the day-to-day responsibilities of the construction project manager.

Covers:
- General conditions of the construction contract
- Contract modifications: change orders and construction change directives
- Negotiations with subcontractors and vendors
- Effective writing: notification and communications
- Dispute resolution: claims and liens

Who should attend: architects, engineers, owner's representatives, project managers.

Mechanical & Electrical Estimating

This *two-day* course teaches attendees how to prepare more accurate and complete mechanical/electrical estimates, avoiding the pitfalls of omission and double-counting, while understanding the composition and rationale within the RSMeans mechanical/electrical database.

Some of what you'll learn:
- The unique way mechanical and electrical systems are interrelated
- M&E estimates—conceptual, planning, budgeting, and bidding stages
- Order of magnitude, square foot, assemblies, and unit price estimating
- Comparative cost analysis of equipment and design alternatives

Who should attend: architects, engineers, facilities managers, mechanical and electrical contractors, and others who need a highly reliable method for developing, understanding, and evaluating mechanical and electrical contracts.

Professional Development

Unit Price Estimating

This interactive *two-day* seminar teaches attendees how to interpret project information and process it into final, detailed estimates with the greatest accuracy level.

The most important credential an estimator can take to the job is the ability to visualize construction and estimate accurately.

Some of what you'll learn:
- Interpreting the design in terms of cost
- The most detailed, time-tested methodology for accurate pricing
- Key cost drivers—material, labor, equipment, staging, and subcontracts
- Understanding direct and indirect costs for accurate job cost accounting and change order management

Who should attend: corporate and government estimators and purchasers, architects, engineers, and others who need to produce accurate project estimates.

RSMeans CostWorks® CD Training

This one-day course helps users become more familiar with the functionality of *RSMeans CostWorks* program. Each menu, icon, screen, and function found in the program is explained in depth. Time is devoted to hands-on estimating exercises.

Some of what you'll learn:
- Searching the database using all navigation methods
- Exporting RSMeans data to your preferred spreadsheet format
- Viewing crews, assembly components, and much more
- Automatically regionalizing the database

This training session requires you to bring a laptop computer to class.

When you register for this course you will receive an outline for your laptop requirements.

Also offering web training for CostWorks CD!

Facilities Est. Using RSMeans CostWorks® CD

Combines hands-on skill building with best estimating practices and real-life problems. Brings you up-to-date with key concepts, and provides tips, pointers, and guidelines to save time and avoid cost oversights and errors.

Some of what you'll learn:
- Estimating process concepts
- Customizing and adapting RSMeans cost data
- Establishing scope of work to account for all known variables
- Budget estimating: when, why, and how
- Site visits: what to look for—what you can't afford to overlook
- How to estimate repair and remodeling variables

This training session requires you to bring a laptop computer to class.

Who should attend: facility managers, architects, engineers, contractors, facility tradespeople, planners, project managers and anyone involved with JOC, SABRE, or IDIQ.

Conceptual Estimating Using RSMeans CostWorks® CD

This *two day* class uses the leading industry data and a powerful software package to develop highly accurate conceptual estimates for your construction projects. All attendees must bring a laptop computer loaded with the current year *Square Foot Model Costs* and the *Assemblies Cost Data* CostWorks titles.

Some of what you'll learn:
- Introduction to conceptual estimating
- Types of conceptual estimates
- Helpful hints
- Order of magnitude estimating
- Square foot estimating
- Assemblies estimating

Who should attend: architects, engineers, contractors, construction estimators, owner's representatives, and anyone looking for an electronic method for performing square foot estimating.

Assessing Scope of Work for Facility Construction Estimating

This *two-day* practical training program addresses the vital importance of understanding the SCOPE of projects in order to produce accurate cost estimates for facility repair and remodeling.

Some of what you'll learn:
- Discussions of site visits, plans/specs, record drawings of facilities, and site-specific lists
- Review of CSI divisions, including means, methods, materials, and the challenges of scoping each topic
- Exercises in SCOPE identification and SCOPE writing for accurate estimating of projects
- Hands-on exercises that require SCOPE, take-off, and pricing

Who should attend: corporate and government estimators, planners, facility managers, and others who need to produce accurate project estimates.

Unit Price Est. Using RSMeans CostWorks® CD

Step-by-step instruction and practice problems to identify and track key cost drivers—material, labor, equipment, staging, and subcontractors—for each specific task. Learn the most detailed, time-tested methodology for accurately "pricing" these variables, their impact on each other and on total cost.

Some of what you'll learn:
- Unit price cost estimating
- Order of magnitude, square foot, and assemblies estimating
- Quantity takeoff
- Direct and indirect construction costs
- Development of contractor's bill rates
- How to use *RSMeans Building Construction Cost Data*

This training session requires you to bring a laptop computer to class.

Who should attend: architects, engineers, corporate and government estimators, facility managers, and government procurement staff.

New Titles - Reference Books

Complete Book of Framing, 2nd Edition

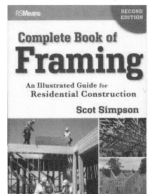

by Scot Simpson

This updated, easy-to-learn guide to rough carpentry and framing is written by an expert with more than thirty years of framing experience. Starting with the basics, this book begins with types of lumber, nails, and what tools are needed, followed by detailed, fully illustrated steps for framing each building element. Framer-Friendly Tips throughout the book show how to get a task done right—and more easily.

$29.95 per copy | 352 pages, softcover | Catalog no. 67353A

Estimating Building Costs, 2nd Edition

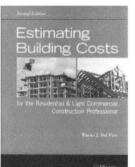

For the Residential & Light Commercial Construction Professional
by Wayne J. DelPico

This book guides readers through the entire estimating process, explaining in detail how to put together a reliable estimate that can be used not only for budgeting, but also for developing a schedule, managing a project, dealing with contingencies, and ultimately making a profit.

Completely revised and updated to reflect the new CSI MasterFormat 2010 system, this practical guide describes estimating techniques for each building system and how to apply them according to the latest industry standards.

$65.00 per copy | 398 pages, softcover | Catalog no. 67343A

Risk Management for Design and Construction

This book introduces risk as a central pillar of project management and shows how a project manager can be prepared for dealing with uncertainty. Written by experts in the field, *Risk Management for Design and Construction* uses clear, straightforward terminology to demystify the concepts of project uncertainty and risk.

Highlights include:

- Integrated cost and schedule risk analysis

- An introduction to a ready-to-use system of analyzing a project's risks and tools to proactively manage risks

- A methodology that was developed and used by the Washington State Department of Transportation

- Case studies and examples on the proper application of principles

- Information about combining value analysis with risk analysis

$125.00 per copy | Over 250 pages, softcover | Catalog no. 67359

How to Estimate with Means Data & CostWorks, 4th Edition

by RSMeans and Saleh A. Mubarak, Ph.D.

This step-by-step guide takes you through all the major construction items with extensive coverage of site work, concrete and masonry, wood and metal framing, doors and windows, and other divisions. The only construction cost estimating handbook that uses the most popular source of construction data, RSMeans, this indispensible guide features access to the instructional version of CostWorks in electronic form, enabling you to practice techniques to solve real-world estimating problems.

$70.00 per copy | 292 pages, softcover | Includes CostWorks CD
Catalog no. 67324C

How Your House Works, 2nd Edition

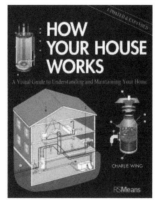

by Charlie Wing

Knowledge of your home's systems helps you control repair and construction costs and makes sure the correct elements are being installed or replaced. This book uncovers the mysteries behind just about every major appliance and building element in your house. See-through, cross-section drawings in full color show you exactly how these things should be put together and how they function, including what to check if they don't work. It just might save you having to call in a professional.

$22.95 per copy | 192 pages, softcover | Catalog no. 67351A

RSMeans Cost Data, Student Edition

This book provides a thorough introduction to cost estimating in a self-contained print and online package. With clear explanations and a hands-on, example-driven approach, it is the ideal reference for students and new professionals.

Features include:

- Commercial and residential construction cost data in print and online formats

- Complete how-to guidance on the essentials of cost estimating

- A supplemental website with plans, problem sets, and a full sample estimate

$99.00 per copy | 512 pages, softcover | Catalog no. 67363

Reference Books

For more information visit the RSMeans website at **www.rsmeans.com**

Value Engineering: Practical Applications

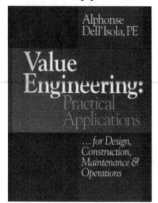

For Design, Construction, Maintenance & Operations
by Alphonse Dell'Isola, PE

A tool for immediate application— for engineers, architects, facility managers, owners, and contractors. Includes making the case for VE—the management briefing; integrating VE into planning, budgeting, and design; conducting life cycle costing; using VE methodology in design review and consultant selection; case studies; VE workbook; and a life cycle costing program on disk.

$79.95 per copy | Over 450 pages, illustrated, softcover | Catalog no. 67319A

The Building Professional's Guide to Contract Documents

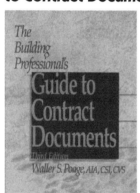

3rd Edition
by Waller S. Poage, AIA, CSI, CVS

A comprehensive reference for owners, design professionals, contractors, and students

- Structure your documents for maximum efficiency.
- Effectively communicate construction requirements.
- Understand the roles and responsibilities of construction professionals.
- Improve methods of project delivery.

$70.00 per copy | 400 pages Diagrams and construction forms, hardcover
Catalog no. 67261A

Cost Planning & Estimating for Facilities Maintenance

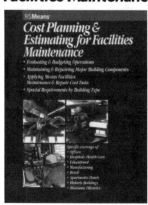

In this unique book, a team of facilities management authorities shares their expertise on:

- Evaluating and budgeting maintenance operations
- Maintaining and repairing key building components
- Applying *RSMeans Facilities Maintenance & Repair Cost Data* to your estimating

Covers special maintenance requirements of the ten major building types

$89.95 per copy | Over 475 pages, hardcover | Catalog no. 67314

Facilities Operations & Engineering Reference

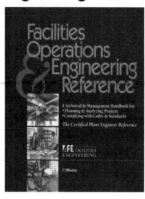

by the Association for Facilities Engineering and RSMeans

An all-in-one technical reference for planning and managing facility projects and solving day-to-day operations problems. Selected as the official certified plant engineer reference, this handbook covers financial analysis, maintenance, HVAC and energy efficiency, and more.

$109.95 per copy | Over 700 pages, illustrated, hardcover | Catalog no. 67318

Green Home Improvement

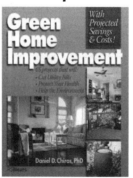

by Daniel D. Chiras, PhD

With energy costs rising and environmental awareness increasing, people are looking to make their homes greener. This book, with 65 projects and actual costs and projected savings help homeowners prioritize their green improvements.

Projects range from simple water savers that cost only a few dollars, to bigger-ticket items such as HVAC systems. With color photos and cost estimates, each project compares options and describes the work involved, the benefits, and the savings.

$34.95 | 320 pages, illustrated, softcover | Catalog no. 67355

Life Cycle Costing for Facilities

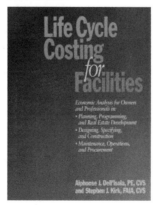

by Alphonse Dell'Isola and Dr. Steven Kirk

Guidance for achieving higher quality design and construction projects at lower costs! Cost-cutting efforts often sacrifice quality to yield the cheapest product. Life cycle costing enables building designers and owners to achieve both. The authors of this book show how LCC can work for a variety of projects — from roads to HVAC upgrades to different types of buildings.

$99.95 per copy | 396 pages, hardcover | Catalog no. 67341

Reference Books

For more information visit the RSMeans website at **www.rsmeans.com**

Interior Home Improvement Costs

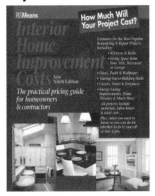

9th Edition

Updated estimates for the most popular remodeling and repair projects—from small, do-it-yourself jobs to major renovations and new construction. Includes: kitchens & baths; new living space from your attic, basement, or garage; new floors, paint, and wallpaper; tearing out or building new walls; closets, stairs, and fireplaces; new energy-saving improvements, home theaters, and more!

$24.95 per copy | 250 pages, illustrated, softcover | Catalog no. 67308E

Exterior Home Improvement Costs

9th Edition

Updated estimates for the most popular remodeling and repair projects—from small, do-it-yourself jobs, to major renovations and new construction. Includes: curb appeal projects—landscaping, patios, porches, driveways, and walkways; new windows and doors; decks, greenhouses, and sunrooms; room additions and garages; roofing, siding, and painting; "green" improvements to save energy & water.

$24.95 per copy | Over 275 pages, illustrated, softcover | Catalog no. 67309E

Builder's Essentials: Plan Reading & Material Takeoff

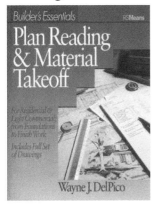

For Residential and Light Commercial Construction
by Wayne J. DelPico

A valuable tool for understanding plans and specs, and accurately calculating material quantities. Step-by-step instructions and takeoff procedures based on a full set of working drawings.

$35.95 per copy | Over 420 pages, softcover | Catalog no. 67307

Means Unit Price Estimating Methods

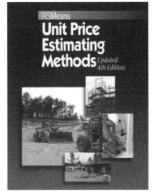

New 4th Edition

This new edition includes up-to-date cost data and estimating examples, updated to reflect changes to the CSI numbering system and new features of RSMeans cost data. It describes the most productive, universally accepted ways to estimate, and uses checklists and forms to illustrate shortcuts and timesavers. A model estimate demonstrates procedures. A new chapter explores computer estimating alternatives.

$65.00 per copy | Over 350 pages, illustrated, softcover | Catalog no. 67303B

Concrete Repair and Maintenance Illustrated

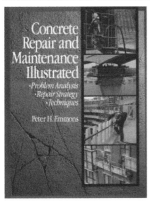

by Peter Emmons

Hundreds of illustrations show users how to analyze, repair, clean, and maintain concrete structures for optimal performance and cost effectiveness. From parking garages to roads and bridges to structural concrete, this comprehensive book describes the causes, effects, and remedies for concrete wear and failure. Invaluable for planning jobs, selecting materials, and training employees, this book is a must-have for concrete specialists, general contractors, facility managers, civil and structural engineers, and architects.

$69.95 per copy | 300 pages, illustrated, softcover | Catalog no. 67146

Total Productive Facilities Management

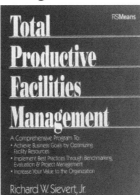

by Richard W. Sievert, Jr.

Today, facilities are viewed as strategic resources. . . elevating the facility manager to the role of asset manager supporting the organization's overall business goals. Now, Richard Sievert Jr., in this well-articulated guidebook, sets forth a new operational standard for the facility manager's emerging role. . . a comprehensive program for managing facilities as a true profit center.

$79.95 per copy | 275 pages, softcover | Catalog no. 67321

Reference Books

For more information visit the RSMeans website at **www.rsmeans.com**

Means Illustrated Construction Dictionary

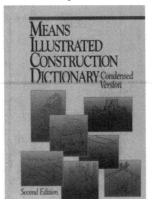

Condensed, 2nd Edition

The best portable dictionary for office or field use—an essential tool for contractors, architects, insurance and real estate personnel, facility managers, homeowners, and anyone who needs quick, clear definitions for construction terms. The second edition has been further enhanced with updates and hundreds of new terms and illustrations . . . in keeping with the most recent developments in the construction industry.

Now with a quick-reference Spanish section. Includes tools and equipment, materials, tasks, and more!

$59.95 per copy | Over 500 pages, softcover | Catalog no. 67282A

Facilities Planning & Relocation

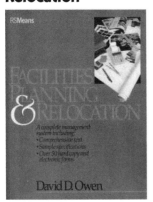

by David D. Owen

A complete system for planning space needs and managing relocations. Includes step-by-step manual, over 50 forms, and extensive reference section on materials and furnishings.

New lower price and user-friendly format.

$89.95 per copy | Over 450 pages, softcover | Catalog no. 67301

Electrical Estimating Methods

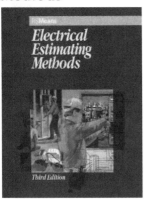

3rd Edition

Expanded edition includes sample estimates and cost information in keeping with the latest version of the CSI MasterFormat and UNIFORMAT II. Complete coverage of fiber optic and uninterruptible power supply electrical systems, broken down by components, and explained in detail. Includes a new chapter on computerized estimating methods. A practical companion to *RSMeans Electrical Cost Data.*

$64.95 per copy | Over 325 pages, hardcover | Catalog no. 67230B

Square Foot & UNIFORMAT Assemblies Estimating Methods

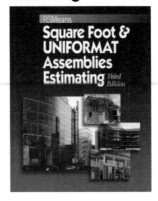

3rd Edition

Develop realistic square foot and assemblies costs for budgeting and construction funding. The new edition features updated guidance on square foot and assemblies estimating using UNIFORMAT II. An essential reference for anyone who performs conceptual estimates.

$69.95 per copy | Over 300 pages, illustrated, softcover | Catalog no. 67145B

Mechanical Estimating Methods

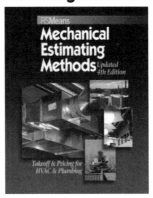

4th Edition

Completely updated, this guide assists you in making a review of plans, specs, and bid packages, with suggestions for takeoff procedures, listings, substitutions, and pre-bid scheduling for all components of HVAC. Includes suggestions for budgeting labor and equipment usage. Compares materials and construction methods to allow you to select the best option.

$64.95 per copy | Over 350 pages, illustrated, softcover | Catalog no. 67294B

Residential & Light Commercial Construction Standards

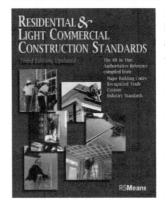

3rd Edition,
Updated by RSMeans and contributing authors

This book provides authoritative requirements and recommendations compiled from leading professional associations, industry publications, and building code organizations. This all-in-one reference helps establish a standard for workmanship, quickly resolve disputes, and avoid defect claims. Updated third edition includes new coverage of green building, seismic, hurricane, and mold-resistant construction.

$59.95 | Over 550 pages, illustrated, softcover | Catalog no. 67322B

Reference Books

For more information visit the RSMeans website at **www.rsmeans.com**

Construction Business Management

by Nick Ganaway

Only 43% of construction firms stay in business after four years. Make sure your company thrives with valuable guidance from a pro with 25 years of success as a commercial contractor. Find out what it takes to build all aspects of a business that is profitable, enjoyable, and enduring. With a bonus chapter on retail construction.

$49.95 per copy | 200 pages, softcover | Catalog no. 67352

Project Scheduling & Management for Construction

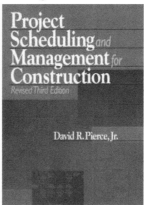

3rd Edition
by David R. Pierce, Jr.

A comprehensive yet easy-to-follow guide to construction project scheduling and control—from vital project management principles through the latest scheduling, tracking, and controlling techniques. The author is a leading authority on scheduling, with years of field and teaching experience at leading academic institutions. Spend a few hours with this book and come away with a solid understanding of this essential management topic.

$64.95 per copy | Over 300 pages, illustrated, hardcover | Catalog no. 67247B

The Practice of Cost Segregation Analysis

by Bruce A. Desrosiers and Wayne J. DelPico

This expert guide walks you through the practice of cost segregation analysis, which enables property owners to defer taxes and benefit from "accelerated cost recovery" through depreciation deductions on assets that are properly identified and classified.

With a glossary of terms, sample cost segregation estimates for various building types, key information resources, and updates via a dedicated website, this book is a critical resource for anyone involved in cost segregation analysis.

$99.95 per copy | Over 225 pages | Catalog no. 67345

Preventive Maintenance for Multi-Family Housing

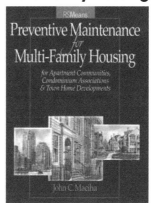

by John C. Maciha

Prepared by one of the nation's leading experts on multi-family housing.

This complete PM system for apartment and condominium communities features expert guidance, checklists for buildings and grounds maintenance tasks and their frequencies, a reusable wall chart to track maintenance, and a dedicated website featuring customizable electronic forms. A must-have for anyone involved with multi-family housing maintenance and upkeep.

$89.95 per copy | 225 pages | Catalog no. 67346

Green Building: Project Planning & Cost Estimating

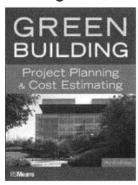

3rd Edition

Since the widely read first edition of this book, green building has gone from a growing trend to a major force in design and construction.

This new edition has been updated with the latest in green building technologies, design concepts, standards, and costs. Full-color with all new case studies—plus a new chapter on commercial real estate.

$99.95 per copy | Over 450 pages, softcover | Catalog no. 67338B

Job Order Contracting Expediting Construction Project Delivery

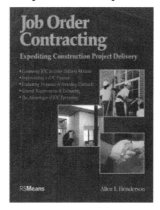

by Allen Henderson

Expert guidance to help you implement JOC—fast becoming the preferred project delivery method for repair and renovation, minor new construction, and maintenance projects in the public sector and in many states and municipalities. The author, a leading JOC expert and practitioner, shows how to:

- Establish a JOC program
- Evaluate proposals and award contracts
- Handle general requirements and estimating
- Partner for maximum benefits

$89.95 per copy | 192 pages, illustrated, hardcover | Catalog no. 67348

Reference Books

For more information visit the RSMeans website at **www.rsmeans.com**

Construction Supervision

This brand new title inspires supervisory excellence with proven tactics and techniques applied by thousands of construction supervisors over the past decade. Recognizing the unique and critical role the supervisor plays in project success, the book's leadership guidelines carve out a practical blueprint for motivating work performance and increasing productivity through effective communication.

Features:

- A unique focus on field supervision and crew management

- Coverage of supervision from the foreman to the superintendent level

- An overview of technical skills whose mastery will build confidence and success for the supervisor

- A detailed view of "soft" management and communication skills

$90.00 per copy | Over 450 pages, softcover | Catalog no. 67358

The Homeowner's Guide to Mold

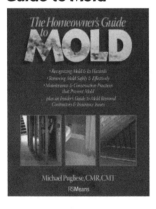

By Michael Pugliese

Expert guidance to protect your health and your home.

Mold, whether caused by leaks, humidity or flooding, is a real health and financial issue—for homeowners and contractors.

This full-color book explains:

- Construction and maintenance practices to prevent mold

- How to inspect for and remove mold

- Mold remediation procedures and costs

- What to do after a flood

- How to deal with insurance companies

$21.95 per copy | 144 pages, softcover | Catalog no. 67344

Landscape Estimating Methods

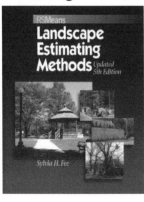

5th Edition

Answers questions about preparing competitive landscape construction estimates, with up-to-date cost estimates and the new MasterFormat classification system. Expanded and revised to address the latest materials and methods, including new coverage on approaches to green building. Includes:

- Step-by-step explanation of the estimating process

- Sample forms and worksheets that save time and prevent errors

$69.95 per copy | Over 350 pages, softcover | Catalog no. 67295C

Building & Renovating Schools

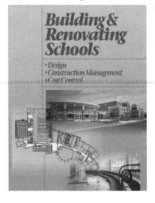

This all-inclusive guide covers every step of the school construction process—from initial planning, needs assessment, and design, right through moving into the new facility. A must-have resource for anyone concerned with new school construction or renovation. With square foot cost models for elementary, middle, and high school facilities, and real-life case studies of recently completed school projects.

The contributors to this book— architects, construction project managers, contractors, and estimators who specialize in school construction— provide start-to-finish, expert guidance on the process.

$99.95 per copy | Over 425 pages, hardcover | Catalog no. 67342

Universal Design Ideas for Style, Comfort & Safety

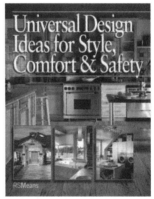

by RSMeans and Lexicon Consulting, Inc.

Incorporating universal design when building or remodeling helps people of any age and physical ability more fully and safely enjoy their living spaces. This book shows how universal design can be artfully blended into the most attractive homes. It discusses specialized products like adjustable countertops and chair lifts, as well as simple ways to enhance a home's safety and comfort. With color photos and expert guidance, every area of the home is covered. Includes budget estimates that give an idea how much projects will cost.

$21.95 | 160 pages, illustrated, softcover | Catalog no. 67354

Plumbing Estimating Methods

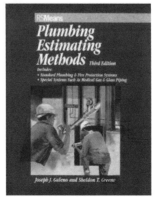

3rd Edition
by Joseph Galeno and Sheldon Greene

Updated and revised! This practical guide walks you through a plumbing estimate, from basic materials and installation methods through change order analysis. *Plumbing Estimating Methods* covers residential, commercial, industrial, and medical systems, and features sample takeoff and estimate forms and detailed illustrations of systems and components.

$59.95 per copy | 330+ pages, softcover | Catalog no. 67283B

Reference Books

For more information visit the RSMeans website at **www.rsmeans.com**

Understanding & Negotiating Construction Contracts

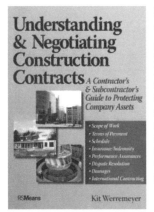

by Kit Werremeyer

Take advantage of the author's 30 years' experience in small-to-large (including international) construction projects. Learn how to identify, understand, and evaluate high risk terms and conditions typically found in all construction contracts—then negotiate to lower or eliminate the risk, improve terms of payment, and reduce exposure to claims and disputes. The author avoids "legalese" and gives real-life examples from actual projects.

$74.95 per copy | 300 pages, softcover | Catalog no. 67350

Means Illustrated Construction Dictionary

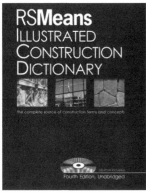

Unabridged 4th Edition, with CD-ROM

Long regarded as the industry's finest, *Means Illustrated Construction Dictionary* is now even better. With nearly 20,000 terms and more than 1,400 illustrations and photos, it is the clear choice for the most comprehensive and current information. The companion CD-ROM that comes with this new edition adds extra features such as: larger graphics and expanded definitions.

$99.95 per copy | Over 790 pages, illust., hardcover | Catalog no. 67292B

RSMeans Estimating Handbook

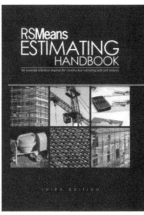

3rd Edition

Widely used in the industry for tasks ranging from routine estimates to special cost analysis projects, this handbook has been completely updated and reorganized with new and expanded technical information.

RSMeans Estimating Handbook will help construction professionals:

- Evaluate architectural plans and specifications
- Prepare accurate quantity takeoffs
- Compare design alternatives and costs
- Perform value engineering
- Double-check estimates and quotes
- Estimate change orders

$99.95 per copy | Over 900 pages, hardcover | Catalog No. 67276B

Spanish/English Construction Dictionary

2nd Edition
by RSMeans and the International Code Council

This expanded edition features thousands of the most common words and useful phrases in the construction industry with easy-to-follow pronunciations in both Spanish and English. Over 800 new terms, phrases, and illustrations have been added. It also features a new stand-alone "Safety & Emergencies" section, with colored pages for quick access.

Unique to this dictionary are the systems illustrations showing the relationship of components in the most common building systems for all major trades.

$23.95 per copy | Over 400 pages | Catalog no. 67327A

The Gypsum Construction Handbook

by USG

An invaluable reference of construction procedures for gypsum drywall, cement board, veneer plaster, and conventional plaster. This new edition includes the newest product developments, installation methods, fire- and sound-rated construction information, illustrated framing-to-finish application instructions, estimating and planning information, and more. Great for architects and engineers; contractors, builders, and dealers; apprentices and training programs; building inspectors and code officials; and anyone interested in gypsum construction. Features information on tools and safety practices, a glossary of construction terms, and a list of important agencies and associations. Also available in Spanish.

$29.95 | Over 575 pages, illustrated, softcover | Catalog no. 67357A

2013 Order Form

For more information visit the RSMeans website at **www.rsmeans.com**

Unit prices according to the latest MasterFormat!

Qty.	Book no.	COST ESTIMATING BOOKS	Unit Price	Total
	60063	Assemblies Cost Data 2013	$305.95	
	60013	Building Construction Cost Data 2013	186.95	
	60113	Concrete & Masonry Cost Data 2013	171.95	
	50143	Construction Cost Indexes 2013 (subscription)	362.00	
	60143A	Construction Cost Index–January 2013	90.50	
	60143B	Construction Cost Index–April 2013	90.50	
	60143C	Construction Cost Index–July 2013	90.50	
	60143D	Construction Cost Index–October 2013	90.50	
	60343	Contr. Pricing Guide: Resid. R & R Costs 2013	39.95	
	60233	Electrical Change Order Cost Data 2013	183.95	
	60033	Electrical Cost Data 2013	188.95	
	60203	Facilities Construction Cost Data 2013	476.95	
	60303	Facilities Maintenance & Repair Cost Data 2013	417.95	
	60553	Green Building Cost Data 2013	155.95	
	60163	Heavy Construction Cost Data 2013	186.95	
	60093	Interior Cost Data 2013	192.95	
	60123	Labor Rates for the Const. Industry 2013	415.95	
	60183	Light Commercial Cost Data 2013	139.95	
	60023	Mechanical Cost Data 2013	183.95	
	60153	Open Shop Building Const. Cost Data 2013	160.95	
	60213	Plumbing Cost Data 2013	188.95	
	60043	Commercial Renovation Cost Data 2013	152.95	
	60173	Residential Cost Data 2013	134.95	
	60283	Site Work & Landscape Cost Data 2013	181.95	
	60053	Square Foot Costs 2013	197.95	
	62012	Yardsticks for Costing (2012)	176.95	
	62013	Yardsticks for Costing (2013)	185.95	
		REFERENCE BOOKS		
	67329	Bldrs Essentials: Best Bus. Practices for Bldrs	29.95	
	67298A	Bldrs Essentials: Framing/Carpentry 2nd Ed.	24.95	
	67298AS	Bldrs Essentials: Framing/Carpentry Spanish	24.95	
	67307	Bldrs Essentials: Plan Reading & Takeoff	35.95	
	67342	Building & Renovating Schools	99.95	
	67261A	The Building Prof. Guide to Contract Documents	70.00	
	67339	Building Security: Strategies & Costs	89.95	
	67353A	Complete Book of Framing, 2nd Ed.	29.95	
	67146	Concrete Repair & Maintenance Illustrated	69.95	
	67352	Construction Business Management	49.95	
	67358	Construction Supervision	90.00	
	67363	Cost Data Student Edition	99.00	
	67314	Cost Planning & Est. for Facil. Maint.	89.95	
	67328A	Designing & Building with the IBC, 2nd Ed.	99.95	
	67230B	Electrical Estimating Methods, 3rd Ed.	64.95	
	67343A	Est. Bldg. Costs for Resi. & Lt. Comm., 2nd Ed.	65.00	
	67276B	Estimating Handbook, 3rd Ed.	99.95	
	67318	Facilities Operations & Engineering Reference	109.95	
	67301	Facilities Planning & Relocation	89.95	
	67338B	Green Building: Proj. Planning & Cost Est., 3rd Ed.	99.95	
	67355	Green Home Improvement	34.95	
	67357A	The Gypsum Construction Handbook	29.95	
	67357S	The Gypsum Construction Handbook (Spanish)	29.95	

Qty.	Book no.	REFERENCE BOOKS (Cont.)	Unit Price	Tota
	67148	Heavy Construction Handbook	99.95	
	67308E	Home Improvement Costs–Int. Projects, 9th Ed.	24.95	
	67309E	Home Improvement Costs–Ext. Projects, 9th Ed.	24.95	
	67344	Homeowner's Guide to Mold	21.95	
	67324C	How to Est.w/Means Data & CostWorks, 4th Ed.	70.00	
	67351A	How Your House Works, 2nd Ed.	22.95	
	67282A	Illustrated Const. Dictionary, Condensed, 2nd Ed.	59.95	
	67292B	Illustrated Const. Dictionary, w/CD-ROM, 4th Ed.	99.95	
	67348	Job Order Contracting	89.95	
	67295C	Landscape Estimating Methods, 5th Ed.	69.95	
	67341	Life Cycle Costing for Facilities	99.95	
	67294B	Mechanical Estimating Methods, 4th Ed.	64.95	
	67283B	Plumbing Estimating Methods, 3rd Ed.	59.95	
	67345	Practice of Cost Segregation Analysis	99.95	
	67346	Preventive Maint. for Multi-Family Housing	89.95	
	67326	Preventive Maint. Guidelines for School Facil.	149.95	
	67247B	Project Scheduling & Manag. for Constr., 3rd Ed.	64.95	
	67322B	Resi. & Light Commercial Const. Stds., 3rd Ed.	59.95	
	67359	Risk Management for Design & Construction	125.00	
	67327A	Spanish/English Construction Dictionary, 2nd Ed.	23.95	
	67145B	Sq. Ft. & Assem. Estimating Methods, 3rd Ed.	69.95	
	67321	Total Productive Facilities Management	79.95	
	67350	Understanding and Negotiating Const. Contracts	74.95	
	67303B	Unit Price Estimating Methods, 4th Ed.	65.00	
	67354	Universal Design	21.95	
	67319A	Value Engineering: Practical Applications	79.95	

MA residents add 6.25% state sales tax _____

Shipping & Handling** _____

Total (U.S. Funds)* _____

Prices are subject to change and are for U.S. delivery only. *Canadian customers may call for current prices. **Shipping & handling charges: Add 7% of total order for check and credit card payments. Add 9% of total order for invoiced orders.

Send order to: ADDV-1000

Name (please print) _____

Company _____

☐ Company

☐ Home Address _____

_____

City/State/Zip _____

Phone # _____ P.O. # _____

(Must accompany all orders being billed)

Mail to: RSMeans, 700 Longwater Drive, Norwell, MA 02061